Evaluation in Planning

The GeoJournal Library

Volume 47

Managing Editors: Herman van der Wusten, University of Amsterdam,
The Netherlands
Olga Gritsai, Russian Academy of Sciences, Moscow,
Russia

Former Series Editor:
Wolf Tietze, Helmstedt, Germany

Editorial Board: Paul Claval, France
R. G. Crane, U.S.A.
Yehuda Gradus, Israel
Risto Laulajainen, Sweden
Gerd Lüttig, Germany
Walther Manshard, Germany
Osamu Nishikawa, Japan
Peter Tyson, South Africa

The titles published in this series are listed at the end of this volume.

Evaluation in Planning

Facing the Challenge of Complexity

edited by

NATHANIEL LICHFIELD
Dalia and Nathaniel Lichfield Associates,
London, U.K.

ANGELA BARBANENTE
School of Environmental Engineering,
Polytechnic University of Bari, Italy

DINO BORRI
Department of Architecture and Urban Planning,
Polytechnic of Bari,
Bari, Italy

ABDUL KHAKEE
Department of Political Science,
Umeå University, Umeå, Sweden

and

ANNA PRAT
Dalia and Nathaniel Lichfield Associates,
London, U.K.

KLUWER ACADEMIC PUBLISHERS
DORDRECHT / BOSTON / LONDON

A C.I.P. Catalogue record for this book is available from the Library of Congress.

ISBN 0-7923-5177-0

Published by Kluwer Academic Publishers,
P.O. Box 17, 3300 AA Dordrecht, The Netherlands.

Sold and distributed in North, Central and South America
by Kluwer Academic Publishers,
101 Philip Drive, Norwell, MA 02061, U.S.A.

In all other countries, sold and distributed
by Kluwer Academic Publishers,
P.O. Box 322, 3300 AH Dordrecht, The Netherlands.

Printed on acid-free paper

Printed in the Netherlands.

TABLE OF CONTENTS

vi

PART III
LINKING PRACTICE TO THEORY

PREFACE

This book is the result of a three day workshop on "Evaluation in theory and practice in spatial planning" held in Ramsey Hall, University College London, in September 1996. Some 30 people from 8 different countries attended and 20 papers were presented. The majority of them now form the basis for this book. This occasion was the third on the topic, the two preceding having taken place in Umeå in June 1992 and in Bari in 1994.

Following these three meetings, we can now say that this small, industrious, international family really enjoy meeting up from time to time at each others places, in the presence of older members and new children, each one presenting his/her own recent experiences. It particularly enjoys exchanging views and arguing about the current state and the future of evaluation in spatial planning (all families have their vices...). It is also pleasing to see these experiences and discussions resulting in a book for those who could not attend and for the broader clan in the field.

Not long time ago, but ages in the accelerated academic time scale, evaluation in planning established its own role and distinct features as an instrument for helping the decision-making process. Now this role and these features are exposed to major challenges. First, the evolution of planning theory has lead to the conception of new planning paradigms, based on theories of complexity and communicative rationality. Second, the planning systems themselves all over Europe and the world seeks to adapt to major political changes and the opening of national boundaries. Central and local relations are changing. It is also frequently held that we may be moving from a democratic model based on representation to one based on participation. Doubtless, whatever our views, the need to make evaluation experiences shareable and communicable between and within countries has increased. This workshop is evidence. Third, evaluation is also challenged by the need to make experiences in different sectoral fields of intervention interchangeable. Above all, the emergence of the environmental and conservation questions poses a series of new problems. How can evaluation in planning adapt to these changes?

The papers in this volume discuss from different viewpoints these open questions. We believe the volume reflects its collective nature in being heterogenous in approach but homogenous in focus.

To present this variety with some order, the volume has being structured in three parts, following an introduction related to the host country. The papers in the first part discuss some relevant theoretical issues. In the second part, methodological questions which incorporate new developments in the field are offered. Finally, in the third part, practical experiences in evaluation in different countries complete the picture. Unfortunately, because of space, we have not attempted to reproduce the lively discussion that has taken place during the workshop. Instead, each contributor reflected it in the redrafting of its paper. The three sectional introduction will guide the reader in the endeavour.

Finally, special thanks are due to the institutions that made all this possible: the Bartlett School of Planning, at University College London, for the accommodation, the Politecnico di Bari for financial help, Dalia & Nathaniel Lichfield Associates for both financial help and organisation and to Petra van Steenbergen at Kluwer for her assistance and patience.

In conclusion, we are happy to say that plans are in progress to carry on this pleasant family tradition in a fourth workshop.

The Editors

Nathaniel Lichfield is an urban and regional planner, specialising in planning and development economics, including in plan and project evaluation. He is Past President of the Royal Town Planning Institute and Professor Emeritus of the Economics of Environmental Planning, University of London. Following posts in local and central government, he founded in 1962 Nathaniel Lichfield and Partners. Left in 1992 to form Dalia and Nathaniel Lichfield Associates. Throughout his career, he has combined practice and academia and published extensively in both fields.

Angela Barbanente studied Urban Planning at the University of Venice. She is researcher at the Bari Polytechnic, School of Engineering for Environment. She has also worked as researcher at the National Research Council (Bari) and as Professor of Territorial Planning (University of Basilicata) and Planning Theory and Urban Planning (Bari Polytechnic). She is author of books and articles on urban rehabilitation, evaluation in spatial planning and development control.

Dino Borri is a civil engineer and professor of Urban Planning Techniques at Bari Polytechnic. He is currently the head of the post-graduate school of Urban and Regional planing and the President of Fondazione Astengo, Istituto Nazionale di Urbanistica, Roma. He has written several books and articles on the application of Artificial Intelligence to spatial planning.

Abdul Khakee has a PhD in Economics and one in Geography. He is currently Professor of Urban Planning at the Department of Political Science at Umeå University. He has been visiting researcher at the University of Delaware, university of Melbourne, Bari Polytechnic and University of Lisbon. He is the author of several books and journal articles on development planning, planning theory, local government and cultural economics.

Anna Prat studied planning and architecture at Turin Polytechnic and regional and urban planning at the London School of Economics. In Italy, she co-founded a non-profit community participation association. She has worked as research assistant at Dalia and Nathaniel Lichfield Associates, London, specialising in evaluation in planning. She is currently working as urban and economic planner at ARUP Economics and Planning, London.

CONTRIBUTORS

Alexander E.R., Faculty of Spatial Science, Public Policy Program, Tel Aviv University, PO Box 39040, Ramat Aviv, Tel Aviv, 69978, Israel.

Barbanente A., Polytechnic of Bari, 2nd School of Engineering, Institute of Engineering for Environment and Territory, Viale del Turismo 8, 74100 Taranto, Italy.

Batty M., Centre For Advanced Spatial Analysis (CASA), University College London, 1-19 Torrington Place, London WC1E 6BT United Kingdom.

Bizzarro F., University of Napoli "Federico II", Department of Conservation of Architectural and Environmental Goods, Via Cesare Battisti 15, 80134 Napoli, Italy.

Borri D., Polytechnic of Bari, Department of Architecture and Town Planning, Via E. Orabona 4, 70125 Bari, Italy.

Clemente M.L., University of Cagliari, Department Engineering of Territory, Piazza d'Armi, 16, 09123 Cagliari, Italy.

Concilio G., PhD in "Evaluation Methods for Integrated Conservation of Architectural, Urban and Environmental Heritage", University of Napoli "Federico II", Via Cesare Battisti 15, 80134 Napoli, Italy.

Corkindale J. 55 Poplar Grove, New Malden Surrey KT3DN United Kingdom.

Fusco Girard L., University of Napoli "Federico II", Department of Conservation of Architectural and Environmental Goods, Via Cesare Battisti 15, 80134 Napoli, Italy.

Glasser H., Foundation for Deep Ecology, 1555 Pacific Avenue, San Francisco, CA 94109, United States.

Hull A., University of Newcastle, Department of Town and Country Planning, Newcastle upon Tyne NE1 7RU, United Kingdom.

Khakee A., Department of Political Science, Umeå University, 901 87 Umeå, Sweden.

Lichfield D., Dalia & Nathaniel Lichfield Associates, 13 Chalcot Gardens, England's Lane, London NW3 4YB, United Kingdom.

Lichfield N., Dalia & Nathaniel Lichfield Associates, 13 Chalcot Gardens, England's Lane, London NW3 4YB, United Kingdom.

Lombardi P.L., School of Architecture, Department of Socio-Economic Analysis, Santa Croce 1957, 30135 Venezia, Italy.

Macchi S., University of Rome "La Sapienza", Department of Architecture and Town Planning, Via Eudossiana, 18, 00184 Roma.

Maciocco G., University of Cagliari, Department Engineering of Territory, Piazza d'Armi, 16, 09123 Cagliari, Italy.

Marchi G., University of Cagliari, Department Engineering of Territory, Piazza d'Armi, 16, 09123 Cagliari, Italy.

Millichap D., Linlaters & Paines, Mitre House, 160 Aldergate Street, EC1A 4LP London, United Kingdom.

Nash C., University of Leeds, Institute for Transport Studies, Leeds LS2 9JT, United Kingdom.

Nijkamp P., Faculty of Economics and Econometrics, Vrije Universiteit Amsterdam, De Boelelaan 1105, 1081 HV Amsterdam, The Netherlands.

Pace F., PhD in "Evaluation Methods in Urban Planning and Architecture Design", University of Reggio Calabria, Salita Melissari - Feo di Vito, 89124 Reggio Calabria, Italy.

Prat A., Dalia & Nathaniel Lichfield Associates, 13 Chalcot Gardens, England's Lane, London NW3 4YB, United Kingdom.

Scandurra E., University of Rome "La Sapienza", Department of Architecture and Town Planning, Via Eudossiana, 18, 00184 Roma.

Selicato F., Polytechnic of Bari, Department of Architecture and Town Planning, Via Edoardo Orabona 4, 70125 Bari, Italy.

Söderbaum P., Department of Economics, Mälardalens Högskola, PO Box 883, Västeras, Sweden.

Torre C., PhD in "Evaluation Methods for Integrated Conservation of Architectural, Urban and Environmental Heritage", University of Napoli "Federico II", Via Cesare Battisti 15, 80134 Napoli, Italy.

Voogd H., Faculty of Spatial Science, University of Gröningen, PO Box 800, 9700 AV, Gröningen, The Netherlands.

INTRODUCTION

TRENDS IN PLANNING EVALUATION: A BRITISH PERSPECTIVE

N. LICHFIELD

1.   Focus

This opening paper has a particular purpose in mind. It is to present a picture of trends in Britain in the topic of the Workshop, in the hope that our discussion of international practice could influence those trends and so help to enrich British theory and practice on spatial evaluation.

In order to comprehend these trends it is necessary to describe the historical and current theory and practice in spatial planning and its evaluation in Britain. That is provided in the companion paper in this volume (Lichfield and Prat). From that review it is apparent that since its inception in 1909, British planning has evolved via a process of continuity, change and paradox. The continuity is seen in the century old attempts by Governments to plan for, and accordingly regulate in the public interest, the market process in the development of land. The change is seen in the continuing shifts of emphasis from the basis of experience, first in the pre-World War II paradigm and then in the post-World War II change in that paradigm. The changes have been related in many ways to the political swing, as the Conservatives and Labour Governments have changed places, each without coalition partners in the British electoral system of "first past the post". The changes have not been linear, except perhaps for the financial solution to the compensation and betterment problem, with three radical proposals by Labour being immediately abandoned by the Conservatives, when the opportunity arose. In this many paradoxes can be found. Relevant examples for this paper can be seen from the influence of the Thatcher Government (1979-1990) on evaluation. While the Thatcherite over-riding criterion of "value for money" was aimed at promoting the private sector, this criterion in itself encouraged the use of the rational approach to evaluation employed by that sector. This was fostered also by privatisation of nationalised industry and the attraction of private finance capital to public sector projects, through the Private Finance Initiative (PFI) for the private financing of public sector projects (Private Finance Panel, 1995). Then, as part of the Government's initiative against regulation, new Government Regulations needed to be justified by producing rigorous assessments of their impacts on business and the resulting costs and benefits to the region (Lichfield, 1996, 309-314). And while the Conservative

N. Lichfield et al. (eds.), Evaluation in Planning, 1–15.
© 1998 *Kluwer Academic Publishers. Printed in the Netherlands.*

Government did not really favour strong development planning it nonetheless introduced in 1991 a mandatory "plan led system", with its greater relative weight to the plan in making decisions. One consequence was the greater attention given to the plan making, with the plan contents being as a result under much greater scrutiny in preparation, and going through more severe tests than before, for example in securing ongoing evaluation.

This historical review of continuity and change, with unexpected paradoxes, provides a useful lead in to the main focus of this paper: an examination of some trends in evaluation in British town planning with an eye to speculation on the future in this field. But it cannot be expected that there will be a simple continuation of the experience of the past fifty years. This arises because of the dramatic change in the complexion of Government in the election of 1997. After twenty years of Conservative Administration, Labour is now back in office, and with a huge majority. Just what this change might mean in practice was clearly a matter of discussion and contention prior to the Election. On this there was great uncertainty, since the Party that has been returned goes under the banner of "New Labour", reflecting a determination to pursue different policies from "Old Labour", which has been identified with old fashioned socialism. Just what that meant in practice was not all clear from the programme of "New Labour"; and the nature of those changes is slowly becoming apparent with the experience of the new Government. Furthermore, the nature of the main opposition, the Conservative Party, is also not clear since they are in the process of major changes in their constitution and policies as a result of the resounding defeat, possibly veering more towards liberalism than before. Furthermore, the dividing line between the policies of the three major Parties (Labour, Conservative and Liberal Democrats) is becoming less clear. For example, new Labour appears to be adopting Conservative economic policies; soon after their defeat, the prediction was that the "New Conservatives" would strongly move to the right but they do not appear to be doing so, and indeed are stealing some of Labour's clothes in their new recognition of the existence of "society" as against the aphorism of Thatcher, that "there is not such thing as society"; the Liberal Democrats remain steadfastly liberal and progressive.

That apart, but evidenced by the landslide victory for Labour, the people of Britain are making their own kind of review of the nature of its contemporary society and the ways in which that could and should be evolving in the 21st century. How this will influence "New Labour", in its avowed vision of modernising the country and speaking for "one nation", as opposed to its traditional base of socialism via the "working classes", still remains to be clarified.

From this background, in the space available, I can only point to a number of themes which could have their bearing on the way that town and country planning and its evaluation will evolve. The themes are varied, and of varying weight in influencing the future path. Thus it is useful to classify them under distinct headings, three in number with sub-headings, although these cannot be taken to present a tidy nor by any means exhaustive categorisation. The first relates to ongoing practice within the field which has penetrated the evaluation process and is likely to continue in that direction. The second relates to ongoing activities which are ancillary to the urban and regional planning system and are exercising considerable influence on it. The final category is concerned with societal change outside the planning system which could have its impact on planning and evaluation theory and practice. Then comes some conclusion on the straws in the wind.

2.   Within the Planning System

It has long been held that planning is for the "people". It therefore cannot be regarded just as a managerial function taking place merely between "the planners" and "the decision takers", both acting without sufficient regard to what has been called "the planned". The reasons for this lie in the ever growing involvement of the public at large in the democratic process, aided, stimulated and also misled by the media; and thereby in planning and its implementation, and also in evaluation which should attempt to appraise the impact of the planning on the people's way of life

(i) Extending Communication

In the pre-war planning system, the "planned" were largely seen as the affected property owners. Changes were formally introduced in post war British planning (Skeffington 1969). In this it was seen that informing the public of the planning process and its decisions was not sufficient; it was necessary also to involve them in the decision making process. In practice this evolved only modestly, and in some cases perfunctorily, and certainly does not meet the need for involving the public as seen today, in "communicative ideology" (Khakee, Voogd this volume). This envisages the setting up of the most complex arrangements for carrying out the communication, in terms for example of those who are to be invited; the style of the discussion; the way issues are identified and filtered; how new policy discourses emerge; and how agreements are reached and monitored.

But, some think, the concept in practice should go wider. It is not only a means of very enriched "public participation" in formulation of plans and the taking of decisions. In addition it augurs the need to abandon the traditional rational model of planning and evaluation for new methods based on discourse and exchange via communication. It implies throwing out the baby with the bath water. The logic here is not so apparent. The wider communication just described will certainly need some rational approach to its comprehension. Why then should it not be linked with an improved means of planning and evaluation? It would then not displace but enrich the traditional approach. Perhaps that is what is conveyed by Healey (1996), a proponent of the communicative, turn in the following:

"The approach outlined in this paper presents strategic spatial planning as a process of facilitating community collaboration in the construction of strategic discourse, in strategic consensus-building […] This approach in some respects revisits the activities of the well-known rational planning process. It involves review of issues (survey), sorting through findings (analysis), exploring the impacts in relation to values (evaluation), inventing and developing new ideas (choice of strategy), and continuous review (monitoring). But these activities are approached in a very different way. They are undertaken interactively, often in parallel rather sequentially; they deal with explicitly in the everyday language of practical life, treating technical language as but one among the many languages to be listened to; as result, the approach extends the reasoning process beyond instrumental rationality, to allow debate in moral and emotive terms. They involve active discursive work by the parties involved, facilitated by planners or other relevant experts, rather than being undertaken by planners themselves. And they are founded on principles of participatory democracy, underpinned by legal 'rights to be heard' and 'inclusionary terms' in which claims for attention must be redeemed, rather than the hierarchical forms of representative democracy.

(ii)   The Public Planning Inquiry

One concept of decision taking through communication, that has evolved in Britain since planning started, is in the form of public inquiries by the relevant Minister into appeals against decisions by local authorities on planning applications, or into representations and objections on a plan prepared by local authorities (Cullingworth and Nadin, 86-88). The inquiry is in front of an Inspector who reports accordingly, where he has no delegated decision taking powers, to the Minister. The structure for this kind of communication is quite stylised in a somewhat pallid echo of the law courts, which stems from the paramount influence of the lawyers in the proceedings. It is not an open free-wheeling discussion but rather an adversarial battleground in which the case of one side is put forward (usually by a legal advocate) and supported by professional experts who are then attacked by the opposing advocate with the help of opposing experts. The process is then reversed. Third parties (those not directly involved in the dispute) are also encouraged to appear and present their evidence. The upshot can be a massive report and decision by the presiding Inspector. There is no appeal against such a decision except on matters of law.

The system has its advantages. It offers a means of ventilating disputes in an open manner under its rules; it offers the Minister the opportunity of influencing policy by the nature of his decisions (not always helpfully); it ensures that planning authorities, developers and their experts need to discipline themselves in the knowledge that their proposal can be open to robust attack.

But there are disadvantages. The procedure is long, drawn out and tedious; the combative stance is not always helpful in elucidating the facts and the realities; and any attempt by one side to present a balanced picture is discredited from the outset as not sufficiently adversarial; the Inspector tends to be a passive as opposed to an active participant; the drawing up of the balance between the parties often lacks rigorous analysis.

Because of this a modification has been suggested which is pertinent to evaluation (Lichfield, N. and D. Lichfield, 1995). In a major inquiry let the issues be presented in advance by an independent expert in a structured way, for example by community impact evaluation. The ensuing discourse of the inquiry, still adversarial, would follow the same structure for the evidence in detail. Thereby the Inspector would be more likely to reach conclusions which are more readily defensible, and parties disaffected by the decision would be clearer as to the reasons.

(iii)   Action Planning

Another manifestation of communication is the extension of public participation beyond reactions *to* the planners to action *with* the planners. This tendency has grown under various names, such as community architecture, community planning, planning jury, as well as action planning, planning weekend and planning for real. While the forms vary in essence, the communicative turn is expanded to result in community led planning and decision making. In effect the planner/public relationship is turned on its head, so that the public become the driving force with the planner becoming the interpreter of their wishes, views etc. In extreme cases, the public becomes in effect the client of the planner, in this way resembling what in the United States is called advocacy planning.

Here again a dilemma arises. Given that the professional planning team is not in the driving seat, the result can tend to be disorganised and not meaningful, in terms of considering rationally the problems, opportunities, constraints, design and evaluation that the professional planning process would take in depth. Furthermore, the result could be so unrealistic that it could hardly be accepted by the planning authority as a basis for a planning decision, or by the developer as a basis for investment or by the banks as a basis for financing.

These weaknesses could be remedied in the following way. The planners as experts (with their elected decision takers) would formulate their proposals along the lines of the conventional planning process. But they would expose *each* of the steps in the process to the involvement of the public in question, in a meaningful as opposed to perfunctory way. Clearly the form of exposure would vary according to the stage in the process. At the outset exploration could be directed towards identification of problems, opportunities and constraints. Later on there would be the exposure of options and later on there would be the exposure of the evaluation as the basis for discussion.

(iv) Public Choice in Decision Making

This does not mean precisely what the term conveys, that is choice by the public at large, but it is nonetheless pertinent here. Rather it is the application of economics to the choices of public bodies, such as political parties, politicians, bureaucrats, interest groups (Buchanan and Tollison 1972; Buchanan, et alia 1978). One aim is to explain choices by such public bodies, which could be used to defend the public at large against the waywardness, at best, of those in political and bureaucratic power. Since the critical decision making in planning rests with such bodies, it follows that there is an area of investigation which could be helpful in scrutinising critically their role in public decision taking and its evaluation.

There is another reason why this contribution can be of importance. In planning, the genesis of public intervention in the market has often been laid at the door of market failure, that is the inability of the market to reflect matters of concern in the public interest, such as externalities, uneven income distribution, etc. An early and somewhat naive assumption followed: that when market failure leads to the need for public intervention, the decisions made by public bodies of the kind described would be necessarily *superior*, since they aimed at remedying the market failure. But as the public choice literature has shown, there can well be failure also in government, bureaucracy and in upholding individual freedoms (Lichfield, N. 1990). The extension of public choice theory and practice in planning could be used to extend evaluation of options beyond the remedy of market failure to addressing also the remedy or potential public failure.

(v) Politics in Local Planning Decisions

As just indicated, the decision taking by politicians is an element in public choice theory. On this, special mention should be made of the increasing politicisation of decisions in local planning in Britain over the last twenty or so years, and its implications for evaluation.

Speaking generally, the planner in Britain used to have a somewhat similar role in relation to the elected members as that of other officials, such as the engineer, medical officer, accountant etc. His professional advice would be considered, and then the decision would be subject to political influence. More recently, in certain authorities, the role have become reversed. Before committee meetings the different parties might meet in caucus to make their decision, in extreme cases independently of the planning advice itself. The significance here is that their evaluation of the case would be heavily influenced by the political as opposed to the professional considerations. Even though these could legitimately be claimed as "in the public interest" (since the political decision makers were the representatives of the victorious electorate), the decision might not be in the public interest as seen in an evaluation, for example, having regard to the impacted community at large, which could spread beyond the particular local administration making the decision.

3. Ancillary to the Urban and Regional Planning System

As will be seen from the companion paper in this volume (Lichfield and Prat) the British urban and regional planning system provides a long standing and deeply penetrating means of intervention by government in the utilisation of land and its development and associated socio-economic activities. As part of this intervention, and inevitably associated with it, are others related to the individual socio-economic activity which involve the use of the land, such as transportation, health, cultural heritage, leisure. These in practice are subsumed in the urban and regional planning process. But outside this have grown up discrete kinds of intervention which could conceivably be incorporated into the urban and regional planning system but have not been, albeit they are pursued in practice alongside. These accordingly can be seen as complementary to the urban and regional planning systems. Five are introduced.

(i) Urban and Regional Policy

Urban and regional planning policy has traditionally been implicit in the urban and regional plans themselves, in relation to the area for which they are prepared. In these, by definition, the various problems and opportunities in those areas are encompassed within the plan policies and proposals, and accordingly are covered in the evaluation process (Lichfield, et alia 1975, pp. 123-289).

But since the 1970's there has been a growing recognition that there are grave social, economic, environmental and cultural problems in our urban areas which require strong pro-active intervention, of the kind that is not provided in the traditional planning process. This has led to the emergence of distinct policy formulation for deprived urban areas (the inner cities) (Hambleton and Thomas, 1995; DOE, 1994a). For them there was in 1994 a consolidation of some twenty Government financial support programmes into a Single Regeneration Budget (SRB). This SRB Challenge Fund is being distributed amongst the competing towns following competitive bids, which are in accord with a centrally devised evaluation of the expected outputs, and potential outcomes (DOE, 1994b; 1995).

(ii) European Regional Development Funds (ERDF)

With the growth of pan-European planning from Brussels, Britain has been drawn into European Regional Policy. Emanating from "social programmes", which focus support on the regions most in need, according to five complementary objectives, Brussels has denoted parts of Britain as eligible for grant. (European Commission 1994) This has established a vertical link between the regions and the Commission which has in the main been implemented alongside the British planning system. The influence here has been powerful, not only because of the generous money grants which have been made available but also for the carrying out of positive developments by means of newly created partnerships. This process has strengthened the socio-economic regional input and also carried with it decided forms of ex ante and ex post evaluation, similar to those for SRB.

(iii) European Planning and Environment Law

Another programme initiated from Brussels is the need to carry out environmental assessment for major projects in parallel with the planning applications. This requirement for methodical environmental assessment, to ensure control over development projects, is based on the rational approach. It requires the evaluation of options in terms of significance for the environment (which could include the socio-economic as well as the natural) as a necessary preliminary to a planning permit. This has profoundly influenced planning practice and evaluation itself (Lichfield, N. 1992).

In consequence of a Directive of 1985 the British Government adopted the necessity in their own Regulations of 1988. And Brussels wishes to see the approach extended beyond projects to plans, programmes and policies. Some extension has already taken place in that Planning Policy Guidance from the Department of the Environment asks for environmental assessment to be made in the preparation of development plans as a framework for the assessment of projects (DOE 1993). There are also likely to be changes in our planning system, in the light of the decided wish in Brussels to harmonise the systems of the different European countries, as evidenced in the draft European Development Perspective (Morphet, J. 1997).

However, while related to the planning system the assessment of projects has been made distinct from it, so that the two are not integrated. Attempts to overcome this for projects have been introduced with the aim of ensuring that in development control the planning decision reflects also the relevant environmental considerations but is not overwhelmed by them (Lichfield, N. 1992; Lichfield, N. and D. Lichfield, 1992).

(iv)   Sustainability

When the United Nations set up the World Commission on Environment and Development they were aiming very high, namely at issues related to the globe as a whole (World Commission, 1987). But it took comparatively little time for the message to percolate down to the national and local levels, namely to aim towards "development that meets the needs of the present without compromising the ability of future generations to meet their own needs". While the concept was clear, it has attracted many different kinds of interpretation for application in practice (Pearce, et alia 1989 App 1).

But this did not stop its adoption by the British Government in 1991 in its first comprehensive White Paper on the environment (DOE et alia 1990), which spread rapidly to the land use planning system namely:

> "The planning system, and the preparation of development plans in particular, can contribute to the objectives of ensuring that development and growth are sustainable. The sum total of decisions in the planning field, as elsewhere, should not deny future generations the best of today's environment."

This was supplemented, following the Rio Conference on the Environment in 1992, by action by a "people central approach" at the local level, in what has been called Local Agenda 21 United Nations Association Sustainable Development Unit, et alia (undated). In this, all local authorities are encouraged to initiate community based programmes for improving the environment, resulting in considerable grass roots activity which has become linked with the local planning process.

In practice, the impact on planning has been marked. In effect the need for "sustainability" has transformed planning in practice, so much so that the term has, in certain quarters, become virtually synonymous with "good planning practice". (Blowers, 1993) Thus it might be doubted whether the practice is precisely following the Brundtland concept, and the definitions of interpretation that have arisen, but there can be no doubt about the repercussions. Since the concept raises the twin major drives relating to development and the environment, which visualises the furtherance of welfare while protecting the environment, there has arisen considerable scope for the introduction for some kind of evaluation to the process.

(v)   The New Public Management

In the Conservative drive towards enlarging the private sector in Government, there has been increasing reliance on management suited to the privatisation, associated regulation and the operations of the market. This in the main has drawn on the conventional Business School, which is oriented to the private sector. In the process public sector management was neglected. This imbalance is in the process of adjustment with what has come to be called the "new public management" (Pollitt, 1990; Dunleavy and Hood, 1994). Associated with the theory and principles there has been an extension in practice, notably in the OECD, where there has been instituted a service for public management (e.g. OECD-PUMA, 1995), and also in the United States, which has resulted in a National Performance Review initiated at Federal level (Government Performance and Results Act (GPRA) 1993). In brief these programmes envisage the introduction of systemic planning at a national level, which involves a strategic plan, performance plan and performance management via programme indicators.

There is evidence that The New Labour Government is making major attempts in this direction. Soon after its election in July 1997, the British Treasury embarked upon a cross departmental public spending review which departed from the traditional financial budgeting process, and sought to review departmental programmes as a basis for financial allocation. Perhaps this is the beginning of moves towards the more comprehensive OECD and United States approaches. Whether this can be so or not, clearly some increase in planning and evaluation, geared towards "new public management" is likely.

4.   From Societal Change

The practice of urban and regional planning is highly sensitive to the kind of society in which it takes place. At the broadest level, countries vary in their political and constitutional philosophy as to the kind of government intervention that is acceptable; a clear contrast is between the mixed and command economies, as evidenced in the different kinds of legislative measures that are required in order to set up the parameters for government intervention in the use and development of land. At the narrower level, in some less fathomable way, changes in public mood, media pressure, lobbying with political parties, for example, will find their way into everyday practice and policy within the current legislation. Pointers in this direction are presented in the contribution to this collection by Millichap who argues for a definition of "community rights" which would be significant in the practice of evaluation. Some current pointers in Britain to societal change of this kind are as follows:

(i)   Communitariasm

Inspired by Amitai Etzioni (1993) this contemporary movement spread from the United States to the British Labour Party prior to its forming the "New Labour" Government in 1997. As a concept it is complex but in essence simple: that individuals in their behaviour should, by choice, have greater regard for their role and place in the community, in terms of their duties to society as well as their individual rights. The resulting "community contract" would be reflected in the evaluation process.

(ii)   Charter 88

From quite another direction there is the call for constitutional change which will again have its impact on the rights and duties of individuals. This flourishes under the name of Charter 88, which describes a powerfully organised movement towards a written constitution, to augment Britain's unwritten constitution, to be embodied in a new (since 1688) Bill of Rights. Steps have already been taken to incorporate in British law the European Courts Convention on Human Rights, which incidentally would avoid the current necessity for individuals and organisations in Britain to take their cases to the European Courts, where there is a conflict between the European and British legal interpretations.

(iii)   Citizen and Consumers Charter

A feature of the early nineties has been the proliferation of government originated charters aimed at the protection of people, both as citizens and also consumers (DOE, 1991). They have multiplied for many activities, as for example customer rights on railway trains, patients rights for hospital treatment, with penalties for the offenders. They thus become part of the code of conduct of society. As such they can be seen as statements of minimal rights for the public at large in various capacities.

Clearly the Charters could have a role in prescribing rights and obligations in relation to planning. They could be seen as providing some kind of measure on the degree to which private rights can be protected or not under the planning system. As such they would clearly enter into the evaluation process.

(iv)   Relational Society

While based on traditional sociological theories (Durkheim 1893) this contemporary expression is of comparatively recent origin. The essence of the message is that, in the ultimate, people's welfare in its many different dimensions is very much tied up in personal relations between individuals in all branches of socio-economic and political life (Baker, ed. 1996). Such relationships are influenced, although not determined by, the physical and socio-economic environment. While these are in general hardly within the orbit of spatial planning, since they depend upon a much finer grain, there is nonetheless some scope for bearing them in mind in planning (Lichfield, N. and D. Lichfield 1996).

(v)   A Stakeholding Society

While again not a new concept, it has attracted a contemporary upsurge of interest owing to its espousal by the Labour Party as part of its policies (Hutton, chapter 7). New Labour considers that society is essentially made up of a variety of interests which have a "stake" in it. This is resisted by upholders of the market philosophy (Hutton, et al 1997)

The perception is that a just society should give the stakeholders the right to be taken into account in relation to any private or government led activity, including urban planning. As such the concept is seen as a special kind of distribution of power. This will involve the reorganisation of societal organisations with a view to the adaptation of institutions which would enable the concerns of the stakeholders to be reflected in ongoing activities and decisions.

The concept has a more direct relationship with planning and evaluation than some of the others. This stems from the fact that over the years the role of stakeholders has become identified in planning. For example, Lichfield (1966b) showed how the definition of community sectors, as production and consumption stakeholders in a proposed mass transit system, was used in a Community Impact Evaluation and so could be employed to galvanise general support for the system.

(vi)   The Audit Explosion

Historically the term "audit" has been applied to the practice of financial review of the annual accounts of public or private sector companies, in order to check financial probaty, accuracy of taxation liability. But more recently the approach has gone beyond this in various dimensions. For example a company would supplement its financial audit by progress in social, safety, environmental and other matters pertaining to their workforce and the local community (Medawar 1978).

This approach has been extended at the national level through the National Audit Office and, for local authorities, the Audit Commission, which looks at particular institutions with a view to comment on their practices with an eye to the public interest. So widespread is this approach that the activity has been described as the Audit Explosion (Power, 1994). This of itself is a stimulus to a greater role for evaluation; for while the normendature is different, the wider audit can be seen as "systematic attempts to incorporate into project appraisal all major dimensions of both economic and social impact, extending beyond the usual concern with internal commercial project viability to look at community costs, benefits and opportunities (Haughton, 1988).

(vii)   Standards in Public Life

From yet quite another source the conduct of local government in statutory planning has been brought under the public gaze. The origins were the setting up in 1994 of the Committee on Standards in Public Life, in consequence of the many instances of low standards by members of Parliament in the House of Commons, in such areas as bribery, corruption, departure from "family values" in individual and social life. This focus has been extended to a wider spectrum of public life, including not only Parliament but also central government (Minister and civil servants), various types of appointed bodies at national and local level, and local government. On this, the Committee stated (Committee on Standards in Public Life, 1997).

> "Planning is clearly a subject which excites strong passions and for good reasons. The planning system frequently creates winners and losers and involves the rights of others over one's property; the financial consequence of a decision may be enormous".

The Committee made a number of recommendations on the conduct of planning. While these were of necessity oriented towards the Committee's Terms of Reference on standards, they nonetheless make their contribution to increasing the rigour and accountability of the conduct of local planning, by members and officers. As such the conclusions and recommendations could enter into the practice of evaluation.

5.   SOME CONCLUSIONS

From the preceding it is possible to draw some conclusions on the possible impact of the above trends on future plan evaluation practice in Britain

Many of the trends can readily be seen as a possible useful enrichment to the context or plan evaluation. For example, public choice requires abandoning an assumption for planning intervention, that *any* result is *necessarily* better than the market which has "failed", and so affect the options adopted for choice; urban and regional policy for the deprived urban areas brings into play new kinds of partnerships and stakeholders; Local Agenda 1921 activities bring grass roots to the fore; the Charters helped to clarify the rights of the "planned"; communitarism and relational theories bring the inter-play between individuals into sharper focus; the stakeholding concepts introduce greater recognition of the key players and actors in decision making in society.

But not all the trends have this promise. For example, both communicative ideology and also community planning, action planning etc., in the full expression which has here been given to them, embody possible dangers from the swamping and dilution of the rational evaluation model in planning through a huge shift of emphasis from the core of the planning system (the planners, the authorities, the machinery for dialogue etc.) to the consumers, who are brought into such powerful play that they could obscure the core purposes. To avoid this therefore, some clarification of the linkage between the two is important, whereby the obscuring and swamping does not take place.

A concluding question is thus the implication of these trends for the traditional planning and evaluation models. Some would support it (e.g. the new public management, the audit explosion); some could be seen as a means of attack on them; and some can be recognised as valuable reactions to ongoing theory and practice which should be taken as warning on the nature and use of the rational models. My own money is on the latter. The greater the complexity the greater the need for rationality, albeit not necessarily of the traditional kind. I am reinforced in this by the recognition that spatial planning has proved to be a robust and flexible instrument which is capable of adaptations, as any comparison of changes in them and practice over the last 50 years can show; the wider planning profession, taken as a whole, have shown themselves able and willing to listen, reflect and adjust, albeit often very slowly. This is because "the planner" is not, in this field of activity, the isolated bureaucrat he is frequently depicted, often as a straw man to be knocked over. He is best seen as an integrating and bargaining player in the total process, with the backing of the multi-disciplinary planning team which brings relevant issues to bear (economics, environment etc.); the stakeholders who are consulted; the different government levels who are entrusted with the democratic decisions, the implementation agencies who are likely to be involved; the impacted public in the relevant community which must participate. This total process is part of evolving society. It is a facet of how the world is painfully learning to organise itself.

It is that process which needs to protected, enriched, improved and strengthened by the intelligent use of evaluation.

References

Baker, N. (ed) (1996) *Building a Relational Society: New Priorities for Public Policy.* Ashgate, Aldershot.

Blowers, A. (ed) (1993) *Planning for a Sustainable Environment. A Report by the Town and Country Planning Association*, Earthscan, London.

Buchanan, J. M. and Tollison, R. D. (eds) (1972) *Theory on Public Choice: Political Applications of Economics,* The University of Michigan Press, Ann Arbour.

Buchanan, J. M. *et al.* (1978) *The Economics of Politics,* Institute of Economic Affairs, London.

Committee on Standards in Public Life (Nolan Committee) (1997) Third Report *Standards of Conduct in Local Government,* HMSO, London.

Cullingworth, J. B. and Nadin, V. (1994) *Town and Country Planning in Britain* 11th Edition, Routledge, London.

DOE (1991) *The Citizens Charter* HMSO, London.

DOE (1993).

DOE (1994a) *Assessing the Impact of Urban Policy,* HMSO, London.

DOE (1994b) *Bidding Guidance: A Guide to Funding Under the Single Regeneration Budget,* HMSO, London.

DOE (1995) *Single Regeneration Budget Challenge Fund. Guidance Note 2: Monitoring and Periodic Review,* HMSO, London.

Dunleavy, P. and Hood, C. (1994) *From Old Administration to New Public Management.*, Public Money and Management **14**, pp. 9-16.

Durkheim, E. (1893) *The Division of Labour in Society.* Translated by G. Simpson (1965), The Free Press, Glencoe.

Etzioni, A. (1993) *The Spirit of Community: The Re-invention of American Society,* Simon and Schuster, London.

European Commission (1994) *Guide to the Community Initiatives 1994-9.* Brussels: The Commission.

Hambleton, R. and Thomas, H. (1995) *Urban Policy Evaluation, Challenge and Change*, London, Chapman.

Haughton, G. (1988) Impact Analysis - The Social Audit Approach., *Project Appraisal* **3**, pp. 21-5.

Healey, P. (1996) The Communicative Turn in Planning Theory and its Implication for Spatial Strategy Foundation. *Environment and Planning B: Planning and Design* **23**, pp. 217-234.

Hutton, W. (1997) *The State We're In.* Jonathan Cape, London.

Hutton, W. *et al* (1997) *Stakeholding and its Critics,* Institute of Economic Affairs, London.

Lichfield N (1990) Land Policy: finding the Right Balance in Government Intervention: An Overview. *Urban Law and Policy.* **3**, pp. 193-203.

Lichfield, N. (1992) The Integration of Environmental Assessment with Development Planning, Part I: Some Principles, *Project Appraisal* **7**, pp. 58-66.

Lichfield, N. (1996a) Community Impact Evaluation - UCL Press, London.

Lichfield, N. (1996b) Social Benefits and Costs to the Community in *Proceedings of Tel Aviv Metropolitan Administration for Mass Transit* (unpublished).

Lichfield, N. (1997) Integrating Environmental Assessment with Development Planning, in Borri, D. *et al* (eds) (1997) *Evaluation Urban-Rural Interplay in Regional Planning,* Kluwer Academic Publishers, Dordrecht.

Lichfield, N. and Lichfield, D. (1992) The Integration of Environmental Assessment and Development Planning, Part 2: Prospect Park, Hillingdon. *Project Appraisal* **7**, pp. 178-85.

Lichfield, N. and Lichfield, D. (1995) Environmental Assessment and the Planning Decision, *Proceedings of the Conference on Environmental Assessment of Airports,* Airfields Environment Trust, London.

Lichfield, N. and Lichfield, D. (1996) Urban Relationships: A Challenge in Town Planning, in Baker, N. (ed) (1996) *op cit,* pp. 217-230.

Lichfield, N., P. Kettle and M Whitbread (1975) *Evaluation in the Planning Process,* Pergamon, Oxford.

Medawar, C. (1978) *The Social Audit Consumer Handbook: A Guide to the Social Responsibilities of Business to the Consumer,* MacMillan, London.

Morphet, J. (1997) Enter the ESDP - Plan Sans Fanfare *Town and Country Planning* **66**, pp. 265-267.

Oecd-Puma (1995) *Governance in Transition: Public Management Reforms in OECD Countries,* OECD, Paris.

Pearce, D. *et al* (1989) *Blueprint for Survival,* Earthscan Press, London, Appendix 1.

Pollitt, C. (1990) *Managerialism and the Public Services. Cuts or Cultural Change in the 1990's,* Blackwell, Oxford.

Power, M. (1994) The Audit Explosion, DEMOS, London.

Private Finance Panel (1995) *Private Opportunity, Public Benefit,* HM Treasury, London.

Secretary of State for the Environment, *et al* (1990) *This Common Inheritance: Britain's Environmental Strategy,* HMSO, London.

Skeffington Committee (1969) Report of the Committee on Public Participation in Planning, HMSO, London.

United Nations Association Sustainable Development Unit *et al* (1996) *Towards Local Sustainability,* The Association, London.

World Commission on Environment and Development (Brundtland) (1987) *Our Common Future* Oxford University Press, Oxford.

PART I

Emerging Issues For
Evaluation Theory

EMERGING ISSUES FOR EVALUATION THEORY
Introduction

A. KHAKEE

The papers in this section reflect two sets of emerging issues in evaluations theory. The papers by Fusco Girard, Söderbaum and Barbanente and her colleagues discuss these issues with the help of substantive issues like cultural and environmental heritage, ecological sustainability and infrastructure development. The papers by Khakee, Voogd and Hull have planning or policy-making approach in their analysis of evaluation theory.

Fusco Girard's paper discusses the procedural aspects of good governance and democratic control in dealing with substantive issues of the conservation of cultural and natural heritage. The only way to take into account all the values that culture and nature represent is through a dialogical and participatory procedure. However, Fusco Girard feels that evaluation matrix methods which account for values can facilitate dialogue between stakcholders. Söderbaum rejects evaluations based on narrow monetary or welfare theoretic perspective. His 'positional' analysis as a interdisciplinary approach to evaluate programme with an environmental dimension emphasizes the integration between valuation, politics and ideology and requires an interplay of all stakeholders. Barbanente and her colleagues also base their analysis on 'value conflicts' which are fundamental in decision making when environmental issues are involved. Their suggestion is to discard instrumental rationality with its quantitative and utilitarian grounds and instead espouse a radically new and wider perspective. Their communicative and dialogical approach is very much in line with the discursive approach based on the critical theory.

Khakee's paper describes the new developments in planning theory towards communicative action and interactive practice and in evaluation theory towards responsive constructivist approach in order to develop and apply a methodology to evaluate planning process. The key point in his approach is that it focuses on qualities of process understood in terms of efficiency and legitimacy rather than just product. Do the new communicative approaches advocated in planning and evaluation imply a fundamental break with the rational paradigm? In order to reply to this question Voogd develops a typology of planning arenas in order to discuss the usefulness of various evaluation methods for consensus building. His conclusion is that with the exception of a few specific evaluation issues, conventional evaluation methods shall have decreasing importance in spatial planning. They shall be increasingly replaced by methods focusing

N. Lichfield et al. (eds.), Evaluation in Planning, 19–23.

on discourse and participation of stakeholders. Hull's paper draws on the political economy approach and social constructivist ontology in order to evaluate the way decisions about the allocation of land for housing are made in England. Her analysis shows the lack of coherence in the British planning system with its narrow and hierarchical sectoral approach which prohibits the participation of new actors who could bring new perspectives and values in the policy process.

A set of central issues for evaluation theory has to do with the impact of critical approach. According to this approach the analysis of a plan or a programme is essentially a political process full of values rather than some kind of scientific pursuit for truth or a rational inquiry. The critical approach has been influenced by Berger and Luckman (1966) who developed a theory that reality is not 'objective', it is socially constructed and by Foucault and Habermas who emphasize the need to analyse politics and policy as modes of discourse which structure reality (Edelman, 1988). These theories have been responsible for increasing misgivings about the dominant - positivist, rational, quantitative - paradigm in social science as a whole. For evaluation theory the whole ethos of performance measurement as a rational approach to improving the performance of the public sector has been questioned (Kelly, 1987). Does this development amount to a shift in the paradigm in planning and policy analysis?

Innes (1995) asserts that planning or policy making theory has moved from the rational planning paradigm to the communicative planning paradigm. Her contention is supported by Lindblom (1990) who talks about a shift in social theory from that of a 'scientifically guided society' to that of a 'probing society'. Evaluation in the rational planning position involved various techniques to measure the relationship of costs to benefits. The objective was to identify an optimal plan or a programme. Even if it was generally recognised that problems were poorly defined, that many goals were formulated only qualitatively, that the relationship between goals and means were poor on account of value uncertainty and scarcity of knowledge and that goal formulation involved a great deal of politics, advocates of the rational approach nevertheless contended that evaluation should try to emulate as far as possible optimization procedure. The focus of interest was in developing different techniques which would satisfy performance requirements which were 'objective' and 'value-free' (Miller, 1984). Thus we find interest focused in developing three categories of techniques: monetary, overview and multicriteria techniques (Faludi and Voogd, 1985). Lichfield (1996) distinguishes 25 different techniques extending from the simple check lists to more comprehensive input-output analysis which provides information on effects, processes and structures in order to bring about a holistic judgement. All these techniques have been dominated by the rational performance measurement ethos.

The communicative paradigm has resulted in an identity crisis for the field of evaluation. Rather than a single rational orientation several alternative approaches to evaluation have sprung up. Kelley (1987) describes three approaches to illustrate the current development: The multiplist approach (Cook, 1985) according to which there is no way of proving what is correct; the evaluation process should involve a pluralistic, multi disciplinary and open exchange of knowledge. The design approach (Bobrow and Dryzek, 1987) favours different ways of looking at problems from the perspective of different frameworks of values and methodology. Frameworks are tools for discussion and critical dialogue rather than techniques to provide answers, facts, costs or benefits. In the responsive constructivist approach (Guba and Lincoln, 1989) evaluation is a form of negotiation involving all stakeholders who are in some way affected by the evaluation of a specific policy. The negotiations result in a set of claims, concerns and issues. There may exist a consensus with regards to certain matters while the others for which there is no such consensus further negotiations are required. Evaluation in this respect resembles an arena to spread experience and perspective from one stakeholder group to another.

This shift in evaluation theory has been taking place at the very time when the ideas derived from managerial evaluation and the urge to quantify have come into vogue in public activities. This has to do with changes which have taken place in the working methods of the public sector. Control with the help of norms and statutory requirements has been gradually replaced by framework legislation. The state and local governance look for new ways of cooperating with private firms, voluntary organizations and other actors in the private sector. The boundary between the public and the private sector has become increasingly blurred. This has led to contradictory tendencies in the field of evaluation - between theory and practice. According to Henkel (1991) this development has led to a tension between evaluation theory which has moved away from positivism and evaluation in practice which has sought to be highly positivist in believing that things can be and should be measured.

Environmental and cultural problems provide an important source of challenge to evaluation theory. These issues can not be approached or resolved by purely quantitative techniques. They involve wider strategic decisions, raise conflicts of values and judgement and there exists a wide range scientific and experiential opinions about these problems (Vogel, 1986; UNESCO, 1995). One response to approaching these problems has been with the help of 'impact assessment'. The history of impact assessment goes back to the 1960s when social indicators came into use as a result of the widespread belief in the capacity of governments to resolve social problems. These expectations were ill-founded and by the 1980s the use of these indicators was largely abandoned (Miles, 1985).

In the 1960s it became compulsory in the USA to use environmental impact assessment (ETA). As a result of the European Union directive of 1985 an EIA is required for all major programmes in the member states. The compliance of this directive varies: EIA is required by law in Germany, the Netherlands and Scandinavian member states, whereas other member states have interpreted the directive more flexibly. There is a wide variety of approaches and methodologies for EIA. They involve the use of both quantitative and qualitative techniques.

One of the central issues involved in the case of the preservation of natural and cultural resources is that these have an indirect value besides their use value. The indirect value include a 'future use value' which means that a person is willing to pay for the option of using these assets in future and an 'existence value' which is expressed by a person's willingness to pay despite the fact that the person in question would not use the resources herself. Evaluation methods based on willingness to pay for the resources have a lot of methodological shortcomings. Environmental and cultural evaluation require the assessment of a large number of non-material and potential benefits. The latter implies that the evaluator's conception of ecology and/or culture has to be extended beyond the range of observable phenomena (Jacobs, 1991). In the case ecological evaluation the concept ecology has to be extended beyond the question of protection versus exploitation to include aspects like the balance in the eco-system, liveable world and sustainable relationship between economic, social and biospheric systems. In the case of culture evaluation has to be extended to include values attributed in the other definitions of culture: 'communication', 'social glue', 'heritage', etc. (Khakee, 1997). Ultimately this does not only mean resolving the question of defining 'ecology' and 'culture' but also establishing the relationship between their various dimensions. The 'trade off' between these dimensions require negotiation, mediation and bargaining (Lichfield, 1996). Impact assessment methodology provides a partial help if distinctions between fact, value and value judgement are kept in mind and the involvement of various stakeholders ensured. Effectivity requirement has to be supplemented by other equally important requirements: justice, legitimacy, mutual understanding, integration of scientific and experiential knowledge and democratic pluralism. It becomes a question of tackling ecological and cultural problems in an open discursive manner. Evaluation becomes a part of the 'transformative learning process'. It is accorded an emancipatory role - to change the society and not just an environmental or a cultural programme.

References

Berger, P.L. and Luckman, T. (1975) *The Social Construction of Reality. A Treatise on the Sociology of Knowledge*, Penguin, Harmondsworth.

Bobrow D.B. and Dryzck, J.S. (1987) *Policy Analysis by Design*, University of Pittsburgh Press, Pittsburgh.

Cook, T.D. (1985) Postpositivist critical multiplism, in Shotland, R.L. and Mark, M.M. (eds.) *Social Sciences and Social Policy*, Sage, Newbury Park.

Edelman, M. (1988) *Constructing the Political Spectacle*, Chicago University Press, Chicago.

Faludi, A. and Voogd, H. (eds) (1985) *Evaluation of Complex Policy Problems.* Delftsche Uitgers Maatschappij, Delft.

Henkel, M. (1991) *Government, Evaluation and Change*, Jessica Kingsley, London.

Innes, J. (1995) Planning theory's emerging paradigm: Communicative action and interactive practice, *Journal of Planning Education and Research* **14**, pp. 183-190

Jacobs, M. (1991) *The Green Economy*, Pluto Press, London.

Kelley, R.M. (1987) The politics of meaning and policy inquiry, Palumbo, D.J. (ed.) (1987) *The Politics of Program Evaluation*, Newbury Park, Sage.

Khakee, A. (1987) *Meeting the challenge of the collaborative state. Potential changes in cultural economic research*, Paper for the Conference for Researchers in Culture, Cultural Research and Cultural Policy, The Royal Swedish Academy of Letters, History and Antiquities, Stockholm.

Lichfield, N. (1996) *Community Impact Evaluation,* UCL Press, London.

Lindblom, C. (1990) *Inquiry and Change: The Troubled Attempt to Understand and Shape Society,* Yale University Press, New Haven.

Miles, I. (1985) *Social Indicators for Human Development,* United Nations University/Frances Pinter, London.

Miller, T.C. (ed) (1984) *Public Sector Performance: A Conceptual Turning Point,* John Hopkins Press, Baltimore.

UNESCO (1995) *Our Creative Diversity. A Report from the World Commission on Culture and Development,* UNESCO, Paris.

Vogel, D. (1986) *National Styles of Regulation: Environmental Policy in Great Britain and the United States,* Cornell University Press, Ithaca.

CONSERVATION OF CULTURAL AND NATURAL HERITAGE

Evaluations for good governance and democratic control

L. FUSCO GIRARD

1. Post-Modern Age, Individualistic Values and Collective Choices

The cultural dimension of sustainable development is often neglected. In the Summit of Rio the ecological dimension of sustainable development has been emphasised. Last year the World Summit in Copenhagen has analysed the social dimension of sustainability (ONU, 1995).

Cultural dimension represents a critical question in pursuing sustainablity. Culture is considered as the whole of meanings, symbols, values, ideas, organisational rules of a society, which are reflected in the way the society itself shapes institutions, uses environment and nature, regulates human relationships.

Nowadays various co-existent cultures in our pluralistic and fragmented society constitute an extremely rich mosaic: e.g. local cultures and the homologating ones, traditional cultures and cultures characterised by new paradigms, scientific and humanistic cultures, the technological and sapiential ones, business and ecological cultures, the homologating and identity-searching ones, etc.

If this mosaic and this fragmentation will grow, we wonder: will it be the source of mutual exchange, of creative innovation or of exasperation of differences? Will it be the source of development or of systemic involution, and then the source of collapse?

Of course, the answer depends first of all on the capability of these different cultures to communicate among them, to make one another fruitful, to recognise inside common elements, and then to co-evolve.

Therefore, the kind of answer depends on the level of communication/dialogue that it is possible to activate among them.

In order to do so, these cultures have to identify reasons which are beyond their specific cultural identity.

The Istanbul Summit (HABITAT II, ONU 1996) is just an example of the possibility to find some common elements of agreement in considering those human rights which are not satisfied in our cities. The human rights to health, to environment, to employment, to housing, etc. are common shared values and their realisation is a common interest element in open communication and dialogue.

N. Lichfield et al. (eds.), Evaluation in Planning, 25–50.

Some values which involve all man's welfare depend on the conservation of cultural and natural capital. They are strong enough to build a dialogue around historic/architectural/environmental values, which is respectful of differences but is also able to build cooperative consensus and to reduce conflicts between cultures.

In this post-modern age which puts away the issue of objective and universal values and turns them into mere expression of individual preferences, the conservation of cultural and environmental assets gives the opportunity to recognise that not only subjective/private values but also collective values exist: not only economic values but also the extra-economic ones, which are independent of utility, have a value: i.e., values in themselves or intrinsic values.

In this paper I would underline that the conservation of natural and manmade-cultural capital may be considered as a cultural challenge to post-modern age and to its fragmentation. It is the opportunity to start in reconstructing sense and exceeding consumer's culture which impoverishes everything by dissolving values which go over the daily horizon.

This culture does not humanise our society, does not reduce fragmentation and conflicts. It cannot build the "new" city; it cannot transform the reality into a more desirable direction.

In this perspective evaluation methods are an useful tool. They concern public institutions, in order to make them work in a more unbiased and effective way, and different stakeholders (in particular the third sector) too, in order to share the construction of collective choices with them, and especially each subject, which should succeed in exceeding his personal utility in respect of his daily choices.

In order to humanise our city, we have to promote a culture which can transform the consumer into a citizen. Our future depends more and more on the quality of our collective decision processes. Those choices depends on culture.

2.   A Post-Economic Culture for the Humanisation of the City

2.1.   THE TYRANNY OF INDIVIDUALISM

At the end of XX century and at the beginning of XXI century we are near a revolution due to re-engineering processes with the always more massive introduction of technological innovations in the production of goods and services, with several social and human impacts, and a potential higher expulsion of labour force (Rifkin, 1995). Unemployment, exclusion, marginalisation, poverty are increasing (ONU, Copenhagen, 1995). The benefits of economic growth favour those who are already included in economic processes, whereas those who were on the borders or out are more and more excluded.

This exclusion and poverty are concentrated in urban/metropolitan areas. This puts new relevant ethical questions added to which comes from the consumption of natural capital due to existing productive and consuming patterns.

The relevance of ethical question (due to social and ecological crisis) is strongly emphasized by current culture.

Current culture is mostly influenced by economics, for which the individual is considered only as a consumer of goods and services.

The view proposed by economics reflects an atomistic idea of reality and of society: the latter is an aggregation of individuals without links, in which each one pursues its specific interest. It erodes or weakens the community sense and deteriorates the environment (Daly and Cobb, 1990).

This culture is characterised by the radical individualism: it does not matter what happens "outside" of the individual, to others and to nature.

The most recent sociological surveys in Western Europe (Arval, 1994; Ester et al., 1994) outline that the individual is more and more characterised by *homo economicus* behaviours, i.e., by behaviours in which individualist values prevail. He is less and less a "citizen", i.e. he is not able to go beyond his own interests in a frame of more general interests.

This culture which considers man as an always more consumer (and always less citizen) subject; for which the bonds of belonging and of community are getting weaker, looks at the market institution as the fundamental tool to connect each person with others.

This kind of culture is the most antithetical for the promotion of a sustainable development. It erodes the trust and fairness between people, the idea of common good and the ethical capital of society, that are necessary for the market to work. It promotes conflicts between groups and individuals.

Obviously conflicts are a "vitality indicator" of a society only if they do not exceed a threshold, beyond which they destabilise the social system.

Reciprocal trust is an essential "ingredient" for economic growth like the conservation of the other forms of capital.

Individualism expresses the acknowledgement of the supremacy of human subject, who becomes the centre of life and society idea. The individual is not obliged to accept the traditional values, being free in evaluating them. These values, in fact, are not "given", but they are evaluated and chosen by considering what he perceives as a good. In our "post-modern time" only individual values exist.

A perspective limited only to its own interest, to personal and particular welfare ("radical individualism") reinforces the *homo economicus* behaviours and leads to more and more social atomisation, fragmentation and disintegration, and to ecological crisis too.

The "tyranny of individualism" (Norgaard, 1994) is an effect, and in its turn a cause of the ecological and social crisis, because it does not catch the reciprocity relationships among individuals and between men and nature. Each element is related to the others within a circular and not linear circuit, which is not only the one of the trading exchange, but it is the reciprocity circuit, as the complex ecosystems functioning shows. Therefore, each part should co-evolve with the others, within a reciprocal change.

A culture oriented by the priority of individualistic values and by the lack of solidarity causes life styles which are in a structural conflict with two fundamental goals of our time: ecological integrity and social justice. The trend towards sustainable development cannot be awarded to sovereignty of the consumer, i.e. to the free market or to "minimal" public institutions.

We need to recognise that reciprocity is a potential regulating rule and to foresee institutions to be able to reflect this widened exchange, richer than those suggested by the market.

Our future depends on the new institutions through which we will be able to reduce ecological crisis and social inequality.

Institutions are the rules for human, social, economic and political exchange (North, 1993). They reflect and produce culture.

We need new institutions which are able to improve the construction and the implementation of public interest, favouring the cooperation between people and improving collective decision processes.

Sustainability does not only concern public institutions, but all subjects. It begins to develop itself in everybody's everyday life, developing the relational values, the sense of identity, due to acknowledgement of rules and of common shared values.

The cultural dimension of sustainability recognises, indeed, that development crisis depends on our mind, values, ideas, culture. It is, first of all, an "inner" man crisis.

It is not possible to entrust the evolution of our society only to economic and technology reasons.

The risks of existing situation seem enormous according to human rights and de-humanisation of society. Re-civilisation process may seem definitely compromised if our society becomes more and more consolidated as the "society of consumers", in which behaviours coincide always more with the *homo economicus* ones.

2.2.   CURRENT CULTURE AS ECONOMIC CULTURE

Today this economic culture offers winning and definitive criteria of private and public choices, because:

- It becomes perfectly integrated with miniaturised technological processes which improve use by each person and new telecommunications technologies, which through computer decentralisation allow an exchange and a global communication (or at least so considered) of each one with the others;
- It exploits the concreteness, the capability to adapt each one to different circumstances from time to time.

- This culture, although it has certainly contributed to the vitality of economic system, to the capability of producing more and more man-made capital, of satisfying wants, of creating employment, of increasing spaces of freedom and from some points of view of democracy too, presents some limits which have to be better outlined. This culture, in fact:

- Devaluates future and implies lack of future. It cancels the possibility of building up any view or perspective. It is the culture of "here and now", which refuses to admit any utopia (Passet, 1979).

- Breaks up the social links between past, present and future generation.

- Breaks up the society, by promoting the more and more extremist individualism, which sociological analyses constantly localise as a structural process (Tchernia, 1995). It is the culture of the "I" without the "We" (Etzioni,1988; Caselli, 1996), i.e. without the community, which reflects herself in the obsolete and "deteriorated" spaces of "We" (squares, streets, public gardens, etc.). It is indeed the culture which reduces the binds of community and the social capital (Daly and Cobb, 1990). It is not involved in reasoning on aims or values. A connection among nihilism, ecological and social crisis exists and it is represented by the linkage between nihilism and economic culture.

- Does not consider the systemic interdependencies of real world, simplifying the complexity: it is a culture as the one offered by the mechanics, i.e. by inanimate sciences, rather than by science of life (Passet, 1979). The notion of balance, of reversibility and the same rationality are in conflict with the ecological and biological view (Tiezzi, 1995).

- Separates people from their environment and community while, instead, each individual grows, evolves within a community that she/he modifies and by which is modified, and that on its turn modifies external environment which is not "given", but is built (Norgaard, 1988).

- Separates each individual in different dimensions, as if each person were not in the same time an economic man, a social man, and an ecological man. It expands to excess the area of trading exchange, which is based on a costs/revenues ratio which becomes the regulating principle of relationships among subjects, by parallely reducing the area of reciprocity and gratuity. It does not respect the integrity of man and of hearth, because it gets the logic and the rationality corresponded on economic calculus, as if only rational choices were from the utility perspective the more convenient ones.

- This culture has eroded the social capital of society.

- Does not civilise society, being linked with having and not being. It does not create social glue; it does not reduce differences and unbalances/conflicts.

- Erodes the natural capital. The consequences of this economic rationality are the more and more negative irrational impacts on social and ecological environment.

2.3.   TOWARDS A POST-ECONOMIC CULTURE FOR THE HUMANISATION OF THE CITY

It seems necessary for the humanisation of city to promote a new culture, which transcends the centrality of economic culture: a post-economic, post-individualistic and post-mechanistic culture.

In a reality in which "facts are less and less certain, playing values are more and more numerous and in competition, stakes become higher and time for choice becomes more and more reduced" (Funtowicz and Ravetz, 1991, p. 210), economic culture should be enriched. It is not able to re-produce community, social cohesion, sense of belonging, acknowledgement of common identity, of mutual interdependence, of sharing the same fundamental organisational rules, i.e. the social capital.

We need to foster cooperative, civil and community values, aiming to reduce the gap between personal and collective choices and, then, the level of conflict among systemic components and to re-produce a "joining force" within the pluralist society (Viederman, 1996).

We need to add the rationality in choosing goals to the instrumental rationality (which is able to choose the means to pursue "given" goals), through communication, dialogue, public comparison of good reasons (Habermas, 1986 ; Sen, 1995).

The real vitality of a social system, i.e. his self-sustainability, depends on the wealth of its values, i.e. on its culture.

The future of our natural and built environment depends on the capability to understand limits of the existing *homo economicus* culture, and to contribute in pursuing a system of shared values which are not only focused on competition, efficiency, domination and expansion, but also on co-operation, reciprocity, responsibility.

It is impossible to identify any perspective of development and of democracy if society keeps on taking only the centrality of exchange and use values in her regulating mechanisms, by neglecting relational values: that is, if it is not recognised the importance of social bonds besides that of consumer goods.

The decline of a society depends on her incapability of improving her knowledge and her capability of replying to sense questions. It is possible to fight the decline by recurring to cultural innovation: by rediscovering the reasons of common living, of coordination of everyone with others in a "collective action", the sense of community, extended also to the ecological system, the awareness of general interests, the feeling of common belonging.

The perspective of sustainable human development introduces new attention to solidarity between man and man, between our and future generation, between man and natural system ; to the ethical dimension.

Ethics is reasoned remark on goals which are of value, because they express a desirable reality, a society view.

Facing with the ethical question of sustainable urban development means asking how it is possible to pursue in physical space general interest, when particular interest becomes more and more able to prevail; how it is possible to go ahead with social construction of the city, when fundamental ingredients, which are constituted by shared relational values, are more and more fainting.

It means asking how it is possible to avoid city becomes the place of conflict of each one against everyone, where only the strongest people win.

Ethical dimension confers a particular quality to act, by ascribing a meaning to action which is linked not only with a subjective utility but also with social utility, being aware of systemic interdependencies which link each part with the whole.

It occurs a cultural revolution, namely a revolution in the assigned priority of different values which orientate choices and behaviours and which not consider economic values as already winning criteria in public and private choices.

This new culture should enlarge time horizons; it should consider each person's role in the social and ecological systems: i.e. the rights of each person in his community and ecological dimension. It should determine a new engagement in recognising, assessing and identifying hierarchies among values; they are not on the same level. There are intrinsic values and instrumental values.

3.   The Post-Modern City, Collective Choices and Urban "Good Governance"

3.1.   THE POST-MODERN CITY AND URBAN SUSTAINABILITY

Post-modern City is the site of pluralities of values, but in practice is the site in which economic culture shapes spaces and relations, identifies bonds and divisions, and becames style of life. Here the results of economic-technological culture are more and more evident.

Post-modern cities are the sites where wealth but also poverty (unemployment, exclusion, marginalisation, social conflicts) are concentrated. The city expresses different values and interests which characterise pluralistic asset of our society. They are the expression of the homogenisation (with the commercial centres which are the same in Zurich, London or Paris, where winning values are those of private consumption). They are simultaneously the place where the "dualities" of post-modern age are more and more evident: i.e., the site of economic integration and of socio-cultural dis-integration. They are the site where the sense of belonging, the feeling of collective interests are more and more lost.

The city expresses in the physical dimension, the relationships between private and public interests, between economic interests and collective values, and the way the conflict among them has - or has not- found its solution (Benevolo, 1993).

The fundamental problem of post-modern cities, with their diversities and fragmentation, is the reconstruction of public spaces, not only in physical but first of all in cultural terms.

The cities are the places where dialogue/communication between different cultures should be tested and improved. In this dialogue, we need to start by facing relevant questions which involve everybody: the acknowledgement of human rights to health, to work, to environment. These are questions which overcome the belonging to a specific culture , as the Rio (1992), the Copenaghen (1995) and the Istanbul (1996) agreements demonstrate. These rights should be concretely realised in the cities.

In particular, conservation, re-production and maintenance of cultural and natural man-made capital are occasion for a dialogue, a communication and an agreement between each one, for reconstructing an idea of common good, and not only to improve urban beauty or scenographic assets for enhancing consumption levels.

Conservation and reproduction of natural and man-made capital stock are a fundamental occasion for building citizens in a post modern city of consumers.

As the wealthiest societies are those where the human/social capitals and man-made/natural capital rate is 2 to 1 (Seregaldin, 1995), so in our more and more urbanised world, the wealth of a city is made not only by the wealth of its man-made and natural capital, but by its human and social capital too. Human-social capital and post-economic culture are strictly linked: they are produced when a post economic culture is promoted.

The human capital is embodied in the skills and knowledge of individuals (Coleman, 1994). The social capital is the capability to co-ordinate one's own actions with those of other individuals in order to achieve common objectives.

Among these four forms of capital, human and social capital plays a critical role. In fact, it explains up to 80 % the conditions of wealth and welfare of a community in advanced societies.

On the contrary, the same percentage concerns the natural capital in backward societies. The incidence of man-made capital is substantially constant in different countries: it does not exceed approximately the 20 % (World Bank, 1995; Scregaldin, 1996).

Nevertheless, in development strategies all attention is always and wherever concentrated on man-made capital.

Also in conservation strategy of cultural and natural assets the above circumstance is neglected. Everything seems turning out through rehabilitation of the largest possible number of cultural assets, in order to increase the income due to tourist expenditure flow, the contribution to local economy, to employment, etc.

Actually, the real contribution of conservation of cultural and natural capital has to be seen in relation to an other perspective which is neither aesthetic nor economic: the capability that this conservation may contribute to production/reproduction of human and social capital.

This is the real source of wealth of a country and of a city. Therefore we have to face the development and conservation issues by starting from this form of capital.

The idea of sustainable development, interpreted as the capacity to maintain "as many opportunities as we ourselves have had" (Seregaldin, 1996), can be transferred from national society to cities.

Cities, and not Nations, create economic wealth (Jacobs, 1969); i.e. cities are the very cause of economic growth in a Nation. To rehabilitate urban areas means to restore/reproduce this four kinds of capital and, first of all, to modify their combination. Sustainability is realised as far as we are able to secure the conservation of man-made, human, natural and social capital. In this way it is possible to supply the required inputs of a "circular production process". The notion of human and social capital is critical for sustainable development.

It is possible to produce human and social capital also through city planning participation activities.

The high demographic density which characterises cities can reduce the per capita production cost of man-made capital, energy resources consumption (due to reduced mobility), per capita quantity of infrastructures for individuals and enterprises, etc. But, on the other hand, these high concentrations erode the natural capital of cities, through air, water and soil pollution - due to waste products - and building on green areas.

Human capital concentrates in cities, where a growing share of people live in a state of underemployment or without job, in conditions of overcrowding, in housing unsuitable because of its quantity/quality of services, and the lack of public services, etc.

Moreover, we have to underline the erosion of social capital in cities, due to the fast disintegration of the traditional rural society and to the loss of identity just following the urbanisation processes. The city, with its way of life and consumption, with its want between supply and demand of public goods and services, with its exclusions/marginalisations, destroys the solidarity among individuals and between individuals and nature.

There is a need to recognise that the future of people and their well-being, or their ill-being, is linked to the future of cities, i.e. to their wealth or their poverty.

The relationship between wealth and poverty depends on the kind of development of cities, and then on their capability to govern the processes in progress well: on capability of their "good governance".

3.2.   URBAN GOOD GOVERNANCE

"Urban good governance" not only means to make the overall benefits of high densities exceed the economic, ecological, social costs. It means that the areas of poverty under the different forms (pollution, inequality, etc.) are to be minimised and, on the contrary, the area of participation of people to collective choices has to be maximised.

We need strong public institutions, able to achieve general interest (to redistribute wealth, to reduce ecological crisis, to re-orient free market), but at the same time, we need to involve all the stakeholders in democratic game.

Good governance is the balanced constitution of strong public authority and large people participation.

Human and social capitals are a fundamental ingredients of good governance and pursuing of general interest.

Good governance is not the good management of public services and infrastructures. It is the capacity to construct general interest in programming-implementing-managing development processes, without being conditioned by particular interests.

Good governance refers to the way in which public decisions are constructed, which should be transparent and opened to participation and to control by a strong civil society. In the city of "good government" public institutions are efficient/effective and strong enough to coordinate all private and social subjects towards general interests and shared goals through "enabling strategies". They are responsible of the results, of their actions, because they are able of a continuous process of evaluation (ex-ante, on going and ex-post) and self-evaluation.

Good governance requires integration among production, conservation and management of the above four kinds of capital and not only of the man-made (monuments, social infrastructures, etc.) and natural capitals of cities (parks, forests, etc.), in order to guarantee a sustainable human development.

3.3.   COLLECTIVE CHOICES CONSTRUCTION FOR CONSERVATION OF NATURAL AND CULTURAL CAPITAL

The above lead to the conclusion that conservation of man-made/cultural capital and natural capital is not sufficient to realise a "good governance" of cities. There is a need to invest in the *civitas* in order to reproduce human and social capital, that are critical capitals to regenerate our cities.

The preservation of the material expression of the culture of a community - represented by its artistic/cultural heritage - should take into account the interdependencies whit human and social capital.

The conservation/reproduction of man-made and natural capitals must become a base to build a "collective consciousness": a process to build human and social capital through interactive communication which produces new values.

Cultural assets allow to interlace sense issues in a period of crisis of sense, or better in a period in which the only sense is conferred to market exchange. They help in the intentional reconstruction of collective memory for identifying a better future. In the society of vertical communication they can become source of horizontal communication and of promotion of a post-economic and post-mechanistic culture.

Conservation should become an activity able to contribute in producing/reproducing human and social capital.

The problem lies not only in conserving stones, but in acquiring awareness that conservation has to be used in order to construct capability of exceeding particular interest and of pursuing general interest.

The real wealth and welfare depend on this form of immaterial capital, which plays a critical role.

Conservation of cultural and natural capital is able to stimulate a process of participation and to produce civil consciousness, because it obliges people to a continuous confrontation among particular and general interest. It links past with present and future time. In this perspective, it plays a strongly educational role.

The problem lies, then, in valorising not only technical dimension of conservation, but especially the educational/formative dimension, which is linked to participation process which can be promoted.

Considering that the production of human and social capital occurs in the system of social or civil economy, it is necessary to involve as much as possible this system in conservation. The third sector, the social enterprise, environmental associations, etc., should be involved in the construction of collective choices on the most compatible use of cultural assets and in their management, aiming to pursue general interests.

Substantially, conservation of cultural man-made capital should be put in a perspective directly linked to the same source of wealth of a society: the promotion of human and social capital.

The involvement of the civil economy system (third sector, volunteering, social and cooperative enterprise, etc.) strongly helps in this direction. In this system the economic culture is overcome, time horizon is enlarged and truth and reciprocity is produced. Civil economy system should be improved for the humanisation and civilisation of our cities.

Conservation has to be seen as an extraordinary event to form the citizen, to transform subjects as simple consumers into citizens which are able to exceed their interest because they are able to act not only on the base of exchange values, but also of relational values.

Conservation of cultural heritage is not only the occasion of physical improvement of urban space for increasing willingness to consumption, but can be also the beginning of regeneration of urban public space in cultural dimension. It is a critical activity in forming cultural values and building civil society.

In other words, if we recognise that the wealth of a city depends on suitable combination of hardware (physical, man-made, natural capital), software (human capital in terms of professional capabilities), and brainware (human capital in terms of education/formation, critical and civil knowledge), we have to stress that conservation becomes an activity more and more linked to brainware and less to hardware.

Civil economic system should be oriented to produce rehabilited public urban spaces , to improve quality of nature and man-made environment.

4.   The New Institutional Frameworks for a "Good Governance" of Cities

4.1.   THE "ENABLING STRATEGIES" AND URBAN PLANNING

The Summit of Istanbul has been the occasion to discuss/analize the way to redesign the strategies of urban governance in order to improve the cities' working and doing so to enable their sustainable human development (ONU, 1996).

This means actual acknowledgment of the right of access to housing and to the different urban services also on the behalf of the poorest people, the acknowledgement of the right of young people to employment, right of everyone to health and to a good quality environment, etc.

The section 12 of the final Statement is the most interesting one because it redesigns the institutional organizational model in order to realize the sustainable human development of human settlements. It proposes the "enabling strategy", which substitutes the "direct production strategy" of urban goods and services (housing, infrastructures, services, land use, employment, etc.) by public institutions.

First of all the "enabling strategy" implies a systemic approach to urban development and not a sectorial one anymore.

It implies the clear acknowledgement of a third important urban system often undervalued and ignored by public institutions: i.e. the civil economic system (made of social enterprises, cooperative economy, voluntary organizations, non-governmental organization, non-profit organizations, etc.). This strategy enhances the role of citizen groups, community organizations, voluntary associations, and ensures more coherence between their actions.

The "enabling strategy" does not imply the mere recourse to market at all in giving up the direct public investment, but it implies the building of rules which allow the market to run efficiently and efficiously also in those sectors where it does not work well. Moreover it means to rebuild a public activity able to govern with consensus and participation of citizen and to arrange the relationship and the cooperation between public-private and private-social in such a way to "sustain" the different undertakings in a coordinate way, in order to actually improve the existing overall conditions.

The "enabling strategy" underlines the capability of management at local level of an exceptional multiplicity of subjects or movements (environmental groups, eic.) bearers of different requests, objectives and values and the capability to orient them towards goals of common and general interest.

This strategy is possible only at local level in a decentralized institutional asset, i.e. with the implementation of subsidiarity principle. This principle implies that the production of public goods and services, the taxation, the private economic sector rules should be decentralized as much as possible, that is at the nearest level possible to the affected subjects, unless there are specific reasons to assign them to higher levels. (Quadro Curzio, 1996). This new organizational system is able to solve the dilemma between strong public institutions and large democratic participation of people to collective choices.

In this perspective cities and metropolitan areas turned into the starting point for a new development: not only the place where a growing share of population lives, where the potentiality (but also the problems) of our society is concentrated, but also the points from which a new, fairer and environmentally respectful economic growth can start, and where an actual dialogue/communication and democracy can be rebuilt.

In front of the crisis of our cities, the proposal of decentralized institutional assets means to recognize the self-organizative potential of society as in bio-ecological natural systems where the complex order is not guaranteed by a centralized management, but by a self-organization asset (Birbracher et al., 1995).

The implications of this new institutional asset are multiple.

First of all, this model is a bottom-up construction with the participation, the agreement, the cooperation, the coordination of actions and activities, through responsibility and not control.

Secondly, this model implies new guarantee mechanisms: all the questions of control are reviewed. Controls related to the results (and not to formal aspects) are necessary not only by higher government levels, but also by each decentralized asset.

These controls and self-controls to guarantee the comprehensive systemic effectiveness are possible through evaluations and self-evaluations process.

Thirdly, this model implies an interactive construction of programmes and projects, by a progressive/incremental approach, through dialogue and communication among actors.

The urban metropolitan areas have become the example of maximum effort of human construction, but also of the clearest failure in the construction of a "good society". It is possible, by starting from them, to rebuild a different development, able to secure the fulfillment of needs of current and future citizens: housing and services also for the poorest ones, employment, health, chances of common life.

The improvement of local government level, as the most adequate to satisfy the housing, services, etc., need of people, puts in a new perspective the rehabilitation of cities, the conservation/reproduction of their man-made and natural capital - i.e. their (also historic) sites, their parks, their services (public care, public health, cultural assetts preservation, education) - for the sustainable development.

The subsidarity principle as regulating principle of the new institutional order implies the relaunching of govern capability expecially with reference to urban planning integrated with employment question and the improvement of the social civil economy. This system can absorb the unemployment from other sectors, providing that it is improved by means of a proper system of financial support, facilities, etc.

The city becomes then the place of production/reproduction not only of natural and man-made capitals, but also and above all, of accumulation of social capital, by enlarging the area of the social civil economy system which plays a new role in satisfying needs (that market is not what is to satisfy) and producing new jobs.

"Enabling strategy" emphasizes molteplicity of actors/groups/singles. It is the collective construction of a general vision through dialogue/communication of different social groups and the capacity to implement it on the basis of some performance rules which identify the results of the projects/plans.

Participation is structurally incorporated in this "enabling strategy", in which the technical approach is only the starting phase of a incremental/progressive process, based on agreements, negotiations, feed-backs, etc. Participation is considered as a constructive relation with the different actors of urban system, to build new interrelation between governed people and government institutions.

Participation can help the collective construction of the city, through the dialogue between different points of view, through the public discussion of different "good reasons" about human rights, etc.

This process can produce new values, not already given at the beginning: these values are less individualistic. People which participate become more aware of their role as members of an interdependent eco-social system: more citizen and less consumer, more tollerant, inclined to conviviality, and to recognize dual utility.

Participation is a learning process through which a new urban culture is built: new cultural, environmental and civic values.

But participation means also evaluating, because evaluation is aknowledgement of the reality confrontation of a situation against another one.

People participate and communicate through evaluations. Participation and evaluation are strictly linked.

"Enabling strategy" is possible only if we are able to improve the local level of government, coherently with the subsidiarity principle, which promotes decentralization at districts level.

It means not to ignore the civil economy system, but to build a patnership among public institutions, private enterprises, and social enterprises in order to cope with the main urban problems.

Actually, it is about to support, sustain, promote but also to orient the effort of a high number, sometimes made of hundreds of organized individuals, groups, institutions. The "enabling strategy" can produce very relevant effects, much bigger than those produced by only one public institution. But the precondition of "enabling strategies" is the existence of a strong civic society, i.e., of strong human and social capital (Putnam, 1993).

This "enabling strategy" affects also the urban planning activity, through which it is possible to realize the objectives of sustainable development in the space. Urban planning becomes a central activity in the perspective of "enabling strategy". It becomes the real test of its concrete implementation. It shifts from control to constructive public dialogue discussion and to communication, realizing the coordination, the promotion and the enabling of the activity of different subjects.

If there exists social capital, this does not mean mediation or negotiation, but first of all, the social construction of a "vision" of the future.

This is possible through a big organizational effort of this multiplicity of subjects, aimed to build a collective action, i.e. aimed to build decision-making processes with the participation of many subjects which bear at the same time different specific values.

4.2. EVALUATION AS A TOOL OF "ENABLING STRATEGIES"

In order to realize the "enabling strategy" in planning, it is necessary to identify new accountable and transparent approaches, specific tools and rules which discipline, from the very beginning, the participation of the different subjects at as decentralized as possible level (round tables, hearings, referendums, public debates enable clearness and transparency in the whole decision-making process).

It is evident that this strategy can work only by means of suitable technical, organizational and institutional tools able to make ex-ante, on going and ex-post evaluation and able to identify the conflicts among the different interests, objectives and values beared by each individual, and to reduce these conflicts at acceptable levels, through processes of constructive dialogue, discussion, communication and negotiation. These conflicts are so much stronger as the community social capital is weaker.

By means of economic, financial, ecological and social evaluations, it is possible to give a basis to subsidiarity principle.

Evaluation is a tool to improve institutional infrastructure at local level and to realize the principles of human sustainable development in urban areas.

Evaluation is always a multidimensional complex exercise which can help to realize enabling strategies in urban planning, by helping to identify priorities among different goals, to search the best sustainable use of a site or of a man-made cultural asset, and to choose how much, where, how conserve/reproduce man made and natural capital.

Evaluation is a difficult process, also when it is related to a single asset.

For example, for a natural asset it is possible to identify different kinds of value: market values, use values, potential use values, existence values. Certain uses or activities can produce environmental damages which can be assessed by considering the above mentioned values.

A careful comparison among existence value, use value and market value can help to identify the more sustainable use of an area or a man-made cultural asset.

In the perspective of "enabling strategy", evaluation is not only a technical exercise which can help to choose among alternatives, but also communicative/dialogue tool which can help to conflict reduction or resolution, to negotiations, etc. and to build human and social capital.

It should be necessary to build a specific balance-sheet matrix which identifies every decision makers, managers, decision-takers, consumers with a special attention to those belonging to the third sector, their objectives and their compatibility with the general objectives, in order to build a process of communication/dialogue and constructive discussion from which strategies of cooperation and co-evolution can arise.

4.3. SOCIAL EVALUATION IN POST-MODERN AGE AS TIME OF EVALUATION

In post-modern age values are only private or individualistic ones: i.e. economic ones. Social values should be built through communication and public discussion (Sen, 1995). Communication is possible through extended evaluation processes. In order to respect the pluralistic composition of our society, the structure of these evaluation processes should be as open as possible and multidimensional, i.e. able to take into account the priorities of all subjects' viewpoints.

Evaluation is not only linked to the prevision of environmental, economic and social impacts of different urban transformation projects on economic, social and environmental systems as technical activity. But it is first of all linked to interpretation of reality, the specific character of a site (genius loci, etc.); to capability to identify a comprehensive or strategic "vision" of site, place, community, to support the communication among the different subjects in order to choose the community relevant goals and derive a rank of the different values; to argue, adduce reasons for certain proposals and, doing so, contribute to build an awareness, critical capacity and responsibility, through circular processes raising (from links, feed-back, etc.).

The robustness of an urban planned transformation of the use/management of natural capital and man-made capital is in the richness and robustness of the evaluations which have been developed. The general objectives of sustainable development are translated into urbanistic rules through evaluations.

By means of evaluations people not only can identify the priority of different values, but take note of a certain context, can learn about it, can better understand the development directions, get critical awareness, produce new ideas and new values, i.e. people produce culture. As long as by means of evaluations it is possible to produce also cultural and civic values. The activity of evaluation contributes to produce human and social capital: a new post-economic/post-mechanistic culture.

In the perspective of "enabling strategies" and goal governance, the activity of evaluation should be regarded not only as a technical activity, characterized by instrumental rationality, but as an activity which is about the constructive discussion of "good reasons" and the understanding between public/private and the civic society, by communicative rationality.

The planning process is an exceptional chance of deciding not only the management of natural capital and man-made capital, but also of creating social capital, i.e. forming subjects who are able to go beyond their own interests, i.e. behave as citizens because they are oriented by relational values and not only by exchange values.

The participation to the planning choices through evaluations develops the formation of a broader viewpoint than the strictly individual one, and helps to make people grow in a community. Unfortunately, it is an underused process to build the social individual, i.e. to train to the evaluation of the Dual (individual and social)

Utility, and to the consideration of every option in the context of the Dual Utility (Etzioni, 1988).

Evaluation can contribute (through values formation from public discussions and social interactions) to build "public spirit", guarantee of democratic control and also of "good governance". On the other hand, it can contribute to build a more robust civic society.

The above makes us state that the conservation of environmental and cultural heritage must become the opportunity to promote the network of horizontal relationships in society, through mechanisms of promotion and stimulus to the cooperative/voluntary/non-profit sector and to participation. In this way, it is possible also to reduce the vulnerability of a community, and to link the questions of good governance to those of democracy and of development.

The communicative function of evaluation does not only interest the analyst and the politician, but all the individuals organized in groups, movements and associations and the stakeholders.

Public institutions should promote the community organisms and the associations belonging to the third sector, which could give a contribution to the achievement of general interests, because these associations aim to strenghten social integration, to make communication easy, to create mutual trust, and to increase the density of horizontal relationships: the social capital produced can help to decrease the costs of interactions, and to multiply sinergies and the average productivity of human capital.

5.   The Different Dimensions of Evaluation and the Production of Human/Social Capital

Evaluation becomes a central activity in management/conservation/reproduction of the four forms of capital (natural, man-made, human and social).

5.1.   THE COMPLEX VALUE OF NATURAL CAPITAL

The planning evaluation as communicative activity starts and performs in a set of technical evaluation of the different forms of natural and man-made capitals. But these evaluations should be "revised", modified and integrated in an open interactive process of public discussion.

The task of a local plan should be to make an evaluation of the characteristics of resources in such a way that proposed use values with planning activities can improve - and not reduce - intrinsic values and at the end overall values - or complex values - of a site. The "complex social value" offers a broader perspective to cope with evaluation problems at the local level of planning. Some relevant issues are: a) evaluation of the inherent value of the land, site or resources; b) comparison of site-functional alternatives involving given impacts with overall characteristics or inherent value; c) choosing the more sustainable use of land, site or resources.

The notion of value in planning/programming for sustainability should be seen not exclusively in economic sense, but in a broader post economic perspective. In the economic approach, the good or service value is traditionally expressed by its market value, which represents an unambiguous "measure" univocally defined by a cardinal number.

I do not want here to cope with the question of evaluation for a sustainable development (Fusco Girard and Nijkamp, 1997). I want to note that the notion of value affecting the sustainability questions is not only linked with the exchange, but with a broader set of values. Following on the perspective opened by Lancaster approach (1971) I want to underline the multidimensional character of value, by introducing a notion of utility which is also social, on the basis of the hierarchical scheme of Maslow (1970), and ecological (Costanza and Tognetti, 1996).

It is convenient to insist on the multidimensional character of evaluations and on the need that adequate procedures are developed in order to make evaluations operational and available.

If we assume with conviction the perspective of sustainable development, we have to make a "comprehensive" or "complex" evaluation of natural and man-made capital.

The limit of the value theory that the traditional economics supplies is evident: it is uninterested in the future generations' demand, in social/communicative dimension, in the ecological dimension (Passet, 1979). It recognizes only the economic man and ignores the social and ecological dimensions of man. Economic approach is not able to capture all the values of an asset (Pearce, 1995).

A more general value theory, able to grasp the social dimension and the ecological dimension, besides the short term interests/utilities, should combine the value in itself, i.e. non-use value or intrinsic value with use value (Daly and Cobb, 1990).

An intrinsic value, independent on the human subject's use, is that related with the self-production of bio-ecological systems (Faber et al., 1995). An autopoietic system has got also another goal which is the supplying of goods and services which supports life in its different forms for other subjects. This heteropoietic aspect emphasizes the use value. The autopoietic capabilities of a bio-ecological system emphasize the ecological value of a good, that is an intrinsic or non-use value: an existence value.

In other terms, the natural factor earth is not only a "resource" because it satisfies human needs, but also because it satisfies the needs of other living organisms pre-existing compared to the man.

The instrumental (or economic, or monetary) value of natural capital is due to the fact that it produces services for the tourism, forestry, recreational, food, fishing, chemical, etc. sectors. The not instrumental value of natural capital is due to the fact that it secures the water and air cycles, the preservation of every living species in their own ecological niches, etc., by means of its autopoietic processes.

Desides these values, It Is possible to ackowledge also a specific value, the "intrinsic" value.

The complex value V_c of an ecological system is a multidimensional profile represented by the socalled total economic value (TEV) and by the intrinsic value (ε): $V_c = (TEV, \varepsilon)$. TEV is use value plus non-use value expressed in monetary terms. The term ε which represents the intrinsic value cannot be expressed in economic terms (Turner, 1992). It is assessed by means of ecological evaluations. These evaluations emphasize the systemic behaviour of every component with respect to the whole and to its evolution. This sistemyc model is not known to public in general. Sometimes, they have been expressed in energy terms (Turner et al., 1993), i.e. identifying the capability of the natural system to catch solar energy through the photosynthesis process.

The complex value provides for a criterion to make sustainable development notion operational in planning choices regarding the land use. It overcomes current economic criteria.

If, for example, the site's use alternatives are those of transformation (development) or those of conservation, the choice criterion refers to the monetary net benefits of transformation ($B_{tr} - C_{tr}$) compared to the complex value of the area. Only if net benefits are higher than complex value V_c (even if this latter is on different scales) the transformation can take place (Fusco Girard and Nijkamp, 1997):

$$[V_{present} (B_{tr} - C_{tr})] > V_c \qquad (1)$$

where is: $\qquad$ $V_c = (TEV, \varepsilon)$
$\qquad$ TEV = "complex" value in monetary terms
$\qquad$ ε = "intrinsic" value in non-monetary (ordinal and/or cardinal) terms

This is what it is about a rule for choosing among alternative functions, that it is not reducible only to one dimension. It implies a multidimensional comparison, including non-monetary components, and then a public participation process, in which economic values are compared with other values through public discussion.

In choosing between different functions for the same site, it will be preferable the combination of functions, which is able to maximize the vector composed by the whole of economic net benefits and the whole of non-economic values.

These "complex" values (Fusco Girard, 1986, 1987; Nijkamp and Coccossis, 1995; Fusco Girard, Nijkamp, 1997)) arise from an autopoietic approach to sustainability and are expressible by means of quanti/qualitative multicriteria anlysis.

Turner (1992) underlines the need to acknowledge to the autopoietic system (and then to its biotic and abiotic components linked by interdependence relationships) a "primary value", that is the very requirement so that the system can supply functions and useful services for the man. We are talking of latent functions, underlying those generally appraised, which express the value of the whole system. It is the main requirement to the carrying out of those heteropoietic activities which identify the "secondary total value".

Therefore, the global value of natural capital, for example, in planning for protected areas can be expressed in economic with contingent evaluation and non-monetary terms together.

Also de Groot (1992) acknowledges the need to integrate economic evaluations with qualitative evaluations which express the existence or intrinsic value(through ordinal evaluations, by means of symbols like +++, ++,+, etc.). Moreover, he introduces the notion of social evaluation in order to take into account the plurality of different viewpoints. This too is expressed by means of ordinal evaluations which integrate the monetary ones.

A further confirmation of the structural limits of economic evaluation and of the need of integration in a multidimensional perspective can be found in Jacobs (1991) who also underlines the impossibility to include the viewpoints of the poorest income classes and of the future generations in the economic evaluations.

Costanza (1991, 1996) suggests a "pluralistic" approach to evaluation which can use the economic and the ecological viewpoints and integrate monetary and non monetary evaluations. Odum (1983) talks of an "holistic approach" which includes cultural, environmental and economic values where complex systems have to be evaluated.

These complex evaluations are a good example of an overcoming post-modern culture, which reduce all values to the same level and to economic dimension.

5.2.   THE COMPLEX SOCIAL VALUE OF MAN-MADE CULTURAL CAPITAL

Sometimes,   it   is   also   possible   to   refer   the   above   to   man-made/ monumental/cultural/artistic capital. These kinds of assetts too have got a complex value which does not correspond to the mere total economic value, but reflects also an intrinsic value.

Cultural assets are not a capital characterized strictly by a bio-ecological vitality. We cannot specifically talk of existence value linked to the bio-ecological vitality of natural ecosystem, in the case of cultural/monumental assetts. But their existence and their use affect the stability and the resilience of the urban ecosystem, because they interact with the living components of the ecosystem, i.e. with the current and future community. Built environment is not only a physical space. It is also a particular system in which man-made capital, human capital, natural capital and social capital interact. Artistic/monumental assets have always been a communication tool able "to put together", to unify, to be elements of social stability.

Man-made/artistic/monumental capital is the element in which today and in future a community is able to recognize itself. It is a source of local identity sense, of integration, of cohesion, of community awareness, of common values, of specificity regarding a homologizing culture spread out by mass media technology. Cultural assetts tell us where we come from, they give us a motherland without which we would be confused statelesses, they help us to recognize our identity. Moreover, they contribute to promote the "horizontal" communication in the era of only "vertical" communication.

Man-made cultural capital is an element which due its diversity contributes in creating the specific "character" of a city. But if it sistematically interacts with the social system, its conservation not only contributes in mantaining the identity, the specificy of a site, but it is also an element which contributes in conserving the identity and the diversity of a community, that is the source of greater interactions, of cohesion, of common values, of opportunities, of stability towards external pressures. It has sometimes an "intrinsic" value (Fusco Girard and Nijkamp, 1997).

We are referring, for example, to the role of some religious monuments, around which a specific and non-repeatable identity is built, i.e., a common feeling that cannot be confused with its economic touristic value is built. This intrinsic value which cannot be valued in monetary terms, and which gives to the cultural asset its vitality, is a source of social capital accumulation. It expresses the genetic heritage of a community, its memory and its specific identity. It can be valued through a dialogic/participative process.

An adequate awareness of the systemic relationship between man-made assets and social life does not probably exist.

We can conclude then acknowledging that cultural/historic/monumental heritage is an element which contributes to the stability and resilience of urban ecosystem. Doing so, it has also got an "intrinsic" value as far as it contributes to produce social capital, i.e. the "glue" which keeps the different subjects of a community together. This "glue" has also got an economic reflection , because it contributes to the "non-economic" conditions of economic development, but cannot be valued in monetary terms.

Sometimes we are able to transform intrinsic values (i.e., artistic value) into use values, i.e. in economics terms, by using techniques from environmental economics, i.e. by environmental economics. But the symbolic value, the artistic value or the historical value of an asset for future generations cannot be expressed in monetary terms. Thus, an economic approach to valuation is only partial (see Turner, 1992). Therefore, we are obliged to consider economic and non-economic values. From this perspective, we speak of a "social complex value" V_c (see Fusco Girard, 1986; Fusco Girard and Nijkamp, 1997): $V_c = (VET, i)$, where "i" is a glue value which reflects the capacity of the system to hold/combine diversities together in a structure.

There is a need in urban planning for a more integrated approach which can refrect various kinds of values. A choice based on the assessment of the complex social value takes into account such different values, and not only the economic ones. Consequently, the leading criterion in urban planning should be to assign to a site its maximun use value which is consistent with its non-use/intrinsic values. Some of these use values could be transformed into market values under specific conditions which have to be identified.

5.3.   THE MANAGEMENT OF COMPLEX VALUES IN PLANNING

The Community Impact Evaluation (Cie) proposed by Lichfield (1989) allows to cope with evaluation problems both in a technical perspective and in a communicative perspective. It is the framework where to put/manage such complex values in planning: use values, non use values, existence values, intrinsic values, and also market values and costs (Lichfield, 1993 ; 1996).

It allows local governments, committed to the conservation, reproduction and management of man-made and natural capitals, to apply the "enabling strategy" starting a constructive discussion among the different public actors and at different government levels, and with the private and the social-private sectors, i.e. coordinating a multiplicity of subjects (even hundreds) with different objectives/interests/values. In order to do so, all the subjects belonging to the third sector, i.e. the neglected social/civil economy system, the ethnic minorities of our multicultural cities (with their particular perception of needs, etc) should be formally included among the producers/users/consumers of the matrix.

The framework of the Cie allows, then, the associations, movements, cooperative firms and groups - even those representing the most neglected subjects - to partecipate to the development choices deducing in an interactive way social values and shared priority, and then evaluating the different alternatives in terms of economic, social, ecological and environmental impacts.

The Cie allows to compare "integrated or complex evaluations", i.e. those combining quantitative monetary elements with qualitative ones. The Cie can be used as a tool to build the consensus among the different stakeholders, that is to build a "shared strategic view" and to derive positive sum strategies.

Moreover, the Cie can allow dynamic iterative processes of evaluation where preferences for the different subjects are not only assumed but also modified during the decision-making process. That is, with this approach, new values can be produced, with respect to those initially "given", in a process of formulation/re-formulation and further re-formulation.

In this way, this evaluation framework support the co-evolution of the different social subjects.

The Cie can be combined with sophisticated multicriteria evaluation techniques, such as the Regime analysis and Evamix, and can allow the development of less sophisticated evaluation, such as (social) financial analysis, etc.

The Cie approach could also be used to organize an Ethical Accounting Statement Balance Sheet which focuses on values which are shared by institutions and groups-stake holders, local society, third sector, etc, and to control the level of the achievement of each value year by year (Bogetoft and Pruzan, 1991). In this perspective, it becomes an instrument for social control for reproduction of human and social capital and for good governance.

6.   Conservation of Natural-Cultural Man-Made Capitals for the Promotion of a
Post-Economic Culture

Problems of post-modern city cannot be solved by improving the condition of physical infrastructures: transportation network, housing heritage, green areas, etc., that certainly make one area more attractive and competitive than the others.

The issue of our cities development is not only linked to aesthetic strategies and market strategies, it is more complex and it refers to integral promotion of human person in her different dimensions. Development involves also the research of new organizative rules for social living. We mean, in particular, strategies of human sustainable development.

It is possible to realize the better urban parks, the more correct restorations of monumental heritage, the more confortable dwellings, but if we do not modify as well latent organizative rules which subject to urban life, economic and social exchange, all this will be unnecessary. In other words, the development issue concerns the structure of our values system. From here it depends the capacity to effectively improve the forms of social living.

Here the hearth of the issue of human sustainable development of our cities lies.

The central issue in a more and more populated and urbanized world, in a context which always more emphasizes the single subjectivity, becomes the coordination of actions of each subject with the others. It means the issue of autonomy of each subject compatible in respect of freedom and rights of others.

We will be more and more forced to elaborate choices and collective actions in order to construct a more desiderable, less conflictual and unbalanced future. (Zamagni, 1995).

It means we are somehow "forced" to research an agreement able to regulate the action of each one with the one of the community through collective action. This process may discend only from intersubjective communication and participation.

The research of a new balance among "I", "We" and Nature, then, is not a technical question, a question of institutional, social, economic, spatial etc. modernization, but it has to be investigated on cultural level, in the shared recognition of general interest notion and, then, in the responsible choice, able to take into account this general interest.

Conservation of cultural and natural assets may help in this perspective of development humanization, if it becomes the starting point to promote a post-economic and post-mechanistic culture; if cultural values can be used in building civil social, in transforming consumers in citizens.

Civil economy enlarges time perspective, promotes a communicative, face to face approach, opens to cooperation because it is based on reciprocity: it produces relational goods. This social capital is critical for "good governance" and democratic control.

Conservation of cultural and natural assets is an activity coherent with the building of a collective shared identity and memory, and the implementation of subsidiary principle.

Evaluations not only produce critical awareness about different values, but they can modify the set of values with public discussions, as already recognized by Knight (1947) and by Buchanan (1954) when they noted that "values are established or validated and recognized through discussion, an activity which is once social, intellectual and creative" and that "individual values can and do change in the process of decision making". This is a fundamental aspect in order to promote the transformation of consumer into citizen.

In our pluralistic society, evaluations are a tool able to promote public dialogue-discussion, coeherently with a partecipative and educational idea of democracy (which is not limited to the mere observance of certain formal rules in the decision-making process). They are a tool for the formation and production of values (Fusco Girard and Nijkamp, 1997) and for deducing priorities of values in our post-modern time.

Social integrated evaluation processes are a good exercise to overcome our post-modern culture, identifying new values and their priorities. They are a tool for "good governance".

From a stricter evaluation point of view, many are the opened problems: how to evaluate the man-made/cultural capital and the natural capital in such a way to take into account every values? How to better value "glue" value (ε) for natural capital? How to value glue value (i) for man-made cultural capital? How to improve the Evaluation matrix as a Values Accounting Statement Balance Sheet, in such a way to better produce common dialogue/communication among institutions, private subjects and social subjects, and reproduction of human and social capital?

References

Arval (1994) Association pour la Recherche sur les Systemes de Valeurs , in *Futuribles*, **200**, pp 5-9.

Benevolo, L. (1993) *Le città nella storia di Europa*, Laterza, Bari.

Birbracher ,C.K., G. Nicolis and P. Schuster (1995), Self-Organization in the Bio-Chemical and Life Sciences, *Report EUR* **16546**, European Commission.

Bogetoft, P. and Pruzan ,P. (1991), *Planning with Multiple Criteria*, North Holland, Amsterdam

Buchanan, J.M. (1954), Social Choice, Democracy and Three Markets, *Journal of Political Economy* vol. **62** (2), pp. 114-123

Caselli, L. (1995), Andare oltre in nome dell'uomo, *Una buona società in cui vivere*, Edizioni Studium, Roma.

Coleman, J. (1994), *Foundations of Social Theory*, Harvard Univ. Press.

Costanza, R. (ed.) (1991), *Ecological Economics: the Science and Management of Sustainability*, Columbia Univ. Press., New York.

Costanza, R. and Tognetti, S. (eds.) (1996), Integrated Adaptive Ecological Economic Modelling and Assessment, *Scientific Committee on Problems of the Environment* (SCOPE), Paris.

Daly H.E. and J.J.Cobb (1990), *For the Common Good*, Beacon Press, Boston.

de Groot R.S. (1992) *Functions of Nature: Evaluation of Nature in Environmental Planning, management and Decision Making*, Wolters-Noordhorff, Groningen.

Ester, P., Halman, L. and de Moor, R. (eds) (1994) *The Individualizing Society*, Tilburg Univ. Press, Tilburg.

Etzioni, A. (1988) *The Moral Dimension. Towards a New Economics*, The Free Press, New York.

Faber, M., Monstetter, R. and .Proops, J.L (1995) "On the Concept of Ecological Economics", *Ecological Economics*, **12**, pp.41-54.

Funtowitcz, S.O. and Ravetz, J.R. (1994) Emergent Complex Systems, *Futures*, n.**6**, vol **26**, pp.210-225.

Fusco Girard, L. (1986) The "complex social value" of the architectural heritage, *Icomos Information* n.**1**, pp.13-22.

Fusco Girard, L. (1987) *Risorse architettoniche e ambientali: valutazioni e strategie di conservazione*, FrancoAngeli, Milano.

Fusco Girard, L. and Nijkamp, P. (1997) *Le valutazioni per lo sviluppo sostenibile della città e del territorio*, FrancoAngeli, Milano.

Habermas, J. (1986) *Teoria dell'agire comunicativo*, Il Mulino, Bologna.

Knight, F. (1947) *Freedom and Reform: Essays in Economic and Social Philosophy*, Indianapolis

Jacobs, J. (1969) *The Economy of Cities* Ramdon House, New York.

Jacobs, J. (1991) *The Green Economy*, Pluto Press, London.

Lancaster, K. (1971) *Consumer Demand: a New Approach*, Columbia University Press, New York.

Lichfield, N. (1989) *Economics of Urban Conservation*, Cambridge university Press, Cambridge.

Lichfield, N. (1993) La conservazione dell'ambiente costruito : verso un valore culturale totale, in L. Fusco Girard (ed.) *Estimo ed economia ambientale : le nuove frontiere nel campo della valutazione*. Angeli, Milano.

Lichfield, N. (1996) *Community Impact Evaluation,* University College Press, London.

Maslow, A.H. (1970) *Motivation and Personality*, Harper & Ron, New York.

Nijkamp P., and H. Coccossis (1995), *Planning for our Cultural Heritage*, London.

Norgaard, R.B. (1988) Sustainable Development: a Co-evolutionary View, *Futures*, n.**12**, pp. 606-620.

Norgaard, R.B. (1994) *Development Betrayed*, Routledge, London.

North, D. (1993), *Institutions, Institutional Change and Economic Performances*, Cambridge University Press, Cambridge.

ONU (1992) *Rio Declaration on Environment and Development*. ONU, Rio Sunmit.

O.N.U. (1995) *Summit mondial pour le developpement social,* ONU, New York.

O.N.U. (1996) *Habitat Agenda*, Conference on Human Settlements (Habitat II), ONU, Istanbul.

Odum, E.P. (1983) *Basi di Ecologia*, Piccin, Padova.

Passet, R. (1979) *L'économique et le vivant*, Payot, Paris.

Pearce, D.W. (1995) *Capturing Global Environmental Value*, Earthscan, London.

Putnam, R. (1993) *Making Democracy Work: Civic Tradition in Modern Italy*, Princeton University Press, Princeton.

Quadro Curzio, A. (1996) *Noi, l'economia e l'Europa*, Il Mulino, Bologna.

Rifkin, J (1995) *La fine del lavoro*, Einaudi, Torino

Sen, A. (1995), Rationality and Social Choice, in *American Economic Review*, vol. **85**, n.1, pp.3-24.

Seragaldin, I. (1996) *Sustainability and Wealth of Nations*, The World Bank, Washington

Tchernia, J. (1995) *La recherche dans le domaine des valeurs*, in *Futuribles*, **200**, pp. 9-24.

Tiezzi, E. (1995) *L'equilibrio*, Cuen, Napoli.

Turner, R.K. (1992) Speculations on Weak and Strong Sustainability, *CSERGE Working Paper* **92-26**, Norwich.

Turner, R.K., C. Folke, M. Green and I. Bateman (1993), Wetland Valuation: Three Case Studies, *CSERGE-Beijer Institute Working Paper*, Norwich.

Viederman S. (1996), Uncivil Society, *The Ecological Economics Bullettin*, n.**1**, vol.**1**, pp.4-6.

World Bank (1995) *Monitoring Environmental Progress*, World Bank, Washington.

Zamagni, S (1995) Soggetti e processi per una progettualità nuova in Italia, *Una buona società in cui vivere*, Edizioni Studium, Roma.

ECONOMICS AND ECOLOGICAL SUSTAINABILITY: AN ACTOR-NETWORK APPROACH TO EVALUATION

P. SÖDERBAUM

As part of mainstream neoclassical economics, current environmental problems are regarded as instances of market failure or possibly government failure. The focus is on trade relationships and on actors in markets. 'Externalities' are defined as impacts on interested parties other than seller and buyer of a commodity and 'market failure' is connected with non-consideration of impacts on third parties. As part of this specific conceptual framework, it is believed that environmental problems can be solved by internalising externalities. When the producer and seller of a commodity act on the basis of 'total social cost', the market mechanism will function smoothly and correctly. Similarly, government failure is connected with market interventions by the state that disturb the correct functioning of the market.

I shall return later to this issue – whether science can point out what are correct or incorrect values for purposes of societal resource allocation. For the moment, I will only contend that there is a role for neoclassical thinking as part of a pluralistic strategy. But there should be room for other perspectives as well, such as alternative perspectives in economics and perspectives emanating from other social sciences, planning sciences included. The current environmental crisis, for instance, may suggest a number of more fundamental issues than those connected with 'optimal levels of pollution control' and other marginal adjustments. In addition to market failure and government failure, there may be failure of world-view, failure of ideology, failure of science (paradigm failure in economics being an example), failure of technology, institutional failure and failure of various organisations, groups and individuals. Even business corporations and environmental organizations may fail.

Thus, neoclassical environmental economists (Pearce and Turner, 1990; Tietenberg, 1992) tend to limit attention to specific issues which some would describe as minor, while overlooking other more basic problems. As I see it, neoclassical economists often act as if they were more interested in protecting the neoclassical paradigm than the environment. Among economists arguing in favour of a broader and more pluralistic approach, where issues of world-view and paradigm are not avoided, Richard Norgaard (1994) can be mentioned here. But similar views have also been expressed by a number of scholars referring to themselves under various labels (social economists, ecological economists, institutional economists, interdisciplinary economists, etc.). Here I will

N. Lichfield et al. (eds.), Evaluation in Planning, 51–71.
© 1998 *Kluwer Academic Publishers. Printed in the Netherlands.*

draw attention to some possibilities connected with such an institutional and interdisciplinarily oriented economics.

1. Political Economics

Economics is unavoidably political economics. Values are always with us in our scholarly activities and in economic analysis:

> "Disinterested research there has never been and can never be. Prior to answers there must be questions. There can be no view except from a viewpoint. In the questions raised and the viewpoint chosen, valuations are implied. Our valuations determine our approaches to a problem, the definition of our concepts, the choice of models, the selection of our observations, the presentation of our conclusions – in fact, the whole pursuit of a study from beginning to end." (Myrdal, 1978, pp.778-779)

It can be added that no scholar (or actor in other roles) is completely isolated from other individuals. We all act in some social, institutional and cultural context and our values are related to those of others. The individual may act opportunistically in relation to specific groups or behave in a way that reflects some degree of integrity.

The project of the 1870s – to develop a 'pure' economics – has largely been a failure. Neoclassical economics (like any other paradigm in the social sciences) is science in some part, but also politics and ideology. Myrdal recommends attempts to be conscious about how values influence scholarly work and also to *openly declare the value premises* of specific studies. For instance, what is the idea of 'development' or 'progress' behind a specific project? In relation to the present subject, approaches to valuation in an urban or rural context, I would like to add the recommendation of *being many-sided in the treatment of values*. In a democracy, there are typically many ideas of progress rather than one and the scholar or other analyst therefore has to refer to various possible valuational standpoints. Some interested parties will typically oppose a construction project, while others are in favour of it. In this situation, there is no value-neutral objective function to refer to. Many are those who would like to adhere to the ideas of constrained optimization of neoclassical economics, for instance in the form of cost- benefit analysis, but such a way of proceeding is simply not compatible with democracy. Here one may even speak of a specific impossibility theorem, the impossibility of 'value-neutral valuation'.

To apply cost-benefit analysis (CBA) or other approaches that reflect an assumption of common societal objectives in a situation characterized by conflict between various interested parties is to support what Peter Self refers to as 'econocracy' (1975). Frank Fischer (1990) uses instead the term 'technocracy' for the tendency to use instruments that increase the power of experts, thereby making politics unnecessary and politicians more or less superfluous. Ezra Mishan, himself an author of text-books on CBA, has admitted that the CBA-instrument can only be used as long as there is a consensus in society about the specific valuation rules or ideological colouring of CBA. Since no such consensus exists, for instance concerning valuation of environmental impacts in relation to other impacts and since no consensus is foreseeable in the near future, the best one can do is to put the CBA manual on a shelf to await some future point in time when the necessary consensus possibly emerges (Mishan, 1980).

2. A Model of the Actor: Political–Economic (Wo-)Man

Not only is economics as a field of study political. Our 'objects' of study, human beings and organisations will also be regarded here as political entities. As part of our present interest in valuation or evaluation, the issue of who is valuing specific development initiatives, arguments or projects cannot be evaded. In this connection, a model of the valuing actor could be useful. Instead of Economic Man assumptions, where man is seen essentially as a consumer and instant optimiser in a machine-like manner, Political-Economic Man or Woman is suggested.

Individuals have many roles in addition to that of being a consumer. In Table 1, reference is made to possible roles R_1, R_2... R_k of the individual. Some roles may be related to markets such as consumer, wage-earner, investor. Our individual may also be a parent, a professional, a member of an environmental organisation, a citizen etc. Connected with the role R_1 is a set of activities, AC_1 and a set of motives M_1. All roles R_1, R_2 ...R_k are somehow kept together by an 'identity' of the individual, while the pattern of activities AC_1, AC_2 .. AC_n form a 'life-style' and all motives or interests M_1, M_2 ... M_m reflect an ethical or 'ideological orientation' of the individual.

While neoclassical economics tends to limit attention to man as a consumer, Table 1 reminds us that there are other relevant roles as well. As an example, the environmental performance of an individual is related to all her roles and not only the one of being a consumer. An individual may value consumer goods and activities connected with consumption but valuation is also part of her/his role as a professional (for instance consultants and bureaucrats in local governments engaged in planning activities) or a citizen.

TABLE 1. Symbolic representation of Political Economic Man/Woman
(Source: Söderbaum, 1993).

Roles		Activities		Motives	
R_1		AC_1		M_1	
R_2		AC_2		M_2	
R_3	'Identity'	AC_3	'Life-style'	M_3	'Ideological orientation'
.		.		.	
.		.		.	
R_k		AC_l		M_n	

In spite of tensions between various motives or interests, the individual is somehow held together by ideas of his or her role or identity in relation to each specific socio-cultural context. Dissonance theory, learning theories and other parts of social psychology are seen as relevant and useful in understanding behaviour. The individual strives for some congruence and balance between roles, activities and interests, and may experience such balance, but incongruence and tensions are equally characteristics of the human existence. Egoistic versus other-related (or community oriented) motives are an example of such tensions. Amitai Etzioni, for instance, has propounded an "I & We Paradigm" (Etzioni, 1988), according to which, the fact that there is a strong ego in each healthy individual is not sufficient reason to denigrate or exclude the social and ethical aspects of human life. An actor plays part in many 'we-contexts' and such relationships involve a number of tensions and ethical issues. The individual is normally embedded in (or locked into) a set of relationships which may be referred to as a network.

As already mentioned, neoclassical theory tends to represent individuals as robot-like, instant optimizers. Institutionalists on the other hand argue that behaviour is largely habitual. Herbert Simon's early argument about selective perception, limited cognitive capacity and search cost are relevant here (Simon, 1945). As humans, we tend to stick to familiar environments and use various rules of thumb to deal with complexity. The development of a habit can be expressed in terms of increases in the probability of a specific behaviour. Emphasis on habitual behaviour does not exclude the possibility of conscious decision making. At times, the individual perceives problems and alternative courses of action. Habits arc reconsidered and behaviour may change. Such decision situations can be discussed in conventional terms of maximizing an objective function, subject to various constraints. In this essay, however, a different and more holistic idea of rationality will be emphasized.

According to neoclassical consumer theory, it is assumed that the consumer maximizes utility, a proposition that hardly can be falsified. Whatever an individual is doing, it can be interpreted as an attempt to maximize utility. However, individual A may consider the wellbeing of others when choosing his or her bundle of commodities, whereas individual B is less concerned about his fellow human beings. Both individuals maximize utility. As part of neoclassical theory, the utility functions of consumers are accepted whatever their ethical or moral implications. Those of us who take environmental issues seriously realize that one has to problematize the way individual A maximizes utility as compared with B. Issues of ethics, ideology and life-style are at the heart of present understandings of environmental problems.

According to Political-Economic Man (Woman) assumptions, complexity is seen as a fundamental fact of life. It is furthermore assumed that the individual is guided by a specific valuational or 'ideological orientation' (cf. Table 1) rather than any desire to maximize utility. 'Ideology' here is used in a broad sense and refers to 'ideas about means–ends relationships' generally, or in relation to specific spheres of activities. Douglas North refers to 'ideology' in a similar sense: "By ideology I mean the subjective perceptions (models, theories) all people possess to explain the world around them." (North, 1990, p. 23)

More recently, ideologies such as 'feminism' and 'ecologism' have been added to liberalism, socialism, etc. among the more general or 'standardized' ideologies (Eccleshall et al., 1994). Six principles of ecologism are suggested, namely sustainability, small is beautiful, reducing the human population, individual responsibility, spiritual reawakening and transition to a Green society. Other interpretations are of course possible and generally one may refer to different versions of liberalism, conservatism, ecologism etc. Combinations can be considered, such as eco-feminism, eco-socialism and so on. It is also possible to speak of ideologies that are more limited in scope for instance 'health care ideologies', emphasizing centralism or decentralism in health care. The debate about environmental issues has made us realize that there are many kinds of environmentalists and that it may therefore be more relevant to speak of different environmental ideologies than mere 'ecologism' (Dobson, 1995). Degree of technological optimism is here a possible distinguishing characteristic.

3.   Ideological Orientation as a Basis for Valuation and Decision Making

As part of CBA and CVM (Contingent Valuation Method), 'valuation' is more or less equivalent to 'monetary valuation'. The previous reasoning in terms of Political Economic Man connects 'valuation' instead with the 'ideological orientation' of individuals as actors in the economy. In this broad sense, valuation refers to discrimination between 'good' and 'bad', between 'goodness' and 'badness'. Such distinctions may certainly be made in monetary terms, but also in non-monetary terms, as when one speaks of a lake or forest being in a 'healthy' condition. For a lake in

Sweden, somebody may argue – on the basis of scientific or other knowledge – that a pH level of 6.5 is preferable to, say 4.5. This exemplifies a 'valuation' that is relevant in relation to environmental issues.

It is assumed here that human beings value things, single activities, patterns of activities (lifestyles), projects, programmes and policies mainly through their ideological orientation. The ideological orientation of individuals is based on knowledge and information of a more or less systematic and complete kind. An individual may deliberately choose to value something exclusively in monetary terms, but this is then a special case of valuation and not the only case. Valuations can be carried out in either non-monetary or monetary terms, or both. I will consider the latter case – where both monetary and non-monetary elements are present – as being the normal one in relation to valuation of projects and programmes. We are then moving in the direction suggested by Mark Sagoff, where ethical, moral, ideological and similar considerations become potentially relevant as part of a decision situation. Sagoff essentially argues that man, for purposes of policy analysis, should not be reduced to the role of being a consumer. He (she) is also a citizen. Sagoff takes a stance:

> "against the use of the efficiency criterion in social regulation, and against the idea that workplace, consumer-product and environmental problems exist largely because "commodities" like environmental pollution, workplace safety, and product safety are not traded in markets. I shall argue, in contrast, that these problems are primarily moral, aesthetic, cultural, and political and that they must be addressed in those terms." (Sagoff, 1988, p. 6)

Again, Sagoff's arguments should not be understood as being exclusively scientific, representing another idea of 'scientific correctness'. The relevance of his arguments is mainly a matter of ideology, in this case his view of democracy as a meta-ideology. He does not accept any exclusive rights for the 'market view of democracy' inherent in cost-benefit analysis.

If ideologies can be used as a basis for valuation, then articulation and analysis of potential ideologies as well as 'ideological debate' become important in relation to environmental and other societal issues (Figure 1). Just as alternatives at a more concrete level, for instance housing projects or alternative area plans can be compared, it can also be beneficial to explicitly discuss and compare 'alternatives at the level of ideologies' or 'ideological orientations'. The taking of steps toward ecological sustainability is partly a matter of changes in ideology. Limiting attention to given alternatives at the more concrete level may not be a wise strategy. Such 'given' alternatives reflect specific ideologies or ideas of means and ends and if one is interested in finding new and in some sense better concrete alternatives, 'ideology development and articulation' represents an additional path to progress (cf. Figure 1).

Modification and reconsideration of ideologies may lead to the perception of new alternatives at the concrete level.

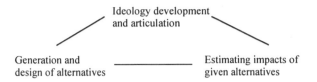

Figure 1. Articulation of ideologies as an integrated part of valuation and impact studies

Our concern for the environment, then makes it important to articulate an environmental ideology or ideological orientation. Many proposals have been made in this direction, the report by the Brundtland Commission being perhaps the best well known. 'Sustainable development' is a development "that meets the needs of the present without compromising the ability of future generations to meet their own needs" (World Commission, 1987, p. 43). The report and the debate that followed paved the way for the 1992 United Nations conference in Rio, where world leaders met to discuss and agree on certain policy principles in relation to biodiversity loss, global warming and other environmental issues. Steps were taken towards making ecology and environment part of the political agenda in many local communities and at national and global levels.

But it should be made clear that the ideological debate about ways of integrating environmental considerations into policy-making did not start with the Brundtland Commission. In the 1970s and 1980s, attempts were made to articulate alternatives to an almost exclusive reliance on conventional indicators such as economic growth in GNP-terms, balance of payments, employment, indices of inflation, etc. Lester Brown was among the early users of the sustainability concept in his discussion of 'sustainable production' and a 'sustainable society' (1981). Among other catchwords, 'qualitative growth' was one of the first to signal a new direction for societal development. It was argued that growth in GNP-terms for some activities is incompatible with environmental goals and therefore "cancerous", while growth in other activities (with connected goods and services) is mainly beneficial. "Eco-development", meaning ecological development, is another catchword, which focuses on impacts upon ecosystems and the natural resource base of future generations. A high degree of local self-reliance, as opposed to further internationalization, was recommended as a strategy leading to improved environmental performance (Sachs, 1984). Attempts to formulate an environmental ethics were made by Goodland and Ledec (1987) among others.

These attempts to articulate ecological sustainability or an environmental ethics will continue. In my own attempt to contribute to this dialogue, 'nondegradation of the natural resource base' as measured in multidimensional positional terms has been emphasized. A set of "ecological imperatives for public policy" were suggested (Söderbaum, 1982), the argument being that this particular ethics is reasonably operational in assessing the long-term environmental performance of a specific ongoing or future activity or course of action. The focus is on non-monetary positions (or states) of the environment at specific points in time. For instance, will a specific development trend or construction project lead to a degradation or improvement in the state of the environment?

4. The Act of Valuing and Making Decisions

In a situation where the traditional idea of optimising by reference to an objective function has been rejected, or pushed into the background as a special case, some alternative view has to be suggested. Such an alternative idea is given by Figure 3. Valuation and decision making can be seen as a matching process between patterns of two kinds. On the one hand are the impact profiles of each considered alternative, on the other, the specific ideological orientations or profiles that are judged to be relevant in a decision situation.

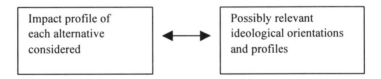

Figure 3. Valuation as a matching process

Instead of the traditional, mainly one-dimensional analysis, a multidimensional analysis is proposed. This is connected with an alternative idea of the decision act and of rationality in terms of 'pattern recognition'. Let us assume that some alternatives for forestry in a given region have already been identified. The analyst could then try to represent each alternative for the growth and development of the forest in visual and other multidimensional terms. Monetary estimates will be part of this picture. In addition, the analyst could try to represent various ideological orientations that seem relevant in relation to current scientific and public debate. 'Ecological forestry' with emphasis on biodiversity and sustainability may represent one such possible ideological orientation and 'traditional forestry', emphasizing monoculture and focusing on the production of cubic metres of timber and monetary profits, might be another. Also various ideas that fall in beween the two extremes, i.e. 'compromise ideologies' might be relevant.

The analyst can formulate conditional conclusions on the basis of his analysis. Assuming that 'ecological forestry', defined in som way, is the relevant ideological orientation, then one of the alternatives, say A2, may be perceived as being closest to this ideology. Some other alternative, say A3, may fit reasonably well into the ideal of traditional forestry, and so on. The actual desision-makers in turn, who normally will differ among themselves with respect to ideological orientation, may find the information supplied by the analyst useful, or not so useful, in making his decision. What counts when the final decision is taken, is of course the decision-maker's own particular ideological orientation, rather than that suggested by the analyst. Similarly, it is the estimates of impacts of the decision maker that counts. But hopefully the preparatory work carried out by the analyst will be useful for those affected and facilitate decision-making.

The term 'pattern recognition' is used in connection with 'image analysis' and visual elements become a natural part (in addition to quantitative estimates) of the information base for decisions. (There is for instance a research tradition related to landscape planning and, more generally, land use changes, which uses drawings, maps, photos, etc. extensively – see many of the contributions in van Lier et al., 1994.) It may be objected that the kind of multidimensional analysis in pattern terms suggested, becomes too complex in relation to the cognitive capabilities of the average decision-maker. My reply to this is that the decision-maker has to take responsibility for his decisions in relation to citizens, those employed, or other stakeholders and if reality is complex, then some part of this complexity has to be reflected in the information base. Simplifying always has a price. If seeking a simple answer were the only consideration, then a dice would serve admirably. I am not saying that CBA is comparable to throwing a dice, but the attempt to capture everything in terms of a 'present value' represents a far-reaching simplification. All kinds of impacts at all times and in relation to all individuals affected are concentrated in one monetary figure.

In relation to the cognitive abilities of human beings, whether university scholars or decision-makers, the latter category exemplified by politicians, I am fairly optimistic. It is true that we as humans have difficulties in memorizing arbitrary numbers, i.e. when we perceive them as rather meaningless. But when meaningful patterns are perceived, things change considerably. Herbert Simon has argued that our strength as humans lies in this ability to recognize patterns such as letters, words and sentences (Simon, 1983, pp. 25-27). I may recognize an individual in the street or the shape of a building, and even in cases where the patterns that I see are incomplete or fragmentary.

5. Economics and Efficiency: A Disaggregated and Ideologically Open-Ended View

The difference between a reductionist and a holistic conception of economics is illustrated in Figure 4. Neoclassical as well as institutional and other economists understand that there are monetary as well as non-monetary impacts in decision situations. The working hypothesis of a reductionist idea of economics, or rather of economic analysis (see upper part of the Figure 4), is that it is fruitful to reduce all kinds of impacts to their monetary equivalent, referring to prices on actual or imagined markets.

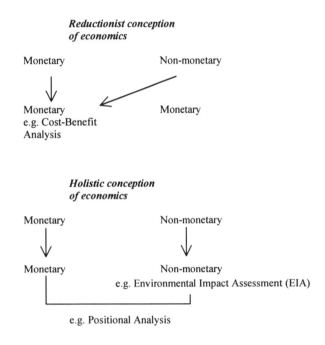

Figure 4. Illustration of the difference between a holistic and a reductionist conception of economics.

The holistic view of economics is instead one of disaggregation. It is believed that different kinds of impacts should be kept separate throughout the analysis (see lower part of the figure). The analyst aims at multidimensional impact profiles, as already indicated.

As examples of reductionist analysis at the societal level, CBA can be given, whereas Positional Analysis (to be described here) is based on a holistic and disaggregated idea of economics. Most versions of Environmental Impact Assessment (EIA) also represent a disaggregated analysis, although limited to environmental impacts. In the case of corporations, or other organizations, monetary cost-revenue analysis is generally part of the picture. Some of the more recent approaches to environmental auditing and reporting (for instance, the Eco-Management and Audit

Scheme, EMAS, of the European Community) exemplify an understanding of the possible advantages of keeping monetary and non-monetary controls separate.

As part of a multidimensional analysis, impacts can be classified according to Table 2. In addition to the previous distinction between monetary and non-monetary impacts, a distinction between 'flow' and 'positional' variables is needed. GNP and the turnover of a forest company are examples of monetary flows (category I), whereas the assets or debts of a country or a corporation exemplify monetary positions, states or 'stocks' (category II). Since qualitative aspects of a situation at a point in time are no less important than the quantitative ones, I prefer the term 'position' to 'stock'. As a concept in economics, stock has mainly been used to refer to homogeneous units of things, such as 'stock of capital'.

It has already been made clear that non-monetary impacts are regarded as being just as 'economic' as the monetary ones. On the non-monetary side, the discharge of mercury from a pulp factory into a nearby lake per year is an example of a non-monetary flow (category III) and the amount of mercury per kilogram of fish caught at a specific place in the lake is a non-monetary position (category IV). Total biomass and the relative presence of various species in a forest are examples of non-monetary positions related more directly to forestry. The health and educational status of an individual are further examples of non-monetary positions.

In a well-known article, "The Economics of the Coming Spaceship Earth", Kenneth Boulding (1966) argues that the welfare of nations and of individuals has more to do with stocks (or non-monetary positions, in my vocabulary) than flows. While Boulding emphasizes the role of stocks in welfare measurement, it should be made clear that all four categories of variables in Table 2 are potentially relevant. They are furthermore interrelated, for instance in the sense that flows of mercury influence positional changes in the mercury content in fish. Depending on the characteristics of mercury as a pollutant and of the lake as an ecosystem, a reduction in flow terms (from, for instance 20 kilograms last year to 15 kilograms this year) may not be enough for an improvement in positional terms to occur. A much smaller amount of yearly pollution may be necessary and even with zero discharges for a number of years, some mercury poisoning will remain. This is the issue of irreversibility, which should, I believe, preferably be described in non-monetary positional terms.

TABLE 2. Four categories of variables in economic analysis

	Flow (referring to a period of time)	Position (referring to a point in time)
Monetary	I	II
Non-monetary	III	IV

What is the meaning of efficiency in relation to the frame of reference suggested? Proponents of CBA themselves realize that an uneven income distribution in society may represent a problem since the 'monetary voting power' of those who own or earn little, may play a negligible role in relation to the voting power of those who own or earn a lot. What they do not seem to understand, however, is that also the efficiency idea that they rely on is controversial. As we have argued, concerning most public policy issues relating to forestry, energy, housing etc., any assumption that there is consensus about one idea of efficiency – more precisely, a set of valuation rules dictated by neoclassical economists as a group – is unrealistic.

In discussing efficiency, one can focus on some ongoing activity or some 'new' activity considered for implementation. Let us assume that we are interested in corn production at some particular place. In terms of Table 2, there are many possibly relevant (non-monetary and monetary) inputs or inflows into the production process and many outflows or outputs. A large number of ratios between outputs and inputs can be constructed, rather than one. Kilograms of corn can be related to hectares of land, as is often done, or one can relate kilograms of corn to work hours. Corn exemplifies a desired output, but there may be undesired outputs as well, such as leaching of pesticides, phosphorus and nitrogen into nearby waters. This means that there are additional possible non-monetary efficiency indicators that refer to environmental performance. (Such indicators may also refer to output–output relationships, as when a desired output is related to an undesired output.) The ratio of one period may be compared with that of the previous period, to see whether things are progressing in some sense. But as already suggested, controlling changes in positional terms is also an important part of the evaluation of an activity. Is the state of the soil or of groundwater improving or not, according to specific indicators? Is the quality of the desired product – corn in our example – improving or not? While positional changes between the beginning and end of a period may be small, over a longer time span, significant changes may occur.

Assuming now that the corn is produced for some distant market, one has to focus on a chain of activities, such as production of the various means (pesticides, fertilizers, etc.) used in corn production, transportation activities (which normally are 'consuming' more non-renewable resources and polluting more, the longer the distance), storage activities, wholesaling activities, consumption, waste disposal and so on.

If a person argues that he or she knows the one and only *correct* way of measuring efficiency from a societal point of view of such a chain of activities, or even of one single activity, then there is reason for scepticism. By overlooking a large number of impacts and only focusing on one ratio, it is of course possible to imagine that one knows when things are improving or deteriorating in efficiency terms. My recommendation is, however, to rely on a large number of indicators and to carry out a many-sided analysis. Rather than assume it away, one has to live with some complexity.

6. Democracy and the Roles of Analyst, Politician, Citizen and Interested Party

Approaches to valuation and planning differ with respect to their relative emphasis on process or product. Traditionally, there has been much emphasis on the final document, for instance a development plan, put forward to interested parties and decision makers. But the interest in process is increasing (Fischer, 1990; Khakee and Eckerberg eds, 1993). If our subject is valuation, then the objects of valuation are not limited to specific alternative courses of action , but may concern the planning process itself. Does the planning process meet specific interpretations of the current rules of the game or, more generally, reasonable standards with respect to various aspects of democracy?

Each approach to decision-making carries with it specific ideas about desired roles for the analyst, politicians or other decision-makers, those affected (i.e. stakeholders or interested parties) and citizens. Table 3 represents an attempt to contrast roles connected with the more traditional approach of CBA and roles that I associate with a more democratic process.

TABLE 3. Roles in public planning: comparison between a traditional vie and an emerging more democratic view.

Actor categories	Traditional view of roles (c.f. CBA)	'Democratic' view of roles
Analyst	Optimal resource allocation referring to an alleged neutral approach. One alternative is recommended (Expert)	Responsiveness and service. Many-sided illumination of decision situation. Conditional conclusions (Facilitator)
Politician	Expected to accept the efficiency idea inherent in the method used and the conclusions reached	Each politician matches his/her ideological profile with the expected impacts of each alternative considered
Interested party	Regarded mainly as a consumer who may be asked to express his/her market preferences. Expected to accept the result of societal optimization.	Interested parties are encouraged to participate in an active, open debate about the planning process and the problems faced
Citizen	Only opinions referring to the role of consumer or other market actor might be regarded as relevant	Participation in debate encouraged, the belief being that this can improve the results of the planning process

As part of CBA, the analyst tends to become an expert in an extreme sense, whereas a more desired role for the analyst would be one of 'facilitator'. The latter role points to the fruitfulness of listening to various interested parties and other actors, for instance what alternatives they see as relevant and how they perceive problems and opportunities. Listening to and documenting the stories or 'narratives' of various actors becomes a natural part of the planning process. Actors in their different roles participate in an interactive learning process. To the extent the analyst is able to listen and learn, he will profit from taking part in the process and hopefully the outcome of the process will improve. As suggested by Forester (1993, p. 76), learning about values (or in the present language, about world-views and ideologies) is an important part of this.

There are other possible roles in addition to those emphasized in Table 3. The analyst may act as a 'devoted analyst' (cf. Mishan, 1982, p. 35), i.e. a person who makes his or her study (and other actions) suit specific power holders, whether they be politicians or business corporations, or some agglomeration of interested parties. Analysts may, furthermore, act as if they were 'one issue politicians', defending specific conservation or development interests, and so on.

7.   A Classification of Approaches to Valuation

This brings us to a classification of approaches to societal project appraisal. Again, such a classification is partly a matter of ideological preferences. Having observed and participated in the methodology debate for some time, I think that 'degree of aggregation' is a relevant distinguishing characteristic between approaches. Nathaniel Lichfield pioneered the attempts to shape approaches of a less aggregated kind (Lichfield, 1964, 1975) by recommending an analysis that clearly separates various interested parties. This represented a departure from conventional cost-benefit analysis (for instance his distinctions between construction period and operational period, his reference to multidimensional thinking in qualitative as well as quantitative terms and his recognition of the ethical issues involved) although in his early writings Lichfield retained a large part of the neoclassical vocabulary, with producers and consumers, etc.

Along these lines approaches may be classified in one of three categories:

A.   Highly aggregated   Example: CBA
B.   Intermediate   Example:Cost-effectiveness analysis
C.   Highly disaggregated   Example: Positional analysis, EIA

While not always achieving this objective, CBA aims at a summation of all impacts in terms of a present value and thus belongs to the highly disaggregated A-category.

Cost-effectiveness (CEA) is rather two-dimensional (in its ambition to find, for instance, the monetary cost of achieving some non-monetary environmental objective) and is therefore placed in some intermediate category. But CEA retains the idea of a single objective function and optimization. Positional Analysis (PA) and Environmental Impact Assessment (Glasson *et al*, 1994) exemplify disaggregated approches. There are many versions of EIA and not all would fall into this category. Similarly, Lichfields more recent Community Impact Analysis, CIA (Lichfield, 1993) – which with respect to impacts is broader in scope than EIA – falls into the C-category with some openings into the intermediate category. Some approaches, for instance multiple criteria approaches are not easily categorized with respect to aggregation. These approaches may retain much of the idea of aggregation, but are rather open with respect to kind of valuational basis for such an aggregation.

8. Positional Analysis

While being different in many ways, positional analysis shares certain features with CBA. Among these is a firm belief in the meaningfulness of systematic study, i.e. analysis in preparing decisions. Other characteristics shared with CBA are identification of possible alternative courses of action, and a systematic comparison of these alternatives. CBA may also start with a multifacetted and multidimensional view of impacts, although the ambition subsequently to reduce such impacts to monetary terms may influence the entire approach from beginning to end.

What is abandoned in PA, on the other hand, is the idea of 'solving' a problem in a way that is correct for purposes of societal resource allocation. Such an objective is too ambitious and, as we have seen, can even negate democracy. Instead, the main purpose of analysis should be one of *illuminating* a decision situation. When no consensus regarding valuational rules can be assumed, one has to refer to two or more valuational standpoints that are possibly relevant to decision makers and those affected, and draw conditional conclusions on the basis of this. Such an approach is more modest than the CBA approach and also transforms the role of the analyst from one of optimizer and expert in an extreme sense, to that of facilitator in a public dialogue. Open-mindedness in relation to all actors and interested parties and many-sidedness in analysis become important criteria of good performance.

Such many-sidedness concerns:

- views of the problem,
- alternatives considered
- impacts of alternatives,
- interests affected, and
- possible valuational standpoints.

Many-sidedness is seen as crucial for a democratic decision process, while one-sidedness in the above senses is connected with manipulation. The extreme form of manipulation is then a study where the problem is perceived only in one way, where only one alternative is considered, only one kind of impact is identified, only one interest taken into consideration and reference is made to only one valuational standpoint. As is already made clear, CBA may be many-sided in some respects but not with respect to valuational standpoint.

While many-sidedness is the ideal, there are normally constraints in terms of search costs and deadlines. Maximizing 'many-sidedness' may lead to confusion, rather than lucidity. In many cases it is, for instance, a good idea to limit the number of alternatives to three or four. Ideally, these alternatives should differ significantly and match the broadness of alternatives advocated by different actors. At a later stage in the analysis, it is possible to return to modifications (marginal adjustments) of each of these alternatives.

Depending upon the kind of decision situation and the social and institutional context, simplified versions of positional analysis can be considered. A more ambitious study would include the following:

(a) Description of the decision situation: historical background; Relationships to other decisions (previous and simultaneous); identification of relevant institutional context (organisational aspects and rules of the game) and interested parties.

(b) Identification of problem/s; reproduction of problem images as perceived and stated by different actors and interested parties.

(c) Design of alternatives and formulation of the problem (choosing a set of options for further consideration); identification of relations between the present and other decision situations and policy options.

(d) Identification of systems that will be affected differently, depending on the alternative chosen.

(e) Identification of impacts (monetary and non-monetary, in flow and positional terms) and comparison of alternatives in relation to these effect dimensions.

(f) Study of possible inertia and irreversibilities in non-monetary positional terms. In what way will first-step alternatives influence future options for different affected parties?

(g) Analysis of activities and interests in relation to the decision situation: identification of activities that will be affected differently depending on the alternative chosen; assumption of goal direction for each activity; activity, together with goal direction, defines an interest; construction of preference ranking from the standpoint of each activity in relation to each alternative.

(h) Analysis of prevailing risks and uncertainties; formulation of possible futures (scenarios).

(i) Summary of information basis for decisions at the two levels of impacts and activities (with associated assumed interests).

(j) Articulation of possible valuational standpoints in terms relevant to the study area (for instance transportation ideologies, energy management ideologies, health care ideologies, etc.)

(k) Conditional conclusions, relating the expected impacts of each alternative to possible futures and valuational standpoints.

Only a few of the above elements of PA will be commented on here. While the CBA analyst tends not to bother much about how present problems came about, this historical aspect is considered important as part of PA. As part of an interest in the process aspects of planning, the institutional context is also described. What are the rules of the game in terms of laws and according to the expectations of various actors and interested parties?

In two senses, PA is based on a multi-perspective philosophy. The analyst is expected to listen to other actors and to the interested parties and to try to interpret their various perspectives and opinions, not only at an early stage, but throughout the planning process. Secondly, since the CBA analyst aims at a specific number, such as a present value for each alternative, he/she must be careful to avoid 'double-counting'. PA, on the other hand, is based on the idea of complementarity between perspectives. The alternatives and their impacts can be elucidated from different angles. Alternatives can be compared at two levels, one referring to impacts and the other to affected activities and hence interests. The two pictures may overlap, but generally complement each other.

'Systems thinking' and positional thinking are essential elements of the PA approach. At an early stage in the analysis and based on some set of alternatives considered, an attempt is made to identify the systems affected. This is a way of moving from single- sector thinking to a consideration of all sectors affected. Systems thinking also represents an incipient attempt to identify differences between alternatives with respect to non-monetary and monetary impacts. The meaning of 'positional thinking' is simply 'to think about impacts in terms of non-monetary positions'. The implementation of one alternative leads to a series of positions for different future points in time. In many cases it is fruitful, however, to see the choice of one alternative as only the first in a series of steps. The choice of one alternative may imply irreversible positional changes or impacts on future options. To illuminate such aspects of future options, decision trees are used (see Söderbaum, 1987). Instead of the usual decision-trees with 'pay-offs' at the end-points as part of conventional game theory or decision theory, impacts are thought of largely as a series of non-montary positions. In relation to planning in urban and rural areas, qualitative change in land use is a relevant example.

An attempt is made as part of PA to illuminate commonalities and conflicts of interest. Just as we were previously looking for systems that were differently affected, depending on alternative chosen, the focus is now on *activities* that are differently affected, depending on the alternative chosen. A number of such human and organizational activities may be relevant, for instance transportation activities between specific geographical points, activities in specific areas such as commerce, forestry and agriculture; water supply, housing and recreation etc. For each activity, a goal direction is *assumed*, which in turn can be used for the ranking of alternatives (in relation to the activity). For transportation, ease of mobility in terms of time and comfort may be chosen as the assumed objective. For those living close to a highway, reducing environmental burden may be a relevant objective, etc.

Table 4 illustrates how activities and alternatives can be related to each other. Should it prove difficult to distinguish between alternatives, this can be indicated by a question mark . There is seldom one alternative which is number one in relation to all activities and hence assumed interests. Conflicts of interest are the norm and such conflicts should be made visible rather than concealed.

TABLE 4. Ranking of alternatives in relation to each activity, with assumed goal direction

Activity	Goal direction	Alternatives		
		A_0	A_1	A_2
AC_1	GD_1	1	2	3
AC_2	GD_2	2	3	1
AC_3	GD_3	2?	3?	1
.				
.				
AC_k	GD_k	2	1	2

By discussing valuational standpoints explicitly, the PA approach may challenge the established ideologies of decision-makers. Expressed in a positive manner, the politician gets an opportunity to modify or reconsider his or her values when faced with possible valuational standpoints and conflicts of interest that are made visible. Such an open discussion of ideology is important, not only for decision-makers and powerful interested parties, but for all affected bodies and citizens in general. This feature of PA, to visualize conflicts of interest and to clarify how activities will be influenced, for instance by choosing A_1 rather than A_0, will not be popular among all actors or interested parties. Not all actors want to know – or like others to know about – the consequences of their actions.

Summarizing all impacts, and more generally the information basis, is an art in itself. Tables where impact profiles of alternatives are compared at the levels of impacts (no illustration given here) and activities (as in Table 4) can be used. A document of 200 pages can be 'reduced' to an executive summary, as exemplified by Brorsson (1995).

Finally, it should be made clear that PA, so far, is a mainly Scandinavian phenomenon. A number of theoretical and applied studies have been presented in Swedish, Norwegian and Finnish, beginning with a PhD thesis (Söderbaum, 1973). Applied studies have been carried out in the fields of road planning, localization of airports, energy systems, urban planning, bedrock quarrying, investments in tourist facilities, forestry and so on. In addition to the study of investment options or future activities, some studies concern past activities, so called 'retrospective PA' (Hillring 1996, Forsberg 1993). Some articles have been published in English (Söderbaum, 1987, 1992) and also a PhD-study at Helsinki University (Leskinen, 1994) on theoretical and practical aspects of road planning in Finland (viewing PA mainly as a language for communication between various actors). Two students have produced a Master's Thesis on the Urrá I Hydro Power project in Colombia (Edlund and Quintero, 1995) and a 'partial PA' has been carried out concerning water pollution in the valley of the Msimbazi River in Tanzania (Mafunda and Navrud, 1995).

9. Conclusions

The political element in economics has been a theme in this essay. Valuation cannot be separated from politics and ideology. The idea of a value-neutral scholar or analyst has to be abandoned. Values and ideological orientations are present not only in our roles as citizens but also as scholars, analysts or other professionals. It is therefore necessary to discuss and handle these ideological issues openly.

Road planning, housing projects or urban planning in general are normally connected with conflicts of interest. Environmental awareness has led to ecologism and other environmental ideologies which may further aggravate conflicts rather than reduce them. A growing number of citizens and professionals reject simplistic ideas of economic growth in GNP-terms. In this situation one has to fall back on democracy as a kind of meta-ideology. Dialogue and conflict management become key concepts. Approaches that refer to correct societal valuations or 'welfare tests' are no longer relevant and may indeed confuse public debate or aggrevate conflict. Only the more disaggregated approaches that leave some room for differences among actors and citizens with respect to ideology seem compatible with democracy.

Positional analysis is suggested as one such disaggregated approach. It is built on Political Economic Man/Woman assumptions, a holistic idea of economics and rationality, systems thinking, positional thinking, analysis of conflicts of interest and conditional conclusions. The idea is to illuminate an issue in a democratic context with the help of interested parties and other actors who may differ with respect to ethics and ideology. In principle, each individual may have his/her particular idea of a welfare test. And such ideas include egoistic as well as other-related considerations.

References

Boulding, K.E. (1966) The Economics of the Coming Spaceship Earth, in: H. Jarret (ed.) *Environmental Quality in a Growing Economy*, John Hopkins, Baltimore, pp 3-14.

Brorsson, K.-Å. (1995) Methodological Development of Positional Analysis and an application to Assjö Water Mill. Sustainable Development in Relation to Environment and Vulnerability (in Swedish). *SLU, Department of Economics, Dissertations 14*, Uppsala..

Brown, L.R. (1981) *Building a Sustainable Society*, Norton, New York.

Dobson, A. (1995) *Green Political Thought*, Routledge, London.

Eccleshall, R. *et al.*. (1994) *Political Ideologies. An introduction*, Routledge, London.

Edlund, J. and Quintero R. (1995) Do Wabura – Farewell to the River! Application of Positional Analysis to the Urrá I hydro power plant in Columbia, *SLU, Dept. of Economics, Report 94*. Uppsala.

Etzioni, A. (1988) *The Moral Dimension. Toward a New Economics*, The Free Press, New York.

Fischer, F. (1990) *Technocracy and the Politics of Expertise*, SAGE, Newbury Park.

Forester, J. (1993) On the Ethics of Planning, in Khakee, A. and Eckerberg K. (eds), pp. 63-90.

Forsberg, G. (1993) Energy systems and ecological sustainability. The municipality of Kil as a case (in Swedish). *SLU, Department of Economics, Report 63*, Uppsala.

Glasson, J., Theerivel, R. and Chadwick, A.(1994) *Introduction to Environmental Impact Assessment. Principles and Procedures, Process, Practice and Prospects*, UCL Press, London.

Goodland R. and Ledec G. (1987) Neoclassical economics and principles of sustainable development. *Ecological Modelling* 38 (1/2), pp. 29-46.

Hillring, B. (1996) Forest Fuel Systems Utilising Tree Sections – System Evaluation and Development of Methodology, *SLU, Faculty of Forestry, Studia Forestalia Suecia No.200*, Uppsala.

Khakee, A. and Eckerberg K. (eds) (1993) Process & Policy Evaluation in Structure Planning, *Swedish Council for Building Research, Report D8:1993*, Stockholm.

Leskinen, A. (1994) Environmental planning as learning: The principles of negotiation, disaggregative decision-making method and parallell organization in developing the Road Administration. *University of Helsinki, Department of Economics and Management, Publications No. 5, Land Use Economics*, Helsinki.

Lichfield, N. (1964) Cost Benefit Analysis in Plan Evaluation *Town Planning Review*, 35 (2), pp.159-169.

Lichfield, N. (1993) Community Impact Analysis in Plan and Project Evaluation, in Khakee, A. *et al.* (eds), pp. 103-131

Lichfield, N. Kettle P. and Whitbread M. (1975) *Evaluation in the Planning Process*, Pergamon Press, Oxford.

Lier, H.N. van, Jaarsma, C.F. Jurgens C.R. and de Buck A.J. (eds) 1994. *Sustainable Land Use Planning*, Elsevier, Amsterdam.

Mafunda, D. and Navrud, S. (1995) Positional Analysis Applied to Water Pollution Problems in Developing Countries, in Dinar, A. and Tusak Loehman E. (eds) *Water Quantity/Quality Management and Conflict Resolution. Institutions, Processes, and Economic Analyses*. Praeger, Westport, pp. 427-437.

Mishan, E.J. (1980) How valid are economic evaluations of allocative changes? *Journal of Economic Issues*, 14 (1), pp. 143-161.

Minhan, D.J. (1983) The non controversy about the rationale of economic evaluation, *Journal of Economic Issues* **16** (1), pp. 29-47.

Myrdal, G. (1978) Institutional Economics, *Journal of Economic Issues*, **12** (4), pp. 771-783.

Norgaard, R.B. (1994) *Development Betrayed. The end of progress and a coevolutionary revisioning of the future*, Routledge, London.

North, D.C. (1990) *Institutions, Institutional Change and Economic Performance*, Cambridge University Press, Cambridge.

Pearce, D.W. and Turner R.K., 1990. *Economics of Natural Resources and the Environment*, Harvester Wheatsheaf, Hempstead.

Sachs, I. (1984) The strategies of eco-development. *Ceres. FAO Review on Agriculture and Development*, No. 17, pp. 17-21.

Sagoff, M. (1988) *The Economy of the Earth. Philosophy, Law, and the Environment*, Cambridge University Press, Cambridge.

Self, P. (1975) *Econocrats and the policy process. The politics and philosophy of Cost-Benefit Analysis*, MacMillan, London.

Simon, H. (1945) *Administrative Behavior*, Free Press, New York.

Simon, H. (1983) *Reason in Human Affairs*, Basil Blackwell, London.

Söderbaum, P. (1973) *Positional Analysis for Decision Making and Planning. An Interdisciplinary Approach to Economic Analysis* (in Swedish), Esselte Studium, Stockholm.

Söderbaum, P. (1982) Ecological imperatives for public policy, *Ceres. FAO Review on Agriculture and Development* **15** (2), pp. 28-30.

Söderbaum, P. (1987) Environmental Management. A Non-Traditional Approach, *Journal of Economic Issues*, **21** (1), pp. 139-165.

Söderbaum, P. (1992) Development: evaluation and decision-making, in Ekins P. and Max-Neef M. (eds), *Real-Life Economics. Understanding Wealth Creation*. Routledge, London, pp 127-144.

Söderbaum, P. (1993) Values, Markets, and Environmental Policy: An Actor-Network Approach, *Journal of Economic Issues* **27** (2), pp. 387-408.

Tietenberg, T. (1992) *Environmental and natural resource economics*, Harper Collins, New York.

World Commission on Environment and Development (1987) *Our Common Future*, Oxford University Press, Oxford.

DEALING WITH ENVIRONMENTAL CONFLICTS IN EVALUATION

Cognitive complexity and scale problems

A. BARBANENTE, D. BORRI, G. CONCILIO, S. MACCHI, E. SCANDURRA[1]

1. Introduction

This contribution discusses the inadequacy of instrumental rationality when dealing with environmental problems, and thus facing complexity and uncertainty characterising both cognitive models and organisations involved. This issue is seen as crucial in environmental planning practice, a field where traditional evaluation processes present many undetermined aspects concerning both latent cognitive structures and problems lacking of solution.

The paper can be divided into four parts. It starts with a review of the paradigm of rationality, arguing the need of a new collocation and definition of evaluation according to the epistemological shift claimed by the new so-called "complexity paradigm" (par. 2). Within this approach, Science and Technique can not be considered any more as neutral, disinterested and unique, and the role of contingency in problems definition and analysis is emphasised, especially with reference to the concept of development, which cannot be seen any more as following a linear evolution and coincident with progress. An example of possible consequences of this shift in perspective is presented (par. 3) with reference to a decision-making process concerning the construction of a road infrastructure: conflicts arising during this process are seen as potential sources of fresh, social based, and unexpected proposals.

The second part examines some problems emerging in environmental planning and decision-making. Two aspects are stressed: that of local/global tensions, which is to be considered as a fundamental issue if we recognise that actions at smaller scales will have effects on global self-regulatory mechanisms following a causal chain that is anything but clear; and that of conflicts arising in environmental planning (par. 4). In a context of major public concern for environmental protection, the relation society-environment should be considered also as a scale-dependent problem, without taking for

---

[1] This contribution is the result of the authors' joint work. Their contributions are as follows: Paragraphs 1, 4, 5 and 6 are by Angela Barbanente, Paragraphs 2 and 3 by Silvia Macchi and Enzo Scandurra, Paragraphs 7 and 8 by Dino Borri. Grazia Concilio participated in the work providing for a number of interesting case-studies of environmental problems setting and solving in multi-agencies.

N. Lichfield et al. (eds.), Evaluation in Planning, 73–95.
© 1998 *Kluwer Academic Publishers. Printed in the Netherlands.*

granted either that involving the general public at the local level or assuming a concept of general interest for environmental actions will imply more environmental sustainable decisions (par. 5); thus, seeking for consensus as the dominant way of dealing with conflicts is also questioned. (par. 6).

Then the paper focuses on theoretical models of organisation in multi-agent environment, emphasising the positive role of endogenous means of knowledge and communication. Non continuous paths of action are viewed as effective and creative in complex domains. Conflict and co-operation issues are analyzed with regards to languages, intended as tools for knowledge expression and transmission (par. 7).

Finally, also referring to recent studies in the field of environmental economics and planning, important unresolved problems of evaluation and decision are stressed, together with the perspective of transformative theory-in-action as an essential direction for further research in the field (par. 8). This implies taking advantage from integrating traditional information and meta-knowledge and heuristics, expert and non-expert knowledge, communicative planning and evaluation, systematic knowledge and creativity, CVM and assessment of individual positions with other more aggregate methods, as well as making evaluation procedures more sensitive to fragile or dissenting voices and developing more friendly and articulated planning and evaluation systems.

2. Origins of the Epistemological Shift in Sciences and their Consequences on Evaluation

Dealing with evaluation includes as a preliminary question to settle the evaluation notion inside the present epistemological change which has led to quit the large and diffuse vision of Science and Technique as "neutral", "disinterested", "unique". This vision, although mostly abandoned by scientists, already finds consensus in the multitude of anonymous practices (which by Foucault are real motors of development and history). We will try to redefine evaluation notion through the change of two paradigms which have engendered such a notion.

The first paradigmatic change concerns the validity and effectiveness of the scientific rationality model proposed by Galileo, Newton, Laplace, Bacon, Descartes and Smith. This largely pervasive model persuaded us on reason omnipotence and ability to dominate the world. The Big Book of Nature was written in intelligible and accessible symbols of mathematics and physics. It gave us a Nature we could control because ruled by deterministic laws. Bacon theorem said "truth and utility are the same thing". Despoiled of its presumed scientificity, this theorem continues to be, in a fuzzier and shiftier way, a prevailing thinking. It is the basis of modern "calculating and utilitarian reason" paradigm, according to which Science would allow people to understand everything, Technique would allow to solve any problem and Market to buy everything.

Today this rationality model has lost its general validity to understand (and change) world. Nature is an equivocal text that can be read in different ways. Cognitive science, ecology, evolutionism, neurophysiology are sketching an history of living that is radically different from the traditional optimising development idea (Ceruti, 1996). Complexity approach is disclosing to us a world that is the kingdom of all what Galileo had left out of his grandiose model: the model which assures us of Mankind magnificent progressive destiny. Present vision of reality is characterised by discontinuity of evolution and history (Gould, 1980), irreversibility of natural process (Prigogine, 1996), systems auto-organisation (Maturana and Varela, 1985), plurality of possible futures (Ceruti, 1996). Sooner we will perceive it, better it will be.

The second change concerns "development" and "progress" notions. Modernity has entrusted to (western) Science and Technique the task to emancipate human society all around the world. But development is not a neutral notion (Barcellona, 1994, 1996a, 1996b; Latouche, 1992, 1995; Morin, 1995). It is an ambivalent notion that has produced and is producing, on the one hand, the foreseen, nice effects that we know and, on the other hand, wicked, unwished, often unexpected effects. This notion, that coincides with world westernization, is responsible of present environmental crisis and man from nature and mind from physis separation.

Modernity abruptly breaks with tradition; it moves us far away from all established kinds of social organisations (Giddens, 1990). It entrusts to "experts" and their technical practices (or technical knowledge) the task to resolve every problem and to build a better world in the name of Development and Progress. But, against any evidence, no one can persist any more in not seeing that, in the global balance of relationships between persons and world, positive values have been equalled and even overcome by negative ones (unemployment, wars, ecological disasters, ...). Physiology of what we call development goes on through constant growth of goods production, in post-fordism rather through productivity growth of time unit, without any manual human work: goods which produce goods, money which produce money.

Then, there are two preliminary questions with which we have to deal in order to redefine evaluation.

First, we must recognise that disinterested knowledge does not exist. It means it is necessary to make explicit the tacit premises, unexpressed points of view, meta-goals and paradigms which underlie "technical knowledge". In fact experts hide besides their knowledge, less or more consciously, a consent to prevailing values system.

Second, we have to wonder if it is possible that a knowledge growing by centuries through nature domination and exploitation can be used for different goals; if it is possible to succeed in modifying this knowledge without abandon its underlying goals.

We refuse "calculating and utilitarian reason" paradigm in order to get a new larger and less contingent perspective of sense, centred on the "social bond", the bond who changes single individuals in "social individuals". In this perspective evaluation is not put aside but placed in a different context. We want to abandon the economic logic with its cost-benefit approach as it does not contribute to strengthen social bond, but rather atomises individuals.

Economic rationality is based on profit calculation (a system is economically rational if it maximises net utility) and its main value is economic efficiency. According to economic rationality, nature is considered as a "resource" to be used, with which different individuals or groups, each having their own interests, compete. Environmental conflicts, engendered by resources competition, are treated as "zero sum" games, where each individual utility implies a loss for other people and it is always possible to identify who gains and who loses. In this merely economic logic, social choice consists in selecting a limited number of individuals who are directly concerned by resources competition and in satisfying their preferences (Dryzec, 1987). Evaluation techniques as cost-benefit analysis deal with these preferences and try to maximise global economic productivity of natural resources within an affordable discontent of concerned individuals. The result is a natural resources use assessment which produces the largest wealth a community can obtain without running the risk of social subversion due to too high individuals discontent.

But this way of making social choices induces also some problems into social systems dynamic.

First, social systems are dissipative like any living system, that means they need some kind of wasting (energy, matter, time, ...). Lynch makes a clear distinction between the waste recognised in economics, by the cost-benefit analysis, and a second type of waste, contemplated by ecology, which he defines as "evolutive" waste. Economics do not allow the separation of the final judgement (a wasteful result) from the process which engendered it (a wasting process), and, on the contrary, recognise a basic similarity between waste and dissipation. Economic waste is "simply inefficiency: time, effort or resources spent without useful results or without extracting the maximum human value possible" (Lynch, 1990). This type of assessment has no sense when applied to ecology (mind, social and natural ecology) where the dissipation processes produce life. "If life and growth are my ultimate value ..., and if dissipation is an essential component of the living and evolving system, it can be judged on the basis of its capacity to sustain this organised growth of complexity. Thus a 'wasteful' event is one which produces discontinuity in biological or cultural growth" (Lynch, 1990). So, aiming at reaching largest economic productivity carries with it the risk to inhibit living processes which are indispensable to humans being and social reproduction.

Second, if economic waste is living processes "gasoline", their "motor" consists in social interaction. Conflicts, which are a form of social interaction, can play a real positive role in social systems if conflicting individuals are relied by a web of strong social bonds. But evaluation techniques are engendered by the wish to reduce social conflicts and they do not care about social bonds; in order to attain this goal, these techniques exclude from social choice process the largest number of persons they can (all who have not direct economic interest in using natural resources) and try to do not discontent the few others. This allows to maintain social order but also inhibits any social innovation and creativity, which need impulse of conflicting interactions as well as co-ordination of social bonds.

3. Promoting Social Innovation and Creativity in Decision-making: An Example

We will try to make explicit what aforementioned by means of an example we get from our national today chronicle. In the last few months Italian public opinion has been engaged in a large debate on the project of widening the Bologna-Firenze motorway. Two fronts oppose each other: "development" supporters, who believe speedy and safe communication availability is the main condition to take part in world-wide network of economic globalization; "environment" supporters, who - really not challenging economic globalization - think it is necessary to reduce its environmental costs. There is a third front too, the silent front of excluded people: those who have no economic interest in globalization; those who will never go from Bologna to Firenze or will never do it if globalization would not oblige them to do it; those who have been despoiled of their mental and social resources by globalization and now are ready to grasp all this project can offer to them, that is a few opportunities of working in the motorway construction and a few minutes saving the only time they really will have to go from Bologna to Firenze. The debate on widening the Bologna-Firenze motorway project goes on according to economic rationality rules: only people who have direct economic interests in market globalization are allowed to participate, strengthening a classical "social bond" rupture due to western development model; "development" versus "environment" supporters conflict does not put out classical economic goals underlying the widening proposal, so that a mere EIA can be naively assumed as able to overcome the conflict without none of the two fronts going to ask for real, deep changes in the proposed project.

What would happen if economic rationality will be abandoned in favour of an "ecological" rationality which supposes that environmental conflicts are "positive sum" games? What would happen if the goal was not to reach an optimal, or even satisfying, allocation of costs and benefits, but rather to co-ordinate all community members in order to obtain a common and indivisible good like the future generations right to life?

First, if we have to assume a real common goal we cannot exclude any community member from social choice process. Economic globalisation can not be a common goal just because it entails social exclusion. If the motorway widening must be considered as a project of public utility - and it is sure that funds to have it will be public - this will be not because of its contribute to such an unequal development model.

Then, if we want to grasp the public utility character of this project we must go back to the original problem that government wishes to resolve by proposing the motorway widening. The problem is that the existing motorway is inadequate to the present traffic demand so that cars flow slowly and there are frequent accidents. It entails high social, economic and environmental costs which are public costs. So the real goal of any public action in this context is not to enlarge the existing motorway but to eliminate these public costs. This new goal admits a large range of solutions as reducing traffic demand by means of activities delocalisation or shifting part of this traffic demand from motorway to railway. The original motorway widening project becomes one of a number of possible solutions.

Now we will try to design a solution following guidelines of what we call ecological rationality whose main value is social bond. "Development" supporters want that a large quantity of goods and persons could flow speedily and safely through all Italy. "Environment" supporters want that it could happen with minimum environmental damage. Are we really sure that the conflict between these two fronts must be lessen or, on the contrary, this conflict goes just to be the source of creative solutions, as for instance a multitude of alternative passages from North to South of Italy shaping a highly connected, redundant network without risk of traffic congestion? The third silent front is still waiting to be compensated for the resource waste stemming in the last decades in Italy from motorways and high speed railways construction and maintenance. Is there any sense in continuing to exclude them when they may suggest this multitude of passages we are looking for just through their territories, passages which can become the source of a multitude of transit economies for them? No living process can afford to waste such a patrimony of creativity.

4.   Local/Global Decision Contexts of Environmental Problems

The idea that following an ecological rationality we are likely to promote social innovation and creativity in decision-making processes leading to choices which allow to reach common, indivisible goods is anything but easy to put in practice. In the followings we start to investigate some questions which seem to emerge as problematic in this context and which are likely to challenge, or even undermine, the affirmation in real world situations of such a different model.

Local-global intertwines characterising current economic and social dynamics induces to reconsider traditional ways of dealing with such problems in planning and evaluation, carrying with it some unavoidable contradictions: on the one hand, local communities are oriented towards finding out peculiar, self-reliant, autonomous paths of development, identifying their needs and opportunities, on the basis of their own cultures and abilities to choose (Friedmann, 1987; Sachs, 1988); on the other hand, it is recognised that is impossible to consider a place as a singular, pure and homogeneous context, exempt from complex and undesirable intrusions coming from globalization (Giddens, 1990; Harvey, 1989; Amin, Thrift, 1993), also due to the rapid growth of information technologies (Rogers, 1986; Castells, 1989). Really, places have not predefinite identities, but are built through the juxtaposition, intersection, articulation of social multiple relations and should be considered as shared spaces, riven with internal tensions and conflicts (Massey, 1991).

Local-global tensions are crucial when dealing with environmental problems, while the origin of environmental issues is mainly global and people now acknowledge more and more the risk of disruption of great planetary regulative mechanisms by human activities, we still lack of local visions and actions consistent with this perspective and with the aforementioned complexity paradigm leading us to consider the single, local point as crucial. For this paradigm a minor, random event, divergent from the 'marked route', may be able to break down the ancient course of events, and actions at smaller scales produce effects on global self-regulatory mechanisms following a causal chain that is anything but clear.

In environmental problems, people seem to attach limited importance to scientific findings (Wildawsky, 1995), due not only to the lack of knowledge on, and understanding of, the complex problems they aim at solving, especially when the global level is concerned (Susskind, 1994), but also to the lack of confidence towards experts in the light of the increasing acknowledgement of their frequently scarce, though rarely exposed, value- and interest-free attitude on this terrain.

In the domain of collective choice and action, these stances, together with the spreading belief that methods and theories based on instrumental rationality have failed to address the problems they were supposed to solve, gave rise to a new kind of communicative and interactive practice based on community involvement in planning discourse which now is more and more supplementing formal governmental decision processes.

In this context, if we accept the idea of self-reliant development as a good basis for dealing with environmental problems, and the associated re-formulation of planners' role as promoting (not only admitting) different possible solutions for local problems as they emerge from community decision settings, even when none of them should be guaranteed by 'scientific blessing', we need to enhance our ability to understand and interpret collective processes as well as related conflicts; and to distinguish, within the community goals, those based on local cultures and traditions and those expressing the pervasive values of modernity. In the following paragraphs we will try to reflect more thoroughly on these issues, also by discussing some cases and experiences.

5. Stances Towards Environmental Issues at the Local Level

When we look at real world planning situations, we find different attitudes towards environmental problems. Scale and place where problems are considered and discussed probably affect the way in which people see them. So, trying to discriminate dominant ways of looking at and scale of dealing with environmental problems may be useful in order to understand how in collective decision settings those problems are formulated and faced, and local/global tensions emerge.

At the local level, on the one hand, rarity characterising increasing quantities of public goods, inducing preservation and prohibition policies, gives rise to conflicts that can easily be associated with the typical land-use conflicts, but that more subtly concern different ideas of development, divergent views about who - among different general publics - should benefit from a protection policy applied to a local environmental good, and different collective uses and users of resources to be allowed. Community involvement, in these cases, is deemed as crucial in order to take more efficient as well as fair decisions.

On the other hand, perceived risks and threats for environment and human health coming from the location of power plants, waste facilities, and other sort of locally undesirable infrastructures, are usually afforded by means of impact assessment techniques aiming at singling out location or design proposals which seem to guarantee the reduction of risks and threats at acceptable levels with calculated uncertainty factors. Conflict solutions in these cases are supposed to converge on magnitude and direction of impacts and the methods in use, which have a technical as well as a consensus building orientation, try to structure information so as to facilitate the comparison among different given options.

The two aforementioned dominant attitudes in facing environmental problems at local level and the associated conflicts hide special kinds of NIMBY syndrome (Lacour, 1994): the idea that catastrophic effects are not imaginable or even possible near to us (where near can assume very different geographical extensions), the unwillingness to give up the advantage of enjoying services and infrastructures located nearby, and the conviction that 'sustainable' ways of life supported by more rich and economically developed regions are unsuitable for disadvantaged (in relative terms) areas and would penalise their local welfare. The risk of these kinds of syndrome may be considered as an impediment for major concern of global environmental problems, and should induce us not to take for granted that outcomes of social interactions and conflicts on these issues involving the general public at the local level will bring with them more environmental sustainable decisions.

What is more, for the reverse side, a concept of general tendency and interest is currently challenged by the new evolutionary paradigms, with the crucial role they assign to local actions and events.

Current strong interactions between local, regional and global decision levels - both in public and private sectors - make these issues to become increasingly relevant in environmental planning, and seem to introduce more general and deep questions regarding the levels at which environmental problems should be tackled and the competencies which should be assigned to different actors, namely World organisations and local administrations, public agencies and communities.

Another problem not yet sufficiently investigated regards the linkage between outcomes of communicative and interactive practices and formal decisions taken by administrative institutions, which also in environmental planning and decision making should be treated not as actors among others or as "simple social mirrors", but as playing a more independent role, and as affecting and not only being affected by society (March, Olsen, 1989). Thus, the question of linking decision process outcomes and choice is not to be seen as limited to the use of techniques and expertise in planning practice, as clearly put by Wildawsky (1979) when stated that politicians tend to make decisions that please salient constituencies, not those that are optimal from analytic view-point, but should be seen as crucial also when problems solution is searched through social communication and interaction.

Finally, in order to build a new normative approach to planning in the light of the sustainability paradigm, we probably need to consider that the society-environment relation may be a scale-dependent problem, with a separate expression of environmental ethics associated with each scale, and avoid the unconstrained extension of a "stewardship concept", suitable for local and - to some extent - regional levels, to the planetary scale, which rather seems to recall a dependency relationship of individuals, society and biogeochemical cycling (Reed and Slaymaker, 1993).

6.   Beyond Consensus Building as Dominant Approach for Dealing with Conflicts

Further investigation into different kinds of conflict and conflict management is worthwhile, as we recognise that they are at the heart of planning and evaluation in multi-agents decision contexts in environmental field and, what is more, they are likely to lead to fresh, unexpected solutions meeting common social goals.

If we roughly distinguish three basic models as available for dealing with such complex policy problems, the policy choice model, which treats disputes as solvable through the application of value-neutral research, the pluralistic model, which assumes the intrinsic multi-values character of decision contexts, and the consensual dispute resolution model (Schön and Rein, 1994), which can be considered as an evolution of the former, it is easy to observe a progressive contamination of the first, still dominant model, on which the most of evaluation techniques are based, with the pluralistic and consensual model. Moreover, basic values of the disputing parties in these models are presumed to be given and constant, and closely linked to, or even identified with, their interests, not considering other components as those related to "recognition conflicts" and "ideological" conflicts (Pizzorno, 1993), which have been increasingly brought to attention by feminist as well as environmental movements, thus becoming more and more relevant in social action and choice.

Moreover, dominant approaches look at conflicts affecting complex decision making processes, as needing to be overcome, if not as negative sides of those processes, not recognising that conflicts may sometimes have beneficial effects, especially when performing constructive functions and leading to creative, unanticipated, problem formulations. Thus, dominant approaches focus on seeking for conflict solutions through agreement building procedures, both when they are specifically built around the use of conflicts as in the theory of games, and when they just acknowledge that conflicts deeply affect decision making and aim at finding ways of managing with them, as in most of the last generation evaluation methods.

Thus, the latters recognise the need to involve communities in decision-making processes, the ongoing and not conclusive character of evaluation, and embrace an orientation towards using the proposed techniques as bases "for community discussion, and the modification of a preferred project option, in order to achieve more general community agreement and support" (Lichfield, 1988), as well as seeing evaluation as a means to promote a broad orientation on the wishes and ideas of social groups in open planning processes, thus stimulating a process of "social will-shaping" (Voodg, 1993). The 'minimum requirement approach' to plan evaluation proposed by Hill and Lomovasky (1980) is even centred around conflict and participation in planning, and adopts a satisfying stance in order to achieve a compromise which satisfies at least minimally the interests of all participants reached through a continuing dialogue among the actors and their negotiations trying to meet minimum requirements.

Actually, that seems to be the path proposed by collaborative and consensus building approaches also in other research fields, from the political and social sciences to Operational Research and, more recently, in multi-agent intelligent support systems (Sycara, 1993), automated systems (Gasser, 1991; Rosenshein et al.), design support systems (Bahler, Dupont, Bowen, 1995).

Susskind and Cruikshank (1987), who pay particular attention to disputes emerging around environmental decisions, distinguish between distributional disputes, concerning environmental standard setting and utility rate setting involving the allocation of tangible benefits and costs, and constitutional disputes, concerning constitutionally guaranteed rights. They propose a negotiated approach for the first, while entrusting the courts with the latter.

But, just as far as environmental problems are concerned, is it so easy to isolate issues not referring to basic values and to avoid dealing with them in order to reach disputes resolution? So, are agreement building procedures always able to help in evaluating and deciding in those cases? And, coming back to the point, are conflicts to be overcome in all situations or are there any situations in which stalemate conflicts may produce fresh, creative, unanticipated ideas and solutions, and consensus building may be likely to put them in the shade? In order to start giving some answers to these questions, we probably need to distinguish between different kind of conflicts as well as different kind of environmental issues which give rise to them.

A possible distinction is among conflicts over urban planning, affecting the environment surrounding it (the human dimension of conflict, the social context of planning, the nation-state context of planning); conflicts in urban planning, involving the elements with which urban planning deals (mainly resource-based conflicts, tangible and intangible); conflicts of urban planning, including methods and procedures as well as social and political issues (Minnery, 1988).

In the light of such an articulation, we can observe that in many local planning situations, particularly when deliberative formal institutional settings are concerned, disputes really deal with procedural questions - who should make decisions, in what fashions - (for example see Grant, 1994), often hiding different visions and interests of the involved players, which a discussion focused on substantive issues would probably surface.

On the contrary, in structured communicative and interactive practices - when using for instance some kinds of "soft system methodology" as those collected by Rosenhead (1989), or evaluation technique - resource-based conflicts tend to emerge, but in a way that usually doesn't put under discussion principles and basic values if decisions directly associated with environmental risks are not involved. In fact, some of those principles and values are taken for granted, also thanks to an apparent terminological consensus on them. A paradigmatic example drawn from our experience in Southern Italy local contexts is that of people attitude towards development: really, they do not discuss it, but divergent goals and visions on specific projects and initiatives, where the attention of participants during planning meetings immediately converges, enlighten value conflicts on, and different definitions of, this central concept. We can observe also a participants attitude for perceiving contingent problems, in particular if linked to the current economic crisis, and a certain difficulty in conceptualising and articulating environmental implications of events or plans (Barbanente, Puglisi, 1996). To let those values and implications emerge would probably mean to address disputes towards a stalemate point. Thus, agreement oriented participation practices are likely to avoid surfacing basic divergence.

As far as environmental problems associated with perceived local risks and threats are concerned, basic values really emerge, and we often face paralysing conflicts.

We can also distinguish between policy disagreement, including disputes in which the parties in contention are able to resolve the question by examining the facts of the situation, and controversies, which are immune to resolution by appeal to the facts (Schön and Rein, 1994). The complexity and uncertainty surrounding environmental issues, and the limited ability of technical experts in supporting their claims on the basis of facts, lead us to conclude that most environmental problems lie in the second set. Thus - following the authors - they will need a reframing of disputants' views, namely a change in their structure of beliefs, perceptions and appreciations, just as required by the vision exposed in par. 2. And it is worth questioning if traditional win-win bargaining where disputants are encouraged to switch their positions in order to build their respect for and understanding of opposing points of view and are helped in inventing win-win options, without changing their basic views, is a suitable procedure in these cases.

7.   Toward a Multi-Agent Decision Theory

If multi-agency means that a number of agents concur in thinking and action processes the still prevailing model of "rational planning" should imply a quasi-virtuous convergence of efforts (definition of aims and goals, analysis and choice of possible actions, implementation, control and feed-back...) which is unthinkable in real multi-agent contexts built on conflict, casualness, and creativity. Furthering and changing this view, the new model of reflective and communicative action locates the cognitive exchange in a sort of role-play, envisaging increasing abilities of devoted persons - here the term is proper, as not only 'specialisation' is required but also emotive devoting to interaction, dialogue, listening - to organise plural and shared knowledge-in-action.

The rational plan model, as rational problem-solving, in the Newell's and Simon's way, draws to the analysis of "rational" behaviour by a single agent. In a broader context, the impossibility of rationally combining individual behaviours into a social dimension has been stated since many years ago by the decision theorists (Arrow, 1951), but the theory of behaving by social sets did not succeed in spanning from political theory and philosophy to planning theory - a theory of knowledge-and-action - in forms plausibly and operationally accounting for the typical transactions of the social domains: in fact, those complex transactions are largely conceived in the rationalistic perspective as quasi-technical "problems" to be solved (where to solve means also to understand) by appropriate "tools" existing independently from the probabilistic interplay of social and environmental events.

To mention Minsky, it is interesting to note that when you study an agent's behaviour in facing very simple tasks you find the concourse, at the same time co-operative and conflicting, of many agents, each of them irrelevant in the absence of others (a part from very limited and partial tasks); only when the agents are globally at work you have a real spreading of knowledge and action per se. Should we say the same about the relationship between persons and society, even if here we acknowledge a substantial difference due to the living feature of the person unity; should we ignore that this person - as a single agent - apparently has unlimited power of understanding phenomena, not different from that showed by society? Is it true that when the single agent behaviour divorces by the social agent (multi-agent) behaviour only modest resources of intelligence materialise?

> "In the same way we would be unable to foresee what will happen in a human community if we know only what each person can individually do; we should also know what is the organisation of persons, that is what relations exist among them. The same holds for any large and complex thing you want to understand. First, we need to know how each single part of that thing functions; second, how each part interacts with the other related parts; third, we need to understand in which way all these local interactions contribute to the implementation of those actions the system, as we see it from outside, performs." (Minsky, 1985).

The mechanistic aspects of Minskian reductionism vs. holism, the wisdom of Minskian interactional view, crashes into the limits of task accomplishment due to "combinatorial explosion"; in the individual human being the interactive functioning of different parts - even in the domain of 'creativity' and ignoring possible self-destructive tendencies - is granted by the basic concurrence of all the sub-agents into the hierarchically superior end of human being preservation and satisfaction; in social groups the interaction is less routinary and foreseeable, because here an analogous hierarchically superior end ruling the different agents can be found only at the level of species, of human community and its life environment, a level hardly - and rarely directly - perceived by senses which fail hence to give useful and unambiguous (even if in aggregate dimension) orientations.

It is interesting, in this light, to look at "abstraction" and "hierarchies" of plans, as ways to navigate in the NP-hard search space of problem solving: an abstract problem space allows to solve the problem at a manageable level before addressing more refined and local problem solutions, using transitional operators among system-problem states fitting conditions ignored in the more abstract problem formulation, substantially based on aggregate and schematic states, each of them corresponding to a number of real states (in this sense, again, Newell and Simon, and Sacerdoti in particular). In fact, problem solving by plan hierarchies and related abstraction hierarchies implies the invariability at the more detailed levels - hierarchically inferior - of those solutions which were found at the more abstract levels - hierarchically superior - according to a property of ordered monotonicity (Knoblock, 1994); this is made possible by the absence of an adaptive interaction among the different levels.

Basically, a hierarchical planner ideally finds a problem solution at a certain level maintaining the structure of that solution while solving the remaining parts of the problem. But if a building project committed to a sole designer can ideally proceed - at least in a rough and aggregate view, looking at fundamental features - from the definition of the basic layout to the definition of structural details and complements, maintaining consistency between previous and later decisions, what should we say about plans whose in-progress development - including implementation - calls for the interaction of different designers and actors who can change dynamically - indeed this being rational in a social, democratic and interactive dimension - the games defined at more abstract levels?

Politics, fundamentally, suggests a solution to these asperities by the different levels of abstraction and decision power affecting social problems, by mechanisms for spurring action through temporal control and renewal of power delegation in the light of a cyclical test of performance in reality.

In uncertain, complex, and dynamic domains the preservation of the structure of solutions moving through the different levels of abstraction cannot be granted. Here, the definition of those different levels of abstraction appears as an indeterminate task and one has to admit that plans need open, arguable, and changeable structure whose basic coherence lies on the fact that those plans cannot be contradicted in their changing qualities and levels due to their lack of structures.

Concurring to this unconventional post-rationalistic view of plans, Winograd and Flores say that the most successful design of an artificial system will occur when people will stop in looking for a global modelling of the concerned knowledge-action domain and will try to "align" the artificial system to the fundamental structure of that domain, allowing for modification and evolution (Winograd and Flores, 1990).

This kind of plan will have a non-monotonic structure: a pattern characterised by high intensity of interaction and cognition, openness to plural efforts, with the sole need of continuous tests of efficacy delegated to a collective agent representing community: a principle of self-conservation open to evolutionary break-downs of social and environmental reality will serve here as a complement and antidote to this total planning relativism.

Expert knowledge and common sense knowledge will interact and fit this *principle of reality*, allowing to go out from the relativistic dilemma: any problem which is approachable on the basis of that knowledge is never solved in isolation and abstraction; in the environmental domain as in the social domain any useful plan is useful when it concurs to a 'locally' - in a space-time sense - useful transformation of reality. The intriguing conceptualisation of usefulness in this context must be afforded by a Pareto-like claim: the advantage of a larger agreement (on the evaluation of usefulness) must not imply the penalty of being forced to register even only one disagreeing voice or entity, human and non-human, about the involved value of usefulness as regards the fundamental issue of conservation and development of species expressed by it. What here is in discussion is a principle of integral morality whose claim potentially paralysing can be only subsumed by a superior mechanism of arguability (here again superior means cognitively broader, more interactive and dynamic and does not mean more expert) which states even against that possible disagreeing individual voice or entity the existence of an ethics and a process of conservation and development operating in favour of it - at least in a global sense, that is beyond the individual self-interest: when this aggregate and 'post-poned' principle of general conservation challenging other individual interests to self-conservation cannot be stated for the uniqueness of the interest involved by the self-conservation instance (the last individual of a disappearing endangered species...), that arguability should imply a principle of caution extended to the non-action possibility.

We integrate, here, a stance of political democracy to an auto-poietic position about living organisations in the way of Maturana and Varela, emphasising the cognitive aspects of the organisation within which the plan and the related task-solving operate:

> "A cognitive system is a system whose organization defines a domain of
> interactions within which the system can act in a relevant way towards
> its conservation, and the cognitive process is the real (the inductive)
> acting or behaving in this domain." (Maturana and Varela 1980).

The reference for this integrated approach here goes to an auto-poietic position which abandons a discrete view of the evolving organisation - a view of "system-states" - adopting a temporal and historical, continuous, view, related to contexts and stories of action, embodying interactive mechanisms inherent to the organisation per se - to its being a living organisation - far more complex than the simple mechanistic feed book of reinforcement as regards external events postulated by the deterministic behaviourism. But, as we will say later, here a problem of linguistic co-operation arises, a problem which confines the existence of the spontaneous "consensual domain" of behaviours stated by Maturana into the narrowness of specific homogeneous organisations (Maturana, 1978). It is clear the distance between this limited domain and the linguistic communication and linguistic interaction whose model we want to introduce and discuss by this paper.

We are claiming here for a cognitive and linguistic re-enforcement that in the applied domain of planning should assume first - as its reference - speech acts grounded on a "demonstrative action" in which actors of the organisation in progress show their abilities to draw on a consensual domain of knowledge-in-action. The more the creation of a common language, in the organisation, will induce - through *demonstrative acting* - an auto-poietic sense among participants the more it will be avoided that the searched cognitive reinforcement remains not understood and maybe even challenged by those participants, overwhelmed by an information flow not understandable from the auto-poietic point of view. Here the dilemma arises of the requirements of demonstrative actions and speech acts relevant for a plurality of actors even when not embodying a mere egoistic local utilitarianism - that is demonstrative and speech acts sensitive to 'other' languages and acts - being at the same time quite evocative and immaterial to keep the language far from a Procustean bed of mere realism, unable to cope with uncertainty, novelty, hard experimentation.

The Heideggerean being-in-the-world as pre-condition of any cognition can constitute a useful reference for the demonstrative action we are speaking about.

Obviously, we have to underline here that planning - spatial planning, too - has shown strong interest in the last two decades to the linguistic, conversational features of planning organisations, for example in the Schön's or Forester's work by importing a large work from the "speech acts" philosophers (Austin, 1962; Searle, 1969, 1979; Habermas, 1979). But this approach seems to fail in considering complex and not simplifiable differences of backgrounds, stories, contexts, staying far from any virtuous agreement.

The rational of both the auto-poietic position and the speech-acts position appear unsuitable to deal with 'real' knowledge and action problems: the auto-poietic plan is not able to incorporate operatively the uncertain and chaotic aspects of physical events, the non intuitive and non opportunistic behaviours of the actors; communication does not solve analogously the mystery and the violence of events substantially autonomous, solitary, temporally and qualitatively unique, which cannot be reduced to mere interaction and communication events.

Moreover, the communicative dimension of the plan - better knowledge into the action through dialogue, listening, and interaction - is based on language sharing. That dimension is therefore meaningless in domains where exogenous events irrupt carrying with them realities not totally manageable in linguistic terms (see the physical environment within the limits of any theory of reality) or still not understandable in their languages.

Although trivial and terrible, the problem is finally of foreseeing and acting for processes undetermined with regards to circumstances both exogenous - processes which develop in opened and dynamic 'systems' in a time configured as irreversible - and endogenous - processes for which it is not possible to compose finality and wishes within thought and action.

Organising the planning multi-agency in such a context means, therefore, to come out from the Procustean approaches of 'optimisation' and catalyst and routine functions carried out by 'single' and 'expert' knowledge - a knowledge mistrusting creativity - to promote more intense and dynamic knowledge environments: quality and quantity of both knowledge and information appear here crucial in their dimensions of plurality, accessibility (i.e. multiplicity of complexity levels), dynamism, possibility of reviewing (free from personal and social constraints), intensity (i.e. availability of information for the problem). These multi-agents environments should be free from formal evaluations inducing cognitive and operational 'restrictions' not emerging from either the free expression of the organisation itself or its free comparison with other organisations; inside these knowledge environments, processes of self-fertilisation and self-conservation, in some way similar to those of other living organisations, should freely emerge.

8.   Unsolved Problems of Planning and Evaluation and Directions for Further Research

If now we turn the attention to planning and evaluation ability to meet the demands for environmentally and socially sustainable actions in the light of the increasing complexity and uncertainty of involved phenomena, a number of unsolved problems of evaluation and decision clearly appear.

The use of meta-analysis as a way to elicit relevant information for problem solving from previous studies or experience, at the same time removing some of the subjectivity from analysis and from forecasting or at least making judgements more transparent, has been recently invoked by van den Bergh and Button (1997) for environmental issues in regional, urban and transport economics; in this sense, meta-analysis could provide -by using a broad range of available techniques -an integrated framework also for evaluation, even if van den Bergh and Button warn against the risk of seeing it as a panacea where social phenomena with their heavy charge of complexity, uncertainty, poor information, are involved making troublesome and vane any effort for the adoption of formal rational calculus.

A first example of the new wave of concepts and methods developed to cope with the complex structures of contemporary economy and society concerns the difficulty in dealing with the socio-economic impact of a given policy in the distributional terms which are typical of mainstream tradition of policy science: recent studies show how conventional socio-economic approaches linking the poverty impact of a crisis to distributional and lack of resources issues are inadequate to deal with a problem which appears primarily linked to subtle relational issues of lack of power and social integration (Smith, 1997).

In fact, this is an issue that apparently the traditional neo-classical economic theory and the correlated decision theory fail to understand: in Smith's study, the individual decision to participate in the labour market refers not only to the typical rational choice between two sharply different uses of time but also to the consideration of a third and alternative kind of time, far long ignored, that is time spent by the woman working at home; Smith points out that in this case the whole structure of rational appraisal and decision of the neo-classical economics of human capital collapsed and that influential economists today link -or substitute- the rational decision with non rational attitudes or behaviour like prejudice, when social and family environments strongly conflictual and based on dominance and power relationships are involved.

It is evident how in those cases both the selection of the range of alternatives to be considered for choice and the selection of the range of choice criteria should be broadened and strongly innovated in a creative direction, including in them also nonrational and nonfunctional dimensions, as yet ignored by approaches substantially inspired by well balanced trade-offs between material entities in traditional economic behavioural contexts.

A second example could concern the equilibrium functions of decision involved by actors and community groups inspired by the economically alternative behaviour of the so-called third sector and social humanistic economics with their propensions for the anachronistic nonmonetary exchanges which neo-classical economists condemn, by the economic self-reliance of the local community, by the reframing of the concepts of time and money and substantially of the concept of value. Pacione, for instance, describing the original market exchange local system, activated in Vancouver in the early 1980s as a solution to the dramatic local economic crisis and the recent diffusion of this model in Glasgow, points out the aspects of dissociation and complementarity which are detectable in the relationship between this informal system -largely based on ethical and moral principles of collective responsibility- and the wider formal economic system including the local community (Pacione, 1997).

Far from being transient returns of strongly ideologised individuals or groups to the archaic economics of the past, these models seem to effectively occupy specific economic, social, and cultural spaces of advanced situations and cities, paralleling structural changes both in production and service systems -with the related labour market- and in lifestyles; the culture of difference helps to appreciate -as in the above mentioned case of rediscovering time and family economics under the push exerted by gender studies- the value of these flexible forms of solidarity and reciprocity in the functioning of complex socio-economic systems which are otherwise highly fragile.

The way in which this highly personalised informal society and economy enter into aggregate contexts of planning seeking for a return to the market must be investigated. In particular we should look at the reframing of economic equilibrium concepts or to the different values of time, a time for developed societies where time is now more abundant than money.

It is interesting, for a discussion on problems of plan implementation and theory/practice relations, to note the tendency of political and administrative machines increasingly affected by deregulation and fragmentation in the last two decades (see the UK case) to coexist with original pushes towards re-centralisation in which what is searched for -more than the traditional strategic approach of top-down government- is the promotion of a virtuous and flexible context with public and private sector cooperating in policy implementation and goal attainment (Newman and Thornley, 1997).

In the new landscapes of political fragmentation interorganisational networks emerge in which decision taking and co-ordination of actions assume original aspects. The recent literature on urban political "regimes", imported from the US, helps to give the right perspective to these configurations of "new governance" (Keating, 1991).

How can we locate rational planning, evaluation, and decision in reticular and horizontal, coalition like, policy contexts, where individual abilities to point out and control policy goal attainment is limited, where self-organising models mostly appear - with their typical informal plans, evaluations, and decisions- and the role of bureaucracy and the need for explicit argumentation and justification for policy action is lowered by increasing privatisation? If, on the one hand, the emerging role of private actors and profit seeking actors leads one to check that the economic dimension could remain crucial in planning and decision making accounts, on the other hand, the increasing role of a non profit third sector and community -as prime actor and addresser of plans- leads one to emphasise language and discourse mechanisms of communicative planning with their basically ethical, conciliative and moral dimension.

The new emphasis on social learning and social capital, as prime requirements of organisational and cultural environments open to co-operative resolution of complex economic and living problems of communities, marginalises the utilitarian rational calculus based on self-interest and focuses on equilibrium and equity to be attained in wider circles and a longer time, on revalorisation of public goods, that is of those goods whose value is lost when self-interest is pursued. A self-organising system can be managed only by self-regulation. From this the increasing distrust for technical expertise, too much detached from common feeling and increasing trust, on the contrary, for relational, intuitive, creative, forms of rationality, showing a strongly innovated concept of value (Wilson, 1997).

Participatory and communicative methods which parallel this new approach and their typical search for direct involving/empowering of all those concerned in policy decisions and actions, evidently also need a radical redesigning of evaluation methods and related indicators and criteria.

In strictly economic terms, the call for evaluations less anchored to narrow and merely budgetary appraisals of costs and benefits it is strengthened by the conversion - definitively reached since the 1980s in local and urban policies and in planning practices of MDC- from the idea of economic development based on the cycle investment-use of resource and labour-selling managed by formal organisations at their different levels (central vs. peripheral) to the idea of alternative development based on more creative and less conventional use of resources, on reduced investment, on active participation of local government and community, on intelligent agreement between public and private actors.

The "Empowerment Zones" program of President Clinton's administration, in the US 1993, merging concepts of "Enterprise Zones" to concepts of local development and decision making, of self-organising communities, of enhancement of the non-profit third sector, indicates both the growth that these ideas are experiencing in recent broadly targeted policies and the new demands which are raised in the face of decisions and evaluations.

Metrick and Weitzman (1996) point out the recurrent asperity of evaluation of policy effectiveness performed through the traditional rational comparison of goals and real outcomes analysing preservation policies for endangered species: we do not have, in fact, acceptable standards for biodiversity on which to anchor analyses and objectives of decision makers; hence, Metrick and Weitzman argue for the need for powerful analytic simplification, and for dealing with relevant and measurable variables, belonging -in the case of endangered species- only to the category of the existence value (within the three potential categories of value involved by the case at hand: market, existence, contribution).

Metrick and Weitzman, moreover, criticise the expenditure decisions of Fish and Wildlife Services (an agency of the Department of the Interior, US, created in the early 1970s to support the 1973 Act on Endangered Species), which are based on cost/effectiveness like appraisals whose costs come from the "reasonably identifiable" expenditures to be faced for preserving individual species; in fact, M&W judge those appraisals as a mere reflection of underlying implicit preferences of decision organisations. They propose a split of competencies between analysers and decision makers in order to prevent the latter from influencing the analyses with their own -more or less conscious- visions about the values of the different species.

The impossibility to attach to biodiversity a value measurable in terms of market prizes, due to the complexity and indivisibility of ecosystems functioning on temporal scales which are beyond human experience or perception, raises the related problems of interest rate of future, of inconsistency (reduction of a complex set of attributes to a single, unidimensional metric of usefulness or money), of the existence of a pure uncertainty (Gowdy, 1997).

The systematic violation of the utility theory for decisions to be taken in complex contexts and the apparent resorting of decision makers -to overcome the knowledge-decision gap- to simplified decision rules, to decision heuristics, which exclude optimisation procedures because of their impossible use in the light of the peculiar nature of information involved, are important to grant regularity and consistency of behaviour: paradoxically, the intuitive decision behaviour induces, in those cases, that is in the case of expert use of decision heuristics , a classical rational behaviour which is not equally attainable by optimising a continuous neo-classical utility function (Mazzotta and Opaluch, 1995, who in this sense restrict the range of validity of the Heiner's well known argumentation).

Linking theory to practice means, in conclusion, a number of different interesting things for the evaluation-in-planning perspective: questioning the dualism theory-practice in the light of a unified perspective of transformative theory-in-action and -on the evaluation side- of adaptive, evolutionary, and cautious behaviour grounded on enhanced intelligent monitoring abilities; integrating different knowledge levels in theory-in-action opportunistically using meta-knowledge and heuristics to fill information blanks and analytical knowledge to deepen insights (on the evaluation side this means to resort to methods processing different forms of knowledge in different but reciprocally consistent ways); integrating expert (formal) and nonexpert (common-sense, informal) knowledge in theory-in-action and in evaluation, in particular when complex environmental and uncertain social and environmental problems -not solvable through established scientific or technical analyses and routines are involved and social knowledge could offer a resilient way to deal with the phenomena at hand; practising communicative planning and evaluation in the social learning perspective, searching from creativity coming from continuous reframing of individual positions and even from conflicts and behaviours radically divergent from routine; exploiting the argumentative potentials of communicative planning and evaluation through the adoption of clear and well debated procedures and criteria; developing nonmonetary evaluation methods to cope with the increasing diffusion -even in the MDC- of informal and moral economy, through a better integration of CVM and assessment of individual positions with other more aggregate methods; making evaluation procedures more sensitive to fragile or dissenting voices, using sensitivity analysis to test robustness of appraisals in the light of radical changes of viewpoints; developing more friendly and articulated planning and evaluation systems which could be consistently used -even through the incorporation of conflicts into the learning mechanism- at the different key levels of decision making, in the emerging crucial dialectics between localism and globalisation.

9.   Conclusions

In the domain of collective action, especially when environmental issues are concerned, and thus complexity and uncertainty characters dominate decision making processes, local/global contradictions and tensions emerge, and value conflicts more than distributional conflicts arise, we envisage a need to drop instrumental rationality and the associated "calculating and utilitarian reason" as a base for evaluation, and to embrace a radically new and wider perspective. The shift in scientific paradigms coming from the "complexity thinking", and the related concepts of processes discontinuity, systems auto-organisation, plural character of possible futures single out some basic frames for dealing with such problems in planning and evaluation.

In the light of the above concepts, anything but adequate seem to be planning approaches based on discrete, sequential decomposition of decision processes which, on the contrary, are continuous in nature and show sudden and unexpected interruptions, stalemates, and breakdowns, due to the uncertain, complex, unforeseeable character of most interactions in multi-agency. Thus, the fundamental assumption in planning that it should be possible to achieve a convergence of goals, actions and choices by producing co-operative solutions at the various steps of plan-making and implementation process has to be reconsidered.

In this paper we started to explore that terrain mainly focusing on conflicts emerging around environmental problems, looking at local community settings as well as to more wide cross-national decision environments, singling out scale-dependent divergencies, and investigating into different kinds of conflict and conflict management.

The idea we upheld is that we need to be deeply embedded in complexity, and to enhance - at the same time reviewing - knowledge resources, in the shape not only of informational flows (that is what really is happening in the current stage of our society), but also of knowledge-in-action, so taking advantage of speech acts based on demonstrative actions. Furthermore, we require to distinguish between what can be assisted and structured by planners evaluation and decision, and what has to remain undetermined in order to avoid inhibiting social innovation and creativity which receive impulse from structural uncertainty of situations and processes, and from conflicting interactions and wide co-ordinated social bonds.

References

Amin, A., Thrift, N. (1993) Globalization, institutional thickness and local prospects, *Revue d'Economie Regionale et Urbaine* 3, pp. 405-427.

Arrow, K. J. (1951) *Social Choice and Individual Values*, Wiley, New York.

Austin, J. L. (1962) *How to Do Things with Words*, Harvard University Press, Cambridge MA.

Bahler D., Dupont, C. and Bowen, J (1995) Mixed quantitative/qualitative method for evaluating compromise solutions to conflicts in collaborative design, *Artificial Intelligence for Engineering, Analysis and Manufacturing* 9, pp. 325-336.

Barbanente, A., Puglisi, M. (in press) Multi-actor knowledge acquisition for sustainable local development, in D. Borri, A. Barbanente (eds.) *Planning for Self-Sustainable Development. Current Issues in Italy,* Kluwer Academic Publishers, Dordrecht.

Barcellona, P. (1994) Critica della ragion tecnica. La pericolosa favola dello sviluppo sostenibile, *Capitalismo, Natura, Socialismo* 2, pp. 72-86.

Barcellona, P. (1996a) Una società da ricostruire, in V.A. *Quale Repubblica?*, Citoyens, Roma.

94 BARBANENTE *ET AL.*

Barcellona, P. (1996b) *L'Individuo Sociale*, Costa & Nolan ed., Genova.

Castells, M. (1989) *The Informational City*, Basil Blackwell, Oxford.

Ceruti, M. (1996) Ambivalenza dell'idea di progresso, in Tiezzi, E. and Marchettini, N. (eds.), *Oltre l'Illuminismo*, CUEN, Napoli.

Cini, M. (1994) *Un Paradiso Perduto*, Feltrinelli, Milano.

Cini, M., (1996) *I Professionisti della Conoscenza, Il Manifesto*, 14.2.1996.

Dryzec, J. (1987) *Rational Ecology*, Basil Blackwell, Oxford

Forester, J. (1989) *Planning in the Face of Power*, Berkeley University Press, Berkeley, Ca

Gasser, L (1991) Social conceptions of knowledge and action: DAI foundations and open systems semantics, *Artificial Intelligence*, 47, 107-138.

Giddens, A. (1990) *The Consequences of Modernity*, Polity Press, Cambridge.

Gould, S.J. (1980) *The Panda's Thumb*, W.W. Norton&Co., New York.

Gowdy, J. M. (1997), The Value of Biodiversity: Markets, Society, and Ecosystems, *Land Economics* **73** (1), pp. 25-41

Habermas, J. (1979) *Communication and the Evolution of Society*, Beacon Press, Boston.

Harvey, D. (1989) *The Condition of Postmodernity*, Blackwell, Oxford.

Hill, M (1985) Can multiple-objective evaluation methods enhance rationality in planning?, in Breheny M. and Hooper, A. (eds) *Rationality in Planning. Critical Essays on the Role of Rationality in Urban and Regional Planning*, Pion, London.

Hill, M. and Lomovasky, C. (1980) *The minimal requirement approach to plan evaluation in participatory planning*, WP-104, Center for Urban and Regional Studies, Technion Institute of Technology, Haifa, Israel.

Keating, M. (1991), *Comparative Urban Politics*, Elgar, Aldershot.

Knoblock, C.A. (1994) Automatically Generating Abstractions for Planning, *Artificial Intelligence* **68**, pp. 243-302.

Lacour, C. (1994) Intégration Environment-Aménagement: fondements théoriques, *Revue d'Economie Regionale et Urbaine* **4**, pp. 537-556.

Latouche, S. (1992) *L'Occidentalisation du monde. Essai sur la signification, la pertée et les limites de l'uniformation plannétaire*, La Découverte, Paris.

Latouche, S. (1995) *La Megamachine. Raison techno-scientifique, raison économique et le mythe du Progrés*, Essais à la mémoire de Jacques Ellul.

Lichfield, N. (1988) The Cost Benefit Approach in Plan Assessment, International Colloquium on Evaluation Methods, in Urban and Regional Planning: Theory and Case Studies, Capri-Napoli, April.

Lynch, K. (1990) *Wasting Away*, Sierra Club, San Francisco.

March, J.G., Olsen, J.P. (1989) *Rediscovering Institutions*, The Free Press, New York.

Massey, D. (1991) The Political Place of Locality Studies, Mimeo, Faculty of Social Sciences, Open University, Milton Keynes.

Maturana, H.R. (1978) Biology of Language: The Epistemology of Reality, in Miller, G.A. and Lenneberg, E. (eds.), *Psychology and Biology of Language and Thought*, Academic Press, New York, pp. 27-64.

Maturana, H.R., Varela F.J. (1985) *The Tree of Knowledge*, New Science Library, Boston.

Maturana, H.R., Varela, F. (1980) *Autopoiesis and Cognition: The Realization of Living*, Reidel, Dordrecht.

Mazzotta, M. J., and J. J. Opaluch (1995), Decision Making When Choices Are Complex, *Land Economics*, **71** (4), pp. 500-15.

Metrick, A., and M. L. Weitzman (1996), Patterns of Behavior in Endangered Species Preservation, *Land Economics* **72** (1), pp 1-16.

Minnery, J.R. (1985) *Conflict Management in Urban Planning*, Gower, Aldershot.

Minsky, M. (1985) *The Society of Mind*, MIT Press, Cambridge Ma.

Morin, E. (1995) *Una politica per l'età planetaria*, in *Pluriverso* **1**.

Newell, A , Simon, H A (1972) *Human Problem Solving*, Prentice Hall, Englewood Cliffs, NJ.

Newman, P. and Thornley, A. (1997) Fragmentation and Centralisation in the Governance of London: Influencing the Urban Policy and Planning Agenda, *Urban Studies*, 34, 7, 967-88

Pacione, M. (1997), Local Exchange Trading Systems as a Response to the Globalisation of Capitalism, *Urban Studies* **34** (8) pp. 1179-99.

Pizzorno, A. (1993) Come pensare il conflitto, in A. Pizzorno, *Le radici della politica assoluta e altri saggi*, Feltrinelli, Milano.

Prigogine, I. (1996) *La Fin des certitudes*, Editions Odile Jacob, Paris.

Reed, M.G. and Slaymaker, O. (1993) Ethics and Sustainability: A Preliminary Perspective, *Environment and Planning A* **25**, pp. 723-739.

Rogers, E.M. (1986) *Communication Technology: The New Media in Sciety*, The Free Press, New York.

Sacerdoti, E.D. (1977) *A Structure for Plans and Behavior*, American Elsevier, New York.

Sachs, I. (1988) *I Nuovi Campi della Pianificazione*, Edizioni Lavoro, Roma (or. title: *Développer les Champs de Planification*, 1984).

Scandurra, E. (1996b) L'urbanistica da disciplina moderna a cultura contemporanea, *Pluriverso*, **3**.

Scandurra, E. (1997) *Città di Terzo Millennio*, La Meridiana, Molfetta

Schön, D. A. (1983) *The Reflective Practitioner. How Professionals Think in Action*, New York, Basic Books.

Schön, D. A. (1987) *Educating the Reflective Practitioner: Toward a New Design for Teaching and Learning in the Professions*, Jossey-Bass, San Francisco, Ca.

Schön, D.A., and Rein, M. (1994) *Frame Reflection. Toward the Resolution of Intractable Policy Controversies*, Basic Books, New York.

Searle, J.R. (1969) *Speech Acts*, Cambridge University Press, Cambridge.

Searle, J.R. (1979) *Expression and Meaning: Studies in the Theory of Speech Acts*, Cambridge University Press, Cambridge.

Smith, Y. (1997), The Houschold, Women's Employment and Social Exclusion, *Urban Studies* **34** (8), pp. 1159-77

Susskind, L., Cruikshank, J. (1987) *Breaking the Impasse. Consensual Approaches to Resolving Public Disputes*, Basic Books, New York.

Susskind, L.E. (1994) *Environmental Diplomacy. Negotiating more Effective Global Arrangements*, Oxford University Press, New York-Oxford.

Sycara, K.P. (1993) Machine learning for intelligent support of conflict resolution, *Decision Support Systems* **10**, pp. 121-136.

Van den Bergh, J. C. J. M., and K. J. Button (1997), Meta-analysis of Environmental Issues in Regional, Urban and Transport Economics, *Urban Studies* **34** (5-6), pp. 927-44

Voogd, H (1993) On the role of will-shaping in planning evaluation, in Borri, D., Khakee,. A. and Lacirignola C. (eds.) *Evaluating Theory-Practice and Urban-Rural Interplay in Planning*, Kluwer Academic Publishers, Dordrecht.

Wildavsky, A. (1979) *The Politics of the Budgetary Process*, Little Brown, Boston.

Wildavsky, A. (1995) *But Is It True?*, Harvard University Press, Cambridge Ma.

Wilson, P. A. (1997), Building Social Capital: A Learning Agenda for the Twenty-first Century, *Urban Studies* **34** (5-6), pp. 745-60

Winograd, T., Flores, F. (1990) *Understanding Computers and Cognition*, Addison-Wesley, Reading, Mass.

THE COMMUNICATIVE TURN IN PLANNING AND EVALUATION

A. KHAKEE

1. Introduction[1]

The interactive and communicative nature of planning is widely recognised by planning theorists and practitioners. It is increasingly argued that the outcomes of planning process are not confined to policies but also include other essential results such as transformative learning (Friedmann, 1987), communicative networks (Forester, 1989), institutional capital (Healey, 1996) and consensus and commitments (Innes, 1994). This development raises compelling questions about which outcomes should be evaluated and how. Traditionally evaluation has been concerned with the assessment of consequences of policies with a view to searching out their comparative advantages and disadvantages (Lichfield, 1996). This type of evaluation is neither entirely adequate nor relevant in the case of communicative planning. Evaluation should provide occasions for ideological and procedural reflections on various results of communicative planning (Faludi and Altes, 1997). Recent research in evaluation moves beyond previous evaluation approaches which focused on rational and systematic measurements and judgements and envisages evaluation as an interactive exploration of claims, concerns and issues among stakeholding groups (Guba and Lincoln, 1989). This advance in evaluation research provides opportunity for extending the assessment of planning process to include even 'communicative outputs'.

There are hardly any studies evaluating communicative planning or policy-making process. Whenever the issue of 'process evaluation' has been raised in evaluation literature, it has been related to the extent to which a particular plan or programme is implemented according to its stated goals or guidelines (Nachmias, 1980). For the most part evaluation is concerned with the appraisal of process outcomes, namely, policy and its constituent programmes (Bryson and Crosby, 1992). More recently in the managerial context, evaluation of people as human resources has received increasing attention (Thompson, 1990). This paper presents a methodology for evaluating such a process.

---

[1] Author's Note. I am indebted to John Forester and Patsy Healey for the excellent comments they made on an earlier draft of this paper. The research is financed by the Swedish Council for Research in Humanities and Social Sciences.

N. Lichfield et al. (eds.), Evaluation in Planning, 97–111.

The model builds on the new communicative planning theory as well as on the theory of responsive constructivist evaluation (the so-called 'fourth generation evaluation').

In a previous paper (Khakee, 199b) the author outlined a methodology of evaluating planning processes based on three major dimensions of planning: the normative that defines the scope and legitimacy of planning; the methodological that specifies how knowledge affects the preparation of plans and the organizational that covers the means of managing and coordinating planning activity. The methodology was outlined in terms of the procedural planning theory and the theoretical positions (incrementalism, advocacy planning, implementation orientation, and strategic choice) which have been put forward as a response to the procedural planning theory (for a discussion of these positions, see, Healey, McDougall and Thomas, 1982). The same framework shall be used in this paper. However the theoretical basis shall be that of communicative planning and responsive constructivist evaluation.

The rest of the paper contains four sections including a conclusion. In the first I very briefly present communicative planning theory and responsive constructivist evaluation, which provide the basis for describing the methodology of evaluating planning processes. This is presented in the second section. A case study evaluating the structure planning process in one Swedish municipality is proffered in the third section.

2.   Theoretical Outline

The theory of communicative planning and responsive constructivist evaluation share some fundamental scientific assumptions. Ontologically, both reject the positivist claim of an objective reality and argue that realities are socially constructed in the community. Epistemologically both refute the notion of subject-object dualism and accept the importance of multiple forms of reason and understanding. Methodologically both dismiss managerialism and techno-rational control and emphasise participatory forms based on inclusionary argumentation.

Communicative planning theory is essentially a theory of planning practice. It describes, interprets and explains what planners do. It is an empirically applicable and defensible theory. But the theory does not stop short of defining, clarifying and elucidating planning practice. It has a strong normative concern as well. It explains what ethically critical planning practice should be. In other words it deciphers the relationship between problems and issues planners experience and deal with in their everyday work and the larger structural shape of the political economy in which they work. A basic feature of the communicative planning theory is the intertwining of the interpretive and normative aspects.

Communicative planning theory attempts to be empirical, interpretive and critical. It owes considerable debt to the critical social theory. Planning is to a large extent an interactive communicative activity where information is presented in a variety of ways. It may "warn others of problems, suggest new ideas, agree to perform certain tasks, argue for particular efforts, report relevant events, offer opinions and comment upon ideas and proposals for action" (Forester, 1982, pp. 278-279). Information presentation is subject to distortion. And the latter is not an accidental problem but often is systematic, structural and institutional and has to be deliberately counteracted. Forester (1980, pp. 69-70) writes that distortions of pretence, misrepresentation, dependency-creation, and ideology are communicative influences with immobilising, depoliticising, and subtly but effectively disabling consequences" and owe to the "contradiction between the disabling communicative power of bureaucratic or capitalistic, undemocratic institutions on the one hand, and the collective enabling power of democratic political criticism, mutual understanding, and self-determined consensus on the other". Thus the communicative planning theory is characterised as the theory of 'progressive planning practice', 'emancipatory way of knowing' and 'inclusionary argumentation'. Healey (1996, p. 219) defines the latter to imply "public reasoning which accepts the contributions of all members of a political community and recognizes the range of way they have of knowing, valuing and giving meaning".

Planning practice is a rich activity which embraces heterogeneous people who have divergent ways to see the world, to identify interests and values, to reason about them and to set about to build consensus and solve conflicts and carry out strategies, policies and proposals agreed upon (Healey, 1996). For further analysis of this rich communicative practice, readers are referred to Fisher and Forester (1993), Forester (1989), Healey (1996 and 1997) and Innes (1995).

Responsive constructivist evaluation represents an alternative to conventional evaluation approaches and is based on two major premises. It is 'responsive' in the sense that its 'parameters and boundaries' are determined by 'interactive, negotiative process' in which all stakeholders have right to put their 'claims, concerns and issues' for consideration. It is constructivist because it rejects the positivist assumptions about 'objective reality' and 'subject-object dualism'. Realities exist as mental constructs and are relative to those who hold them. Knowledge and knower are part of the same subjective reality. Constructivism is hermeneutical and dialectical in the sense that it involves identifying, comparing and describing the various constructions that exist (Guba and Lincoln, 1989).

This form of evaluation represents what Guba and Lincoln (ibid) characterise as the 'fourth generation evaluation' approach. The first three generations of evaluation - belonging to conventional evaluation thinking - were 'measurement-oriented', 'description-oriented' and 'judgement-oriented' respectively. Guba and Lincoln (ibid, pp. 57-63) put forward five major reasons for discarding the conventional methods. They do not contemplate direct work with stakeholders and their knowledge constructions. They are "tied to the verification mode and do not deal with discovery processes". They strip "contextual factors rather than taking them into account". They aim at 'generalisation' rather than unique aspects of every situation. The 'value-free posture' assumed in the conventional methods is "logically disjunctive with evaluation's goal of making value judgements". Thus the three earlier generations of evaluation suffer from a "tendency toward managerialism", "failure to accommodate value-pluralism" and "overcommitment to the scientific paradigm of inquiry" (ibid, pp. 32-35).

In responsive constructivist evaluation stakeholders' 'claims', 'concerns' and 'issues' act as methodological generators in an interactive process where the end-product is not "a set of conclusions, recommendations or value judgements, but rather an agenda for negotiation of those claims, concerns and issues that have not been resolved in the hermeneutic dialectic exchanges" (ibid, p. 13). Guba and Lincoln (ibid, pp. 40-41) identify three general categories of stakeholders: 1)the agents who produce, use and implement the evaluand, 2)the beneficiaries who profit in some way from the use of the evaluand, and 3)the victims who are negatively affected by the use of the evaluand. For further discussion of this theory readers are referred to Guba (1990), Guba and Lincoln (1987, 1989).

This short outline of the two theories shows the similarities between communicative planning theory and responsive constructivist evaluation. However there are some differences. The major difference is that the fourth generation evaluation lacks a sense of encompassing politics that needs to be anticipated and then responded to. A key aspect of the communicative planning theory is its emphasis on different kinds of politically rational responses varying with political boundedness of different kinds (see, for example, Forester, 1989). Another difference relates to the conception of reality which according to communicative planning theory can not be fully understood since there is a multiplicity of causes and effects; constructivism emphasises relativism: realities exist as mental constructs and are relative to those who hold them. Thirdly, communicative theory proposes the elimination of false consciousness and facilitates and participates in social transformation. Constructivism on the other hand identifies, compares and describes the various constructions that exist (see, further, Forester, 1996 and Guba, 1990).

3.   A Methodology for Evaluating Planning Process

The methodology developed here is based on three dimensions of planning which I have defined above. The normative analysis of the planning process focuses on three aspects: public interest, the scope and legitimacy of planning and public involvement in the process. The methodological dimension includes the nature of knowledge, the link between learning, deciding and acting, and management of uncertainty and wicked problems. The organisational aspects include the organising of an inclusionary institutional arena, diffusion of planning culture within the planning organisation and allocation of planning resources.

3.1. NORMATIVE ASPECTS

The terms of discourse, the possibility of influencing the conditions which enable citizens to participate, act and organise effectively and ultimately the likelihood, even if it is very small, to influence the structure of power in society depend on the planners' and other stakeholders' conception of community interests and what they involve. The concept 'public interest' has been severely criticised by the proponents of pluralism who mainly argue against the contention about the existence of a predetermined and unique public interest. Despite this, public planning can not be justified without reference to community interest(s). Krumholz and Forester (1990, p. 253) tell us that the concept is "enjoying a resurgence of interest (in the United States) as a result of the reawakening of public philosophy and classically republican political theories". Moreover, Krumholz's appeal to the public interest broaches two questions: "First, is it possible to evaluate planning efforts without some particular conception of the public good? Second, when does the trickle-down theory of private gain promoting public well-being simply not work - not in economic theory but in public practice?" (ibid, p. 253).

Planning works between systematic pressures of development and the concerns and fears of community residents. It is performed under political constraints and political mandate. The terms of interactive acts depend on what are the limits of empowerment for planning and the ambit of legitimacy to speak and act in the name of community interests. In short it is a question of the scope of planning which determines principles of the planning discourse. A crucial source of planners' power is the control of information. Forester (1982) talks about four types of constraints on the access to and the use of information: 'necessary' and 'unnecessary' constraints and 'ad hoc' and 'systematic' constraints. Necessary constraints are as a result of the division of labour, knowledge and expertise while unnecessary constraints are socially or politically contingent on relations of custom, status or power. Ad hoc constraints are as a result of random disturbances whereas systematic constraints are rooted in political and economic structures. A careful distinction and understanding of these constraints and an analysis of misinformation and communicative distortion allow actors "to anticipate and respond practically to misinforming or distorting communicative influences" and to develop "emancipatory, politically informed and guided (planning) practice" (ibid, p. 71).

An essential aspect of 'interactive acts' is 'inclusionary discourse' which recognises not only multiple claims made for policy attention but also different forms in which claims are made. In order to make the discourse as inclusionary (and thereby as democratic) as possible several conditions have to be met: The discourse draws upon the knowledge and understanding, the values and capacities of all stakeholders within any political community. Agreement on complex issues and consensus-building about strategies should involve all people concerned. The mapping of the stakeholders is an important task. The community is both spatially-based as well as stakeholder-based (Healey, 1996). The interactive practice should build collaborative working relations, create co-operative work culture and dialogue across cultural differences. In short planning should be based on an 'inclusionary ethic' if it has to receive democratic support. Healey (1996, p. 224) defines 'inclusionary ethic' as emphasising "a moral duty to ask, as arenas are being set up, who are members of the political community, how are they to get access to the arena in such a way that their points of view can be appreciated as well as their voices heard, and how can they have a stake in the process throughout. This means moving beyond simple conceptions of distributive justice (everyone has equal standing) to a recognition of diversity (all groupings of people should have equal ability to put over their views)". In a way 'inclusionary ethic' is linked to Habermas' 'communicative rationality' with the latter's emphasis on legitimacy. However, Habermas' ideal speech situation is not adequate in itself to evaluate a communicative process.

3.2. EPISTEMOLOGICAL ASPECTS

Systematic analysis (technical and economic knowledge) constitutes only a tiny part of a communicative planning process which depends more on qualitative interpretive inquiry than on logical deductive analysis. Planners face three types of problems: substantive problems which require knowledge and skills about specific planning items, process problems demand skills for responding to the pressures of among other things, politics, and emotive problems which call for interpersonal skills and the capacity to listen and respect differences. The use of these three types of knowledge depends upon which interested parties are represented at particular instances in the planning process. Thorgmorton (1991) distinguishes between three broad audiences: professionals (which include planners, other policy experts and private developers - my clarification), politicians, and lay advocates (community at large). He emphasizes the varying use of figures of speech, arguments and other language devices depending on who the audience is. The role of rhetorics varies in the discourse of science, politics and lay advocacy. Rhetorics is defined both as "persuasive discourse within a community" as well as "honest argument directed at an audience" (Thorgmorton, 1991, pp. 154-155). Much of the important knowledge includes stories, myths and implicit understanding which need to be examined and evaluated.

A critical issue in planning is to link knowledge to action. How and under what circumstances does knowledge affect decisions? How should the planning process be orchestrated in order to maximise the accumulation and use of knowledge. Innes (1995, p. 185) writes that in an action-oriented policy-making process learning, deciding, and acting can not be distinguished. "The linear, stepwise process, assumed by the model of instrumental rationality, where policymakers set goals and ask questions, and experts and planners answer them, simply (does) not apply." Planners exercise considerable power and discretion to frame problems, inform various stakeholders and call attention to one aspect or another. The theory of critical practice stipulates provisions for good listening, confronting systematic misinformation, negotiating conflicts and building up consensus. These together with tentative codes concerning social learning help to guide the interplay between learning, deciding and acting.

Planning requires people to "imagine" future events. The systematic rational model relied for this purpose on forecasts, projections and predictions. An underlying hypothesis was that most uncertainty facing planners and decisionmakers was of a quantitative nature and that the assumptions regarding past development provided a relevant basis for preparing forecasts, projections and predictions. Since the mid-1970s it has become increasingly evident that uncertainty is no longer only 'quantitative' ("where possible futures are known, but their probability distribution is unknown") but also 'qualitative' ("where the very shape of possible futures is not known") (Dror, 1986, pp. 248-249). Several methodological dichotomies have been called to attention by decision theorists: 1) problems are not 'tame' but 'wicked' and defy any clear-cut definition (Rittle and Webber, 1973); 2) problems are not independent of each other but interact with one another and are often 'messy' and 'confusing' and defy technical solutions (Schön, 1989); 3) problems are not only 'technical' amenable to expert solutions but are 'practical' for which at best some general statement of purpose can be achieved (Ravetz, 1971); 4) it is a misconception to expect that the methods capable of putting a man on the moon can be used to solve the socioeconomic problems of inner city (Nelson, 1974). This is described as the 'moon-ghetto' metaphor and refers to Vice-President Hubert Humphrey's statement in the US Senate in 1968, "[t]he techniques that are going to put a man on the moon are going to be exactly the techniques that we are going to need to clean up our cities".

In response to this new problem-situation decision theorists have proposed several approaches to understand the nature of unsolvable problems - the so-called 'problem structuring methods'. Rosenhead (1989) provides an account of six problem structuring methods. These are 'strategic options development and analysis', 'soft systems methodology', 'strategic choice', 'robustness analysis', 'metagame analysis' and 'hypergame analysis'. The main features of these methods are: participation, problem construction, iterative progress, cyclical mode of decision making, partial commitment and alternative perspectives (Rosenhead, 1989). In short features which are identifiable in communicative planning theory.

3.3. ORGANIZATIONAL ASPECTS

Interactive, communicative planning does not mean that a group of planners, decisionmakers and other stakeholders are simply thrown together. It is a question of reorganising the group into a more deliberative political body. Forester (1996) suggests that the process is not as 'anarchic' as the 'garbage can model' of decision-making proposes. In the latter, choices, problems, solutions and participants are just thrown together into garbage cans. Interactive planning resembles an 'organised anarchy' or as

Bryson and Crosby (1992) call it 'a shared-power, interorganizational, interinstitutional environment'. They give several characteristics of the shared-power environment: goals are inconsistent and pluralistic, positions of various stakeholders change (sometimes quickly), conflict is seen as a legitimate and expected part of the free discourse, information is used and withheld strategically, considerable disagreement about action-outcome relationships, and disorderly decision process. Thus the 'organising' of an 'inclusionary' institutional arena comprises of identifying roles, keeping anticipatory (strategic and normative) from the routine concerns of decision-making, keeping the institutional arena as 'widespread' as possible, maintaining the balance between various networks, actors and relationships.

A central theme in communicative planning theory is 'organizational learning'. (Some times also referred to as 'institutional' or 'community' learning. In the latter case it is learning among lots of organizations and even less organised interests in the society). Forester (1996) proposes a 'transformative' theory of social learning which means that not only arguments change in dialogues and negotiations but also participants. Forester derives his theory from the theories of organization development advanced by Argyris and Schön (1978) and Bennis (1976). The focus in the latter theories was on individual, small group and organization whereas communicative action theorists' 'inclusionary institutional arena' involves even community. A diffusion of critical planning ethics (in the Habermasian sense) through the group, organization and community is necessary if participants were to create 'shared institutional capital' which can continue to prevail even when a particular planning task is done. It helps to institutionalize co-ordinated action in the long run. Innes (1994, p. 12) distinguishes between three types of shared capital: social capital in the form of "personal networks and trust", intellectual capital in the form of "both of common data sets, indicators, and descriptions of problematic conditions, and of mutual understanding among players of each others' needs and situation" and political capital as "new political alliances".

In interactive, communicative planning practice, planners are involved in community concerns, politics and public decision making. Both Forester (1989) and Innes (1995) propose that appreciating communicative rationality, envisioning the ambiguous demands of institutional contexts, gaining understanding of the identities of the participants in the planning process; in short, comprehending the 'inclusionary' institutional arena require new forms of planning education. Surely this applies equally to the resources planners require. The conventional systematic planning emphasized the need for resources for scientific inquiry (forecasts, predictions and projections). In interactive planning practice resources are needed in order to reduce distortion of information. For example, what Forester (1982) calls an 'inevitable' source of information is due to the division of labour and thereby knowledge. It is enhanced with the increasing specialization in the planning profession. The problem can only be resolved by proper input and distribution of planning personnel and the continuing education of planners. Similarly specific educational strategies have to be adopted in order to overcome the uneven position of various stakeholders in the planning process.

Allocation of resources for different purposes is an important aspect of improving communicative planning activities.

4.   Evaluation of Structure Planning

Marks *kommun*, with a population of just over 33 000 inhabitants, is a small semi-urban municipality in western Sweden. The community is well-known in Sweden for its entrepreneurial spirit. People are flexible and inventive and often have two jobs in order to make ends meet. Characteristic for the population is the widespread interest in environment, nature conservation and the preservation of cultural relics. For these and other purposes there are well-developed club activities. There are more than 200 associations and societies. The most well-known, besides several sports clubs, are the nature conservancy associations, folklore associations, 'Living Countryside' movement and the 'Whole of Sweden Shall Survive' movement. Marks *kommun* is also renowned for its solid 'planning tradition' and unusually large competence in this field. Knowledgeable and engaged politicians are actively involved in planning and environmental issues and are willing to pursue them politically. Party conflicts are rare. Politicians try to achieve as broad agreements as possible and cooperate across the party lines. Local government officials in different departments are enthusiastic and cooperative. There are well-developed networks and routines for exchanging information and viewpoints. There is thus a large 'shared capital - social, intellectual and political' in Marks *kommun* in the form of 'networks of communication', 'mutual understanding' and 'alliances and agreements'. (For a more detailed description of the contextual features and institutional arena of Marks *kommun*, see, Khakee, 1994a and 1997).

The following evaluation of the structure planning process is based on a number of in-depth interviews with the planners and politicians responsible for preparing the structure plan. The 1987 planning legislation waived the previous requirement of the state approval of structure plans (previously they were called 'master plans' and were limited to municipalities' urban territory). The legislation also emphasised the local character of the plan because it only requires the approval of the elected municipal council and that the municipalities are free to develop the plan as they like except that they have to pay attention to national environmental and cultural interests.

4.1. NORMATIVE ASPECTS

An open discourse was necessary for developing a 'robust' plan which would otherwise end up as 'decorative' report in the bookshelf. An inclusionary discourse was a 'proxy' for 'comprehensive attention. However, the fact that Marks *kommun* was small helped the planners to make issues more comprehensible. According to Bryson and Crosby

(1992) smaller scale makes situation more comprehensible intellectually and manageable processually. One of the problems was to avoid 'incrementalism' so that differences of opinion and conflicts were given legitimate and bona fide role. According to Forester (1989) Lindblom's incrementalism hangs on his (magnificently optimistic) assumption that every affected interest has its 'watchdog'. Conflicts were regarded as 'productive' and various techniques of group facilitation extending from a daily cup of coffee together to public meeting and workshops were applied (see, Fog *et al.* 1992, about the rules and procedures on conflict processes in Swedish land-use planning).

Stakeholders viewed issues differently. For example, Marks *kommun* has a very decentralised physical structure with one major and several small urban centres. There is a strong public pressure to preserve the identity, social and environmental quality of all urban places (some with only 20 inhabitants - my remark) as well as to ensure contact network between them. The people seem to be interested in working with "relational resources" and developing them. Planners and local government officials perceive the issue from resource and environmental point of view, favouring concentration of resources to a few centres. The public associations put forward the historical tradition of small communities and 'living countryside'. The politicians favoured the least possible number of constraints. Delineating major points of conflicts, creating a discourse among the stakeholders to know what 'the fighting' was about and illustrating the consequences of various alternative policies in a down-to-earth manner helped to build up a consensus for 'balanced development'.

Stakeholders had different notions with regards to the scope of planning and the scope of the structure plan. It became necessary to make clear what use the plan shall have. The discourse took up in turn the plan's educational, symbolic and regulatory role. For example there was settlement pressure in the north west area of the municipality, adjacent to the City of Göteborg. The educational aspect was to make clear the consequences of unrestricted settlement policy. The regulatory aspects were discussed with regards to environmental restrictions owing to the proximity to the Göteborg International Airport and eventual possibility of technological industry wanting to establish the neighbourhood of the airport. Finally the symbolic aspects were summarized in the concept of 'preparedness' to show that the local governance had a plan for the area although the households' and industries' location decisions were outside its reach.

Identification of the stakeholders was not an issue in Marks kommun with its rich network of associations and clubs which had fairly good contacts with planners and politicians. The issue was to involve the members of these associations and especially those citizens who were not active in these associations in a planning process which often took up topics of overall nature, which the citizens could not identify with. This is a problematic issue which has not so far been discussed in the communicative planning literature. One suggestion was to let citizen participation at certain strategic points during the planning process when planners could take up all the controversial issues and disagreements and 'provoke' the citizens to react. Many planners feared that this might put local government in a 'defensive' position. The method adopted was to encourage participation through 'demonstration effect'. Citizens had previously participated in the campaign in connection with 'Whole Sweden Shall Survive' movement. This was a nationwide movement in order to counteract the persistent depopulation of rural Sweden. Planners encouraged the citizens to show the same enthusiasm in structure planning process which in fact succeeded especially with regards to nature conservation issues.

4.2. EPISTEMOLOGICAL ASPECTS

The task of preparing the text of the structure plan was assigned to a consultant. Her job was to prepare proposal about every section of the plan in the course of a discourse primarily with local government officials and politicians but also seeking the advice of other stakeholders on specific topics. The consultant faced three problems. One issue was to sort out the knowledge provided from various sources so that only 'essential' knowledge was presented in the draft for each section. Since all affected parties had access to the information synthesised by the consultant there was hardly any danger of 'backroom deal making'. Back-room deal-making rarely occurs in the case of structure planning which aims at outlining overall strategies for development. Such deal-making is common at the implementation level i.e. in the course of preparing development plans. The other problem was to keep the 'technical-systematic' knowledge at a reasonable level so that the participants did not get lost in details. Too much emphasis on such knowledge often depended on local government officials' and politicians' reluctance to make overall judgements. The third epistemological issue was to obtain a proper balance between various types of knowledge: technical, economic, social and ecological and at the same time make sure that one or the other type of knowledge did not constrain the major aim of the effort, namely to develop broad development strategies or visions.

There was a time table for the structure planning process which included identifying planning issues, carrying out investigations and analyses, preparing proposals, consultations, revisions, and finally preparing the plan text for approval by the local council. But the timetable only provided bench marks for carrying out various tasks within a certain time period. The actual procedure was very much a case of reiteration: knowledge - discourse - proposals - discourse - new knowledge - proposals and so on. Formally local overall planning was very much the work of a single person, often the town planning architect who asked various local government offices to supply relevant information. The only discourse took place when he (often it was 'he') sent out the plan proposal for comments. Structure planning on the other hand was very much a continuous dialogue involving many groups meeting and listening to one another. There is a big difference between acting as purveyors of information and commenting on a proposal, and creating knowledge in a the course of a discourse. Moreover, the process led to the formation of different relationships, better understanding of one another and of local issues.

Preparing a long term plan of various issues involves much uncertainty and wicked problems. A major problem is who decides what these issues are. According to communicative planning theory, the discourse community makes these decisions. The prevention of acidification of land and water resources provided one such example. It was a major issue in structure planning. However, 90% of the problem was caused by sources outside the municipality. Mapping the sources of acidification and likely changes was one way to appreciate the uncertainty surrounding this issue. Co-operating with other municipalities and county authorities was another way to handle the uncertainty. Analysing the values among the stakeholders was a third way to find out the 'willingness to pay' for measures to reduce the problem. Flexibility was another device in meeting uncertainty rather than indulging in futile exercise to reduce it through investigations. There was a certain wariness among stakeholders with regards to discussing the issue of 'urban structure' which had been debated in Marks *kommun* over a long period. The *kommun* is made up of one central place and six smaller urban centres of about equal size. Attempts to get these small centres co-operate with regards to the location of services (schools, sporthalls, etc) had not succeeded. 'Small is beautiful' and other arguments were put forward by the small communities renowned for their spirit of 'self-reliance'. The discourse resulted against making projections about the future 'urban structure' and for retaining flexibility so that adjustments could be made following changes in or outside the municipality.

4.3. ORGANIZATIONAL ASPECTS

Several factors contributed towards creating an 'inclusionary arena' on Marks *kommun*. The new planning legislation emphasised that the structure plan should be a 'local' plan. Marks *kommun*'s 'association model' came in to use since the rich network of public association was used to obtain viewpoints and suggestions from various associations. The Swedish system of consulting all organisations entitled to have an opinion on a public matter is sometimes self-deceptive in providing a feeling that public decision making is participatory. Often very few active persons get involved in this process. In Marks *kommun* some of the association specialising in nature conservancy were invited in the planning discourse in order to enhance public participation. The other requirement was that the structure plan covers the entire municipal territory leading to negotiations between urban-rural, agricultural-recreational and other sets of interests. The planning legislation did not require the implementation of the plan. But in many *kommuns* including Marks *kommun* local governance stated how the plan shall work. This involved dialogue between a large number of development and conservation interests. The Nature Resources Act provided new responsibility for local governments with regard to environmental protection and conservation. This enhanced the role of the

Environment Protection Board and green interests in a community where there was a strong public engagement in conservation issues. Finally the growing awareness of 'place-making' and regional competition in the new political geography of Europe compelled Marks *kommun* into a new arena involving various regional organisations and interests.

As mentioned above, the substantial social, intellectual and political capital in Marks kommun made the structure planning process into such an interactive and participative process. It enabled the consultant to 'draw out' the plan from this shared capital. It enabled her to explore her interpretations with the participants and see that she had got it right. There was a strong commitment on the part of various stakeholders to ensure that the plan should not be just 'a bunch of papers in the drawer'. Four aspects of 'planning culture' paved the way for a successful process. First, planning was used as a way of communication. Second, it formed an important dimension for all local government work. Third, it cemented the work of various boards and departments. Finally, the 'public association model' as a model of citizen participation explicitly utilized by the planners was an important component of this process.

Even if there was an overall satisfaction with the way the structure planning process was carried out, several specific programmatic suggestions were made to do a better job next time. The foremost desire was more resources in informing the public and also to provide training in negotiation and mediation. Both Forester (1989) and Susskind and Cruichshank (1987) emphasize the need for such training so that the 'lay advocates' understand the difference between 'compromising' and 'exploiting', between 'splitting' and 'trading across priorities' so that 'joint gains' can be realised. The local government officials and even politicians showed the desire for special training in dealing with ecological issues in planning, in making use of social knowledge other than in investigating the need of social services.

5. Discussion

With the increasing recognition of the communicative nature of planning, greater interest has been focused on the planning process rather than the outcomes of planning (plans, policies and decisions). While traditional evaluation has attempted to assess the success of planning in terms of judging the outcomes against certain prescribed criteria (objectives or goals), in the evaluation of the planning process the normative component is much less explicit. Nevertheless an evaluation of the process is necessary in order to improve its performance in terms of its ability to enhance such aspects as transformative learning, communicative networks, institutional capital, consensus and commitments.

The methodology presented in this paper attempts to take up some aspects of the planning process categorised according to the normative, methodological and organizational dimensions. The fundamental issue the methodology tried to deal with is if it is possible improve the way the process is organized with respect to several central aspects in planning:

- defining the community interest(s), determining the scope and legitimacy of planning and specifying the role of public participation.

- identifying the role of different types of knowledge and the relation between knowledge and action, managing uncertainty and dealing with wicked problems

- ensuring the diffusion of planning culture within the planning organisation, enhancing the institutional capital and maximising the use of planning resources.

The methodology helps us to develop a probing set of questions in order to open up planners' and politicians' thoughtfulness and reflections on their experiences. For each of the three dimensions - normative, methodological and organizational - planners can relate as they did in the case of Marks *kommun* to what had been expected and unexpected. Looking backwards planners in Marks *kommun* suggested several things which would improve the process of reaching consensus, determining the scope of planning, obtaining stakeholder support and enhancing the shared capital. Of course, the distinctive character of Marks *kommun*, with its rich planning culture, provided fertile ground for employing the communicative approach in both planning and evaluation. Nevertheless the case study does show the need to think more coherently about assessing planning processes. Ultimately the usefulness of this methodology (and for that matter any other evaluation scheme) can only be determined by its application in a broader comparative framework.

References

Argyris, C. and Schön, D.A. (1978) Organizational Learning: A Theory of Action Perspective, Addison-Wesley, Reading, Mass.

Bennis, W. G. et al. (1976) The Planning of Change, Holt, Renehart and Winston, New York.

Bryson, J.M. and Crosby, B.C. (1992) Leadership for the Common Good, Jossey-Bass Publishers, San Francisco.

Dror, Y. (1986) Planning as fuzzy gambling: A radical perspective of coping with uncertainty, in Morely, D. and Shachar, A. (eds) Planning in Turbulence, The Magnes Press, Jerusalem.

Faludi, F. and Altes, W.K. (1997) Evaluating communicative planning, in Borri, D., Khakee, A. and Lacirignola, C. Evaluating Theory-practice and Urban-Rural Interplay in Planning, Kluwer Academic Publishers, Dordrecht.

Fisher, F. and Forester, J. (eds) (1993) The Argumentative Turn in Policy Analysis and Planning, Duke University Press, Durham.

Fog, H., Bröchner, J., Törnqvist, A. and Åström, K. (1992) Mark, politik och rätt (Land, Politics and Justice), Swedish Council for Building Research, Stockholm.

Forester, J. (1980) Critical theory and planning practice, *APA Journal* 46, pp. 275-206.

Forester, J. (1982) Planning in the face of power, *APA Journal* **48**, pp. 67-80.

Forester, J. (1989) Planning in the Face of Power, University of California Press, Berkeley.

Forester, J. (1996) Beyond dialogue to transformative learning. How deliberative rituals encourage political judgements in community planning processes, in Esquith, S. (ed.) Democratic Dialogues: Theories and Practices. Poznan Studies in the Philosophy of Sciences and Humanities, Rodopi: University of Poznan.

Friedmann, J. (1987) Planning in the Public Domain: From Knowledge to Action, Princeton University Press, Princeton.

Guba, E.G. (ed.) (1990) The Paradigm Dialog, Sage, London.

Guba, E.G. and Lincoln, Y.S. (1987) The countenances of fourth-generation evaluation: Description, judgement and negotiation, in Palumbo, D.J. (ed.) The Politics of Program Evaluation, Sage, Newbury Park.

Guba, E.G. and Lincoln, Y.S. (1989) Fourth Generation Evaluation, Sage, London.

Healey, P. (1993) The communicative turn in planning theory and its implications for spatial strategy formation, in Fisher, F. and Forester, J. (eds) The Argumentative Turn in Policy Analysis and Planning, Duke University Press, Durham.

Healey, P. (1996) The communicative turn in planning theory and its implications for spatial strategy formation, *Environment and Planning B* **23**, pp. 217-234.

Healey, P., McDoughall, G. and Thomas, M.J. (1982) Planning Theory: Prospects for the 1980s, Pergamon Oxford.

Innes, J.E. (1994) Planning Through Consensus Building: A New View of the Comprehensive Planning Ideal (Working Paper 626), University of California, Institute of Urban and Regional Development, Berkeley.

Innes, J.E. (1995) Planning theory's emerging paradigm: Communicative action and interactive practice. *Journal of Planning Education and Research* **14**, pp. 183-190.

Khakee, A. (1994a) Innovation in Development Plan-Making in Sweden: The Case of Marks Kommun. Paper for the Second Workshop on Innovation in Development Plan-Making in Europe (March 18-20), Faculteit der Beleidwetenschapen, Katholiche Universiteit, Nijmegen.

Khakee, A. (1994b) A methodology for assessing structure planning process *Environment and Planning B* **21**, pp. 441-451.

Khakee, A. (1997) Working in a democratic culture: structure planning in Marks kommun, in Healey, P. et al. (eds.) Making Strategic Spatial Plans. Innovation in Europe. UCL Press, London.

Krumholz, N. and Forester, J. (1990) Making Equity Planning Work. Temple University Press, Philadelphia.

Lichfield, N. (1996) Community Impact Evaluation, UCL Press, London.

Nachmias, D. (ed.) (1980) The Practice of Policy Evaluation, St. Martins Press, New York.

Nelson, R.R. (1974) Intellectualizing about the moon-ghetto metaphor: A study of the current malaise of rational analysis of social problems, *Policy Sciences* **5**, pp. 375-414.

Ravetz, J.R. (1971) Scientific Knowledge and its Social Problems, Oxford University Press, Oxford.

Rittel, H.W.J. and Webber, M.M. (1973) Dilemmas in a general theory of planning, *Policy Sciences* **4**, pp. 155-169.

Rosenhead, J. (ed.)(1989) Rational Analysis for a Problematic World, John Wiley, Chichester

Schön, D.A. (1987) Educating the Reflective Practitioner: Towards a New Design for Teaching and Learning in the Professions, Jossey-Bass Publishers, San Francisco.

Susskind, L. and Cruichshank, J. (1987) Breaking the Impasse, Basic Books, New York.

Thompson, J.L. (1990) Strategic Management Awarenes and Change, Chapman & Hall, London.

Thorgmorton, J.A. (1991) The rhetorics of policy analysis *Policy Sciences* **24**, pp. 153-179.

THE COMMUNICATIVE IDEOLOGY AND *EX ANTE* PLANNING EVALUATION

H. VOOGD

1. Introduction

Public planning in Western Europe has changed considerably in the last decade. The economic and political climate has provided public authorities with a new set of goals and expectations. These include the assumption that new large-scale infrastructure and urban revitalisation projects are essential for preserving the current level of welfare in the rat race with other municipalities/regions/countries. At the same time public authorities are faced with generally diminished resources. In reaction many authorities have found it necessary to use marketing and communication techniques to attract private investments (see Ashworth & Voogd, 1990, 1994). Similar communication techniques are also being used to oppress counter-forces of those with different opinions. Academic planners in Western Europe, however, are more and more convinced that planning should be a process of facilitating community collaboration for consensus-building (e.g. see Balducci & Fareri, 1996; Healey *et al.*, 1995; Voogd & Woltjer, 1995; Woltjer, 1996a).

New approaches are now being advocated, and sometimes also followed, that suggest a fundamental break with the planning methodology of the past. Traditional professional expertise seems to be losing ground. According to Healey (1996) the planning community therefore needs to engage in vigorous debate and research on the forms and methodologies of this new situation.

The purpose of this paper is to investigate the consequences of the emerging situation for the future use in spatial planning of so-called *ex ante evaluation methods*. These are methods for comparing the characteristics of various choice-possibilities in an explicit and systematic manner. Many so-called 'formal' methods have been developed in the last twenty-five years to support this task. How useful are these methods, which focus on the quality of decisions, in an emerging planning practice primarily focusing on the quality of decision-making?

The structure of the paper is as follows. In order to understand the changing context of planning practice, the rise of the 'communicative ideology' will first be briefly discussed. In addition a typology is given of different planning arenas that may occur.

N. Lichfield et al. (eds.), Evaluation in Planning, 113–126.

This typology is further used for discussing the usefulness of evaluation methods. The paper finishes with some concluding remarks.

2.   The Rise of the Communicative Ideology

The relations between participants of the political process have been changed over the past *decennia*. The growing welfare and increasing individualism have disillusioned many citizens with political parties. The *Representative Democracy* (cf. Korsten, 1978) is more and more under fire. The rise in general levels of education has made many citizens less slavish followers of 'party views' and the number of floating voters has increased considerably. Political parties have had to adapt to this situation. By using the full arsenal of marketing techniques they try to attract the necessary voters. Abstract ideologies are no longer seen as appropriate for selling 'political products'. A fundamental discussion is no 'news' and not suitable for television. The competitive struggle for the voters in this - for political parties - 'post-ideological age' is now done by focusing on fragmented wishes of the electorate and by creating a favourable public image for the party's spokesman. This is a game of 'Old Maid': the most important task of politicians seems to be avoiding mistakes that may harm their public image and at the same time pointing at 'obvious' mistakes by others. Hence, public opinion has become a very dominant factor in political decision-making. Political debate is reduced to exchanging a couple of 'one liners', suitable for broadcasting at 'prime time news'.

Politicians are on their guard. While society is becoming more and more complex, they are taking care not to burn their fingers by complex societal problems. For such problems often involve unpopular measures and this may harm their career. This results in a paradoxical situation that if politicians go their own way based on ideological motives, they are accused of not listening to their voters, and if they follow public opinion they become impotent. In both cases traditional 'party politics' is losing ground. Hence, there is talk of 'a gap' between citizens and politicians. This gap can be seen as the main cause of the crippled functioning of democracy.

This change in political climate has also affected planning theory and - more important - spatial planning in practice. It is fascinating to see that many 'neo-marxist' ideas about communicative planning from the roaring sixties and early seventies re-appear in recent planning literature. However, for many of my academic generation the idealistic opinion of Habermas (1973, 191) that *'Den kommunikativ angelegten Planungstheorien liegt ein Begriff von praktischer Rationalität zu grunde, der am Paradigma willensbildender Diskurse gewonnen werden kan'* could not compete with the challenges that in those days were offered by the seemingly more realistic development of computer-assisted methods and theories based on scientific rationality as presented by, for instance, Chadwick (1971) and Lichfield *et al.* (1975). It was very difficult in the 1970s to imagine governmental influence reduced to a 'referee' in a game, where according to 'Habermasian criteria' all players should be fair and square and able to play the game. Clearly, the rebuilding of the country after World War II was seen by many Dutch planners as a proven success of a strong public planning, for a large extent based on the principles of rational planning (see also Faludi & Van der Valk, 1994).

Obviously, this experience is not the same for countries like the USA, which traditionally have a decentralised, more liberal, planning system, which is 'dominated by working class realists with a low regard for missionaries' (cf. Dyckman, 1961). Also in the 'pre-Habermas period', i.e. before his major work had been translated into English, much planning-oriented literature was published in the USA that stressed the importance of social pluralism and bargaining (e.g. see Dahl & Lindblom, 1953). This is not surprising for a country where democracy appears to have degenerated to 'Hollywood show level' and where legal bribery exists since interest groups actually can buy political attention and political favours. For European outsiders it is fascinating to see how 'neo-marxist' ideas are linked with this 'capitalist' market democracy. It has resulted in many new ideas about communicative planning - for example see Fischer & Forester (1993); other good overviews are provided by Sager (1994) and Healey (1996).

The widespread renewed introduction in Western Europe of, what I prefer to call the *communicative ideology* in public planning, was only possible because of the fundamental societal changes as discussed above. The growing social complexity needed a new - but simple - philosophy by which people come to terms with the world around them. This is an ideology, being pervasive sets of ideas, beliefs and images that groups employ to make the world more intelligible to themselves. Hall (1977) asserts that an ideology only operates by being openly embedded in commonsense wisdom. It is commonsense wisdom in the Netherlands, and probably elsewhere, that public discussions between political parties are more and more replaced by discussions between interest groups. Discussions, that are often fed by - or based on - one-sided research outcomes and normative expert views.   Representative democracy is clearly changing into *participatory democracy*.

The magic word for narrowing down the gap between 'citizens and politicians', and embraced by all actors in this play, is called *communication*. Evidently, 'good communication' is a goal that is giving everybody warm feelings given the 'inclusionary ethic which underpins the approach' (cf. Healey, 1996). Political parties, governmental bodies, interest groups, now all stress the importance of communication, leaving the innocent citizen with an avalanche of 'news letters' and invitations for 'information evenings' and 'open days'. Promotion, persuasion and propaganda have been discovered as communication tools. Marketing has become an ordinary public planning concept, but also other institutional groups are using its techniques (Ashworth & Voogd, 1990, 1994).

Clearly, the communicative ideology has its limitations. Although planning practice in the last decade certainly has moved in a direction that vaguely resemble some 'neo-marxist' ideas of the sixties, the Habermasian dream of 'discourses' based on 'fair play' is still an unattainable ideal. What is left is an uncertain world, where facts seem to be replaced by values. Often values of those who have the money and the power, whose local egoism seems more important than principles of sustainable development. Will there be a place left in this world for old-fashioned planning expertise, such as the clarification of consequences of different choices and planning options by means of *ex ante evaluation methods* that follow the logic of scientific rationality? In answer to this question, in the next section 'this world' will be first further elaborated, by distinguishing a number of different planning arenas. In addition what is precisely meant by *ex ante* evaluation methods will be outlined. These methods are then be compared with the various arenas.

3.   A Typology of Planning Arenas

If we consider the communicative ideology in relation to the variety of situations that occur in spatial planning, a large number of *planning arenas* can be distinguished. By planning arena is meant a configuration of actors that are involved in a product of planning. Since it is realistic to assume that each planning arena has its specific characteristics, a judgement about the usefulness of *ex ante* evaluation methods can only be given in relation to these characteristics.

For clarity the following limited set of criteria will be considered here for defining different planning arenas:

1)   The territorial level of planning

Local planning has an entirely different nature than regional planning. Due to its close distance to the actual users of space, the local level is much more open for consensus-building approaches than the regional level. Evidently, almost all examples in literature about communicative planning deal with the local level. Regional practice has also its variations, since the regional level itself is multi-leveled again, e.g. province (or county), state, European Union.

2)   The level of legal regulations

For a large number of well-defined situations strict legal regulations exist that define both the procedure to be followed as well as the format of the resulting products. An example is the Environmental Impact Statement, that is required for a given type of project. Also other Acts, such as in the Netherlands the Environmental Management Act and the Housing Act, provide constraints that will restrict freedom of planning. However, there are also situations where such legal constraints are negligible and/or avoidable.

3)   The power structure of actors

It is difficult to make a straightforward classification for this criterion given the complexity of social power structures. However, for the purpose of this analysis a distinction is made between a hierarchical, i.e. 'top-down', power structure and a mixed power structure. A hierarchical power structure implies that a higher level authority is able to empower its wishes on a lower level. In a mixed power structure there may be one or more dominant actors, but they are unable to exercise full power over other actors.

4)   The level of integration of planning

Planning can be focused on one administrative sector, such as housing, recreation or traffic infrastructure, but it can also be comprehensive, i.e. focusing at integrated developments.

5)   The level of abstraction of planning

A distinction is made between the strategic level and the operational level. The strategic level is operating with a long-term perspective, whereas the perspective of the operational level is implementation-oriented.

Based on these criteria, in Table 1 a typology of planning arenas is given. It illustrates the complexity of spatial communicative planning, because the simple assessment in two categories of each criterion already results in 32 different arenas. Obviously, this amount will exponentially increase if more variables are taken into account. On the other hand, the theoretically derived arenas of Table 1 are not equally important. The presence of certain arenas will, because of national regulations and cultural differences, certainly differ per country. For this reason, further comments will be solely based on the Dutch situation (see also Faludi & Van der Valk, 1994).

4.   A Typology of *Ex Ante* Evaluation Methods

The current methodological 'state of the art' of *ex ante* plan and project evaluation is the outcome of developments in various disciplinary and scientific areas. Selective overviews can be found in, for example, the following books: Lichfield, Kettle & Whitbread (1975), Nijkamp (1980), Kmietowicz & Pearman (1981), Voogd (1983), Fusco Girard (1987), Shofield (1987), Nijkamp & Voogd (1989), Shefer & Voogd (1990), Nijkamp, Rietveld & Voogd (1990), Janssen (1992), Lichfield (1996).

Already before World-War II various attempts have been made to perform a systematic evaluation of intended government policies (e.g. see Nijkamp *et al.*, 1990). 'Cost-benefit analysis' have long been in many countries the preferred methodology for *ex ante* evaluation. This emphasis on *monetary evaluation methods* gradually changed in the sixties, thanks to the influential publications of Lichfield (1970) about the 'planning balance sheet' and Hill (1968) about the 'goals-achievement matrix'. Their pioneering work has had an important impact on a generation of planners. It was not so much the technical 'sophistication' of their *descriptive overview methods*, but more their 'power of conviction' and 'transparency' in reducing a choice-problem into manageable judgement criteria - goals and objectives - and impact ratings.

TABLE 1. A typology of planning arenas for spatial planning

territorial level	regulations level	power structure	integration level	abstraction level	#
local	formal product and process regulations	hierarchical	sectoral	strategic	1
				operational	2
			comprehensive	strategic	3
				operational	4
		mixed	sectoral	strategic	5
				operational	6
			comprehensive	strategic	7
				operational	8
	relaxed formal regulations	hierarchical	sectoral	strategic	9
				operational	10
			comprehensive	strategic	11
				operational	12
		mixed	sectoral	strategic	13
				operational	14
			comprehensive	strategic	15
				operational	16
regional	formal product and process regulations	hierarchical	sectoral	strategic	17
				operational	18
			comprehensive	strategic	19
				operational	20
		mixed	sectoral	strategic	21
				operational	22
			comprehensive	strategic	23
				operational	24
	relaxed formal regulations	hierarchical	sectoral	strategic	25
				operational	26
			comprehensive	strategic	27
				operational	28
		mixed	sectoral	strategic	29
				operational	30
			comprehensive	strategic	31
				operational	32

A very interesting aspect from the point of view of communicative planning is that in the sixties and early seventies the idea existed that an 'aggregation' of impact scores should not be done by planners, but that it primarily should be a political task. However, this did not work very well. The vast amount of information from a spatial impact analysis always raised questions like 'what do the planners/consultants recommend?'. This gave the impetus for the application of arithmetic *multicriteria evaluation methods* in planning practice, that were able to provide such a recommendation (Voogd, 1997).

The critique of these kinds of approach in the sixties and seventies mainly focused on the technocratic method of their use. Especially the use of arbitrary numerical weights, the fixation on 'hard', i.e. numerically measurable, criteria, and the 'optimization' characteristics were good targets for criticism (see Chadwick, 1971; Lichfield *et al.*, 1975). This critique evoked an avalanche of new multicriteria methods, that were capable of using both 'soft' and 'hard', as well as 'mixed' impact data (e.g. see Nijkamp *et al.*, 1990).

Another, more recent, line of work is in the field of *Decision Support Systems (DSS)* (e.g. see Janssen, 1990, 1992). The post-war advances in computer technology have favoured the introduction of computer-based choice models in spatial planning. These DSS-models are especially advocated for less structured choice situations. However, a DSS approach is not in contrast to multicriteria methods, but rather complementary. Specific man-machine interfaces should create a 'learning process', so that the 'decision-maker' is growing towards the 'best' choice. In Voogd (1985) the interactive learning approach in public planning has been critized because of its conceptual simplicity. For instance, the assumption that a pluriform society can be represented by 'one' individual decision-maker, interacting with his or her computer screen, is in reality extremely naive. Nevertheless, the technical possibilities of a DSS-approach are very interesting and for 'routine' decisions certainly attractive. Unfortunately, there are not so many 'routine' decisions in spatial planning.

5.   Evaluation Methods Versus Planning Arenas

There is no properly documented empirical knowledge yet about the usefulness of various explicit evaluation methods for consensus building. The evaluative remarks in this section are therefore only based on more than twenty years of personal experiences of this author with practical applications of these methods.

Monetary evaluation methods, descriptive overview methods, multicriteria methods and DSS-approaches are all based on the assumption that the impacts of a policy proposal can be assessed for all relevant variables of the proposal. They differ in the way this assessment is done and in the way the results are presented. However, in the literature about these methods it is usually assumed that the impacts can be assessed by experts, whether in a qualitative sense or quantitatively. Some DSS-approaches enable the use of opinions rather than empirical data as input for such an assessment. But in all cases the point of departure is the application of scientific logic for measuring the effects.

If a communicative ideology is pursued, however, it may very well be that impacts are subject of debate and that consensus about the size of the impacts is sought. Hence, '*professional data*' will be replaced by '*negotiated data*'. In principle, all categories of evaluation methods can deal with this type of information. Whether the planner, involved in performing the evaluation, can live with it is another issue, but it should be no problem for the communicative ideology since he or she can communicate it ...

A big problem for the application of many evaluation methods within the framework of the communicative ideology is the *lack of transparency*. All methods are in principle professional tools, designed as an aid for skilled planners. And since participants in a communicative planning process do not need to obtain any certificate before participating, let alone be academic planners, professionals have to come down from their ivory towers. It will depend on the kind of planning arena, whether they can stop 'half-way' down the tower, or whether they have to sink 'to the bottom'. In the latter situation, the only evaluation method that will be acceptable as a vehicle for discussion is no doubt a simple overview method, i.e. an evaluation matrix where options are mutually compared by means of a number of criteria. Any aggregation of this matrix may be considered to be 'too difficult'. If the elevator of the ivory tower can stop 'half-way', i.e. in arenas where only planning and political professionals participate, it is essential that the *concept* of the method can be clearly explained[1]. In addition, the acceptance of the method in this arena will depend on its credibility. This depends among others on: has the method been used elsewhere, has it been accredited by independent experts, who is applying the method and under what circumstances, and - above all - are the results of the method flexible enough to allow different - political - interpretations? Looking at the planning arenas of Table 1, it is difficult to give a straightforward appraisal of the suitability of *ex ante* evaluation methods. However, it is possible to highlight per component a preferred situation for systematic evaluation methods. This in done in Table 2, where the dark compartments represent a high probability that systematic evaluation methods might be useful.

The assessment of the components of the planning arenas in Table 2 is based on experiences in The Netherlands, but it is probably rather universal. Since it is virtually impossible for regional authorities (i.e. provinces, national government) to have a 'discourse' with *all* relevant and interested groups, there will always have to produce documents or plans that account for the preferred policy actions. In other words, always alternatives have to be presented and also a reasoning why a particular alternative is most preferred. Evidently, this will never be done by just pointing at 'fruitful meetings', 'inspiring conversations' and 'deepening discourses' among - usually professional -

---

[1] Looking at successful methods in the past, it seems reasonable to suggest that a method will be successful, i.e. become 'fashionable', if the structure is very simple and easy to grasp, while the appearance looks very complex. This enables less talented brains, once they understand the simple structure, to flaunt their acquired wisdom.

representatives of regional interest groups, civil servants and political representatives! There will always be a need, and hence an attempt to justify such decisions by applying 'traditional scientific methods'. *Ex ante* evaluation methods, such as briefly described in the previous section, belong to this category.

TABLE 2. Suitable characteristics of a planning arena - in grey - for explicit evaluation.

territorial level	regulations level	power structure	integration level	abstraction level	#
local	formal product and process regulations	hierarchical	sectoral	strategic	1
				operational	2
			comprehensive	strategic	3
				operational	4
		mixed	sectoral	strategic	5
				operational	6
			comprehensive	strategic	7
				operational	8
	relaxed formal regulations	hierarchical	sectoral	strategic	9
				operational	10
			comprehensive	strategic	11
				operational	12
		mixed	sectoral	strategic	13
				operational	14
			comprehensive	strategic	15
				operational	16
regional	formal product and process regulations	hierarchical	sectoral	strategic	17
				operational	18
			comprehensive	strategic	19
				operational	20
		mixed	sectoral	strategic	21
				operational	22
			comprehensive	strategic	23
				operational	24
	relaxed formal regulations	hierarchical	sectoral	strategic	25
				operational	26
			comprehensive	strategic	27
				operational	28
		mixed	sectoral	strategic	29
				operational	30
			comprehensive	strategic	31
				operational	32

At the local level the relation of authorities with citizens is much more direct. Evidently, you can not convince citizens that a change in his or her neighbourhood is necessary by just pointing at cost-benefit ratio's or evaluation rankings. Practice teaches that at the local level consensus can only be reached by following an 'open planning process', i.e. by starting a 'discourse' with all interest groups concerned. Traditional evaluation methods can hardly fulfill a role in this process, perhaps with the exception of 'overview methods' for structuring a discussion.

The use of evaluation methods is also determined by existing formal - legal - regulations, notably the *Environmental Impact Assessment (EIA)*. In 1985 a general directive for EIA has been adopted by the Council of the European Union (EU). Consequently, as of 1988, a ruling on environmental impact assessment must be applied in all EU countries (e.g. see Arts, 1994). The formal application of EIA is confined to those decisions in the field of physical, infrastructure and economic planning which are likely to have the most detrimental impact on the environment. The cases in which an EIA has to be carried out are listed in a general administrative order. The Dutch legislation on EIA has appeared to be very influential in the way *ex ante* evaluation has been performed in Dutch planning practice in the eighties and early nineties. This legislation provides strict rules regarding the way alternatives should be distinguished and evaluated, however without providing strict methodological guidelines. Many EIA-studies in The Netherlands use some kind of multicriteria evaluation (see Mooren, 1996). However, although EIA has a central position in Dutch planning, it hardly played any role in actual decision-making. The most obvious example is the expansion of Schiphol Airport. The Dutch Government made its decision about the location of a new runway, even *before* the EIA was available and formally published (see also Voogd, 1987).

Especially the planning of infrastructure projects, such as railways and highways, is in the Netherlands based on a hierarchically organised power structure. It clearly represents all characteristics of a 'top-down' approach (see also Niekerk & Arts, 1996). An hierarchical top-down approach is often too rigid to fit well to societal dynamics (Niekerk & Voogd, 1996). However, as Dutch practice illustrates, professional impact assessments like EIA, can relatively easily be linked with such an approach but sofar they hardly solved actual problems related to capricious decision-making because of autonomous behaviour of authorities and societal groups. According to the communicative ideology, a bottom-up approach would be much better able to include the criteria and needs of local actors. This would imply the recognition of a mixed power structure. However, it is questionable whether explicit *ex ante* evaluation methods can play a proper role in this context. The most probable stage for such analysis is after local authorities have got an agreement about the projects to pursue.

Sectoral planning is usually associated with a strong emphasis on 'technical issues'. Although the consequences of many technical choices can be very well subject of a public debate, this is very often only possible after a proper 'translation' of the underlying technical details. Multicriteria methods can be, and actually are used for this purpose. This is not only a Dutch experience, but also in the USA - the political market economy pre-eminently - this can be witnessed (e.g. see Maimone, 1994). It is surprising, given the multidimensional complexity of plans that aim at integration of various sectoral perspectives, that in the Netherlands multicriteria methods have been hardly used in local and regional comprehensive planning, such as land-use planning.

Strategic planning is a matter of designing possible long-term perspectives and creating the necessary commitments on a preferred strategy. A limitation of the communicative ideology in respect to strategic planning is that abstract issues seldomly raise enough public attention for a balanced 'discourse'. For instance, strategic decisions in the field of infrastructure are in the Netherlands - and probably also in many other countries - often made without proper public and political discussions about the choices and their consequences. A major reason is the fact that such decisions are usually too vague, too abstract, to be properly valued. Only in the operational stage, participants really start to question the desirability of strategies (Niekerk & Voogd, 1996). Evaluation methods can be useful in strategic planning for visualizing the consequences of various long-term strategies, and they are for this reason often used, but they are certainly not able to overcome this fundamental problem.

6.   Some Concluding Remarks

The analysis in this paper of the usefulness of evaluation methods in various planning arenas illustrate the fact that nowadays most planning processes in the Netherlands refrain from applying systematic evaluation methods. It can be concluded from Table 2 that out of 32 theoretically distinguished planning arenas only 1 arena, viz. number 17, has all suitable properties for an optimal use of explicit *ex ante* evaluation methods. In other words, by far most planning arenas have one or more characteristics that favour an approach based on the communicative ideology. If we combine this observation with the general observation in section two that the society is moving from a representation democracy to a participation democracy, then the general conclusion can be drawn that in the next decade evaluation methods will have a limited use in spatial planning. This does not imply that planners refrain from these methods but it strongly depends on the field of planning whether analytical methods, and hence also evaluation methods, are being used. In the Netherlands evaluation methods are mainly applied in sectoral planning fields, viz. environmental planning, mineral planning, infrastructure planning and water management. These fields still operate from what nowadays might be called a 'classical' planning paradigm based on the appreciation of professional knowledge. In urban and regional planning, viz. urban design, urban renewal, strategic regional planning, etc., evaluation methods are hardly popular, although sometimes an evaluation matrix is used.

The future will probably be for new evaluation methods that focus on a permanent discussion between, or among, the parties concerned. By means of such methods all essential moments of choice should be emphasized and brought into discussion. Already in the 1970s informal evaluation procedures have been developed, for instance by Bleiker *et al.* (1971) and Manheim *et al.* (1974), for making choices in concert with, and by mutual arrangement with, all parties involved. More recently, a constructivist paradigm for evaluation is presented by Guba & Lincoln (1989), which should offer empowerment and enfranchisement of stakeholders, as well as an action orientation that defines a course to be followed.

However, it should not be neglected that such participation approaches, based on the communicative ideology, are also subject to a number of handicaps. A summary of these of many years ago is still valid (Voogd, 1983, 19):

- In many cases it will be highly problematic to place great demands, for a relatively long period, on the time of the various participants. This implies that it may be very difficult to find the proper persons (e.g. see also Woltjer, 1996b).

- Only few of the potential participants have the ability to disclose their often dormant views.

- Not everyone wants to nor can participate actively. But everyone has the right to know for what reasons choices are made or not made. However, if decisions are made in such deliberation meetings, the value system underlying the choices might remain unknown to the outside world.

- The law in some countries, including the Netherlands, does not allow decision-making without paying attention to legal arrangements in case of petitions, and time for disposition, and such.

- The role of the expert and of expertise will change, which may affect the ultimate quality of the decisions.

These handicaps can be expanded and elaborated, but they will never become efficacious for refraining from consensus-building. It will be a major challenge for planners in Western Europe to cope with these handicaps and at the same time resist the current tendency to follow the US 'socio-political market approach' in all respects.

References

Arts, E.J.M.M. (1994) Environmental Impact Assessment: from ex ante to ex post evaluation, in H. Voogd (ed), *Issues in Environmental Planning*, Pion, London, pp. 145-163.

Ashworth, G.J. and Voogd, H. (1990) *Selling the city: Marketing Approaches in Public Sector Urban Planning*, Belhaven, London/New York.

Ashworth, G.J. and Voogd, H. (1994) Marketing and Place Promotion, in: J.R. Gold & S.V. Ward (eds), *Place Promotion*, Wiley, Chichester/New York, pp. 39-52.

Balducci, A. and Fareri, P. (1996) Consensus building as a strategy to cope with planning problems at different territorial levels: examples from the Italian case study, Dipartimento di Scienze del Territorio, Politecnico di Milano, Milano, Paper presented at the 1996 ACSP-AESOP Joint International Congress in Toronto.

Barde, J.P. and Pearce, D. (eds) (1991) *Valuing the Environment*, Earthscan, London.

Bennema, S.J., 't Hoen, H., van Setten, A., Voogd, H. (1985) Studying Gravel Extraction through Multicriteria Analysis, in A. Faludi, H. Voogd (eds), *Evaluation of Complex Policy Problems*, Delftsche Uitgevers Maatschappij, Delft, pp. 179-188.

Bleiker, H., Suhrbier, J.H., Manheim, M.L. (1971) Community Interaction as an Integral Part of the Highway Decision-Making Process, *Highway Research Record*, No. 356, pp. 125-149.

Buffet, P., Greeny, J.P., Marc, M., Sussman, B. (1967) Peut-on Choisir en Tenant Compte de Critères Multiples?, *Metra*, **6**(2), pp. 283-316.

Chadwick, G. (1971) *A Systems View of Planning*, Pergamon, Oxford.

Dahl, R.A. and Lindblom, C.E. (1953) *Politics, Economics and Welfare - Planning and Politico-Economic Systems Resolved into Basic Social Processes*, University of Chicago Press, Chicago.

Dyckman, J.W. (1961) What Makes Planners Plan?, *Journal of the American Institute of Planners* **27**, pp. 163-170.

Faludi, A. and van der Valk A. (1994) *Rule and Order: Dutch Planning Doctrine in the Twentieth Century*, Kluwer, Dordrecht.

Faludi, A. and Voogd H. (eds) (1985) *Evaluation of Complex Policy Problems*, Delftsche Uitgevers Maatschappij, Delft.

Fischer, F. and Forester, J. (eds) (1993) *The argumentative turn in policy analysis and planning*, Duke University Press, Durham/London.

Fusco Girard, L. (1987) *Risorse architettoniche e culturali: valutazioni e strategie di conservazione*, Franco Angeli, Milano.

Guba, E.G. and Lincoln, Y.S. (1989) *Fourth Generation Evaluation*, Sage Publications, Newbury Park.

Habermas, J. (1973) *Legitimationsprobleme im Spätkapitalismus*, Suhrkamp, Frankfurt am Main.

Hall, S. (1977) Culture, the media and the 'ideological' effect, In: J. Curran, M. Curevitch, J. Wollacott (eds), *Mass Communication and Society*, Edward Arnold/Open University Press, London.

Healey, P. (1996) The communicative turn in planning theory and its implications for spatial strategy formation, *Environment and Planning B* **23**, pp. 217-234.

Healey, P., Purdue, M., Ennis, F. (1995) *Negotiating Development: Rationales and Practice for Development Obligations and Planning Gain*, E & FN Spon, London.

Hill, M. (1968) A Goals-Achievement Matrix for Evaluating Alternative Plans, *Journal of the Americal Institute of Planners*, **34**(1) pp. 19-28.

Janssen, R. (1990) A Support System for Environmental Decisions, In: D. Shefer & H. Voogd (eds), *Evaluation Methods for Urban and Regional Plans*, Pion Ltd, London, pp. 159-174.

Janssen, R. (1992) *Multiobjective decision support for environmental management*, Kluwer, Dordrecht.

Khakee, A. and Eckerberg, K. (eds) (1993) *Process and Policy Evaluation in Structure Planning*, Swedish Council for Building Research, Stockholm.

Kmietowicz, Z.W., and Pearman, D.A. (1981) *Decision Theory and Incomplete Knowledge*, Gower, Aldershot.

Korsten, A.F.A. (1978) *Het Spraakmakende Bestuur*, VUGA, The Hague.

Lichfield, N. (1970) Evaluation Methodology of Urban and Regional Plans: a Review, *Regional Studies* **4**, pp. 151-165.

Lichfield, N. (1996) *Community Impact Evaluation*, UCL Press, London.

Lichfield, N., P. Kettle and M. Whitbread, (1975) *Evaluation in the Planning Process*, Pergamon Press, Oxford.

Lichfield, N., Kettle, P. and Whitbread, M. (1975) *Evaluation in the Planning Process*, Pergamon Press, Oxford.

Maimone, M. (1994) Computer-assisted Environmental Planning for Groundwater Management, in H. Voogd, *Issues in Environmental Planning*, Pion, London, pp. 100-113.

Manheim, M.L. (1974) Reaching Decisions about Technological Projects with Social Consequences: a normative model, in De Neuville & Marks (eds), *Systems Planning and Design*, Prentice Hall, pp. 178-196.

Massam, B.H. (1993) *The Right Place: Shared Responsibility and the Location of Public Facilities*, Longman, Harlow.

Mooren, R. (1996) Appels of peren: alternatieven vergelijken, in R. Bonte, P. Leroy, R. Mooren (eds), *Milieu-effectrapportage*, Aeneas, Best, pp. 59-78.

Niekerk, F., and Voogd, H. (1996) Impact Assessment for Infrastructure Planning: Some Dutch Dilemma's, Faculty of Spatial Sciences, University of Groningen, Groningen, Paper presented at the 1996 ACSP-AESOP Joint International Congress in Toronto.

Niekerk, F., and Arts J. (1996) Impact Assessments in Dutch Infrastructure Planning: Towards a Better Timing and Integration? *Proceedings Conference Integrating Environmental Assessment and Socio-Economic Appraisal in the Development Process*, Vol. 2, Development and Project Planning Centre, University of Bradford, UK.

Nijkamp, P. (1980) *Environmental Policy Analysis*, Wiley, Chichester.

Nijkamp, P., Rietveld, P., Voogd, H. (1990) *Multicriteria evaluation in physical planning*, North Holland Publ., Amsterdam/New York.

Oelen, U.H. and Struiksma, N. (1994) *Puzzelen met beleid*. Samsom H.D. Tjeenk Willink, Alphen aan de Rijn.

Reiner, T.A. (1990) Choices and choice theory revisited. In: D. Shefer, H. Voogd (eds), *Evaluation Methods for Urban and Regional Plans*, Pion, London, pp. 65-78.

Roy, B.(1972) Décision avec Critères Multiples, *Metra* II, pp. 121-151.

Sager, T. (1994) *Communicative Planning Theory*, Avebury, Aldershot.

Seo F., and Sakawa M. (1988) *Multiple Criteria Decision Analysis in Regional Planning: Concepts, Methods and Applications*, D. Reidel Publishing Co., Dordrecht.

Shefer, D. and Voogd H. (eds) (1990) *Evaluation Methods for Urban and Regional Plans*, Pion, London.

Shofield, J.A., (1987) *Cost-Benefit Analysis in Urban & Regional Planning*, Allen & Unwin, London.

Susskind, L. and Cruikshank, J. (1987) *Consensual approaches to resolving public disputes*, Basic Books, New York.

Thomas, H. (ed) (1994) *Values in Planning*, Avebury, Aldershot

Voogd, H. (1983) *Multicriteria Evaluation for Urban and Regional Planning*, Pion, London

Voogd, H. (1985) Prescriptive Analysis in Planning, *Environment & Planning B* **12**, pp. 303-312.

Voogd, H. (1997) The changing role of evaluation methods in a changing planning environment: some Dutch experiences, *European Planning Studies* **5**(2), 257-266.

Voogd, H. and Woltjer, J. (1995) Besluitvorming via onderhandelen: een oud gebruik in een nieuw perspectief, *Beleidsanalyse*, 95-4, pp. 12-21.

Woltjer, J. (1996a) Consensus-building in Infrastructure Planning, Faculty of Spatial Sciences, University of Groningen, Groningen, Paper presented at the 1996 ACSP-AESOP Joint International Congress in Toronto.

Woltjer, J. (1996b) Consensus-building voor infrastructuurplanning: case studies, deelrapport onderzoek Pi, Ministerie van Verkeer en Waterstaat/FRW-Rijksuniversiteit Groningen.

REGULATION THROUGH THE DEVELOPMENT PLAN
An evaluation

A. D. HULL

1. Introduction

This paper evaluates the way decisions on the allocation of land for housing are made in England drawing on the literature charting the changing style of governance and institutional structure in Britain. In this respect it attempts to link into two of the sub-themes of the Umea and Bari conferences, notably the *theory-practice interplay* using a *case study approach*. The paper takes the practice issue of projected household growth and the spatial implications of absorbing that growth and evaluates the debates that take place within the formal arenas of the British land use planning system. Particular attention is paid to how the political setting is structured through the imposition of values, and how the policy agenda is foreclosed through the detailed specification of criteria and procedures. The North East of England (Figure 1) is used as the study area to identify the effect on local government agency of this regulatory style. Research for this paper is drawn from work commissioned by the Town and Country Planning Association (Hull, 1996a) and consultancy work the author has been engaged in for the Housebuilders Federation in the review of regional planning guidance for the North East.

Conceptually the analysis draws on several theoretical strands. The analysis is firstly informed by a political economy approach which identifies the power and resources available to those who have a stake in the issue area and which seeks to throw light on how these 'cards' are used both in the more visible arenas of interaction and behind the scenes (Fainstein and Fainstein, 1983; Harvey, 1982; Whelan et al, 1994). The second analytical strand employs a social constructionist ontology which seeks to show how an actor's systems of meaning or understanding of the context is revealed through their discourse and actions. Finally the case study approach links both these approaches in focussing on institutional interaction in the formal arenas of the land use planning system in the North East of England. The North East of England provides an interesting case study since it is a region with a distinctive identity where there are strong driving forces to extract some form of regional governance from any incoming Labour government, if one should be elected. Secondly, the region as part of the review of regional planning guidance is formulating advice to central government, through the

N. Lichfield et al. (eds.), Evaluation in Planning, 127–142.

local government associations, on the spatial structuring decisions which need to be made to steer the region into the twenty first century. The analysis focusses on the institutional dynamics to highlight the extent to which policy goals can be transformed through adjustments to policy delivery mechanisms.

2.   Policy Context: Changing Values and Institutional Settings

Central-local government relationships have taken a stormy passage since the mid-1970s in Britain - initially revolving around reduced resource flows from the centre following the International Monetary Fund agreement and more latterly by an attack from the centre on the powers and responsibilities of the elected local state. Central government dependency on the cooperation of local government for policy implementation has slowly been broken down since the late 1970s with the introduction of new players as well as a redefinition of the responsibilities of the existing elected agencies, which has impacted on local discretion and working relationships. This transformation of the workings of the local state has principally been brought about by Conservative governments through imposing new political values and a restructuring of the political setting and mode of operation (Thornley, 1991). The post-war welfare contract was based on a *"fairly robust balance of responsibility for the growth and management of urban areas"* between central and local government (Harding, 1996:2). Local authorities took the bulk of delivery responsibilities in field of social consumption - education, housing, social services - with financing and broad direction decided by central government. Production related interventions, as now, were the product of decisions taken outside the ambit of sub-national government.

"The Thatcher governments, combining neo-liberal economics with a somewhat authoritarian approach to centres of opposition, destroyed this balance. At the most general level, Conservative governments attempted to improve national economic competitiveness through tight monetary policy, public expenditure cuts, scaling down the size and functions of the public sector, increasing the role of the private sector in delivering public and quasi-public goods and liberating the 'entrepreneurial spirit' wherever and whenever it could be found [...] Specifically with regard to economic development, urban authorities found that (a) as soon as they discovered legislative loopholes through which to develop alternative strategies, they were closed down by central government, and (b) central government, supported to some extent by the private sector, assembled a new policy infrastructure outside, and largely beyond the influence of, local government" (Newman & Thornley, 1996:2-3).

The rhetoric and the rationale for state regulation is driven now more by internal economic reasons, to remove monopoly and prevent market failure, than for external social reasons to control market behaviour to meet questions of distributional fairness and environmental impact (Francis, 1993). The market mechanism, on the other hand, is now seen as a tool to both solve public sector 'inefficiencies' and social allocation problems.

One of the important requirements of this game plan was to remove all opponents who could threaten this value transfusion. The Labour controlled strategic policy-making power bases in the metropolitan County Councils were the first to be abolished in 1986. Existing advisory bodies were transmogrified through government appointees with widened powers to carry out city wide services. New agencies have been created to bypass the elected local authorities and to implement new initiatives. There is now a patchwork quilt of fragmented small agencies, with little or no local democratic involvement, creating a very complex institutional picture. Through piecemeal institutional restructuring a whole area of policy making has been removed from the local authority 'formal' agenda.

Many of the policy changes impacting on first the English and now the Welsh and Scottish planning systems have not involved explicit programmes but adjustments to the ways existing activities are carried out. Procedural rules, consultees, and the requisite weight to be attached to decision criteria have been specified for certain policy sectors such as housing land allocation, retail impact assessment, minerals requirements, and road scheme assessments. Local state services have been transferred to newly created agencies (eg Urban Development Corporations, Local Authority Waste Disposal Companies) and national-level governmental agencies amalgamated. These can be seen as 'meta-policy' adjustments to the programme shell that are concerned with means more than ends but which nevertheless imply fundamental goals and values in that as the policy delivery system has been restructured, the policy itself has been transformed. Newman and Thornley's (1996) account of the planning process in London encapsulates the tenor of these changes:

> "In the 1980s the planning objectives of the boroughs which did not fit government priorities were either by-passed or overruled. Now, in a substantially weakened position the boroughs have been allowed back into the decision-making arenas but on the government's terms. All are now agreed on the primacy of economic development objectives, social objectives have been recast. The emphasis on 'social cohesion' in the London Pride Prospectus reflects a recognition of the problems of many boroughs but also the philosophy of social responsibility espoused by Business in the Community and perhaps most importantly a necessary acknowledgement of the priorities of the European Commission who may fund more London projects" (Newman & Thornley, 1996:8).

2.1.   FRAGMENTED INSTITUTIONS AND VISIONS

Part of the price to be paid by the planning system has been the fragmentation of the planning service and a dearth of strategic policies and a failure to address the 'wicked' problems (Rittel and Webber, 1973). A hastily conceived gesture by the government has been to upgrade the regulatory status of the local development plan so ensuring that the plan consultation arenas would be used both as fora to debate the appropriateness of emerging local planning strategies and to bring key stakeholders together to 'own' and implement a spatially-coordinated, centrally-directed, local vision. This process is centrally directed because the Secretary of State for the Environment reserves the right to intervene at any point. The significance of section 54a of the 1990 Town and Country Plannning Act, which introduced this presumption in favour of development which accords with the development plan, is still being debated. This presumption sits unfavourably alongside the presumption in favour of development per se, in the British planning system. Even in the case where the development plan contains a presumption against the proposal, the decision is taken out of the local plan arena to be decided by the Secretary of State for the Environment or an appointed inspector. In this 'court' there has been a strong presumption in favour of development. Land use conflicts are now being debated at length at the local level and new skills of negotiation and compromise are being learned by local authorities.

A second imperative for recent Conservative governments has been to release the entrepreneurial flair which is seemingly inherent in all of us. On the one hand this was to be realised through lifting the burden of minor and petty land use regulations, on the other hand a steady tantalising drip of 'new' resources, top-sliced from existing public sector budgets and more recently lottery funds, was to unleash new working arrangements and a cross fertilisation of ideas between public and market-oriented actors. The government has ensured compliance with both market values and stakeholder collaboration through performance criteria and competition for funds. The bid documents aim to lever-in private sector resources on the back of redirected public sector capital and revenue monies around a vision of the future 'regeneration' of a narrowly defined locale. It only takes several of these non-statutory area 'planning' frameworks from successful bids to be in place at any one time to add to the complexity of the planning system, particularly for a metropolis the size of London: In terms of plans, London is covered by regional and strategic planning guidance and other government planning policy guidance, a public-private prospectus from London Pride, the advisory documents of the London Planning Advisory Committee, and an Assisted Area and Objective 2 framework. Various parts of the capital are covered by sub regional plans such as that for the Thames Gateway, an unofficial list of priority areas as seen by the Government Office in London, Unitary Development Plans - some approved, some not, the London Urban Development Corporation's planning frameworks, detailed development plans of City Challenge and Single Regeneration Budget projects, and a host of promotional documents coming from, for example, the

Private Finance Initiative, the new inward Investment Agency and the Training Education Council (Newman &Thornley, 1996). Though there is some evidence of the overlapping membership of some of these agencies, the system lacks one body with the power and responsibility to coordinate, despite the rhetoric of the enabling local authority.

Coordination hinges on the procedures laid down by central government for the preparation of Regional planning guidance notes, Structure plans, and Local plans. To aid coordination of the proliferating agencies the policy process is simplified by a focus on individual sectors and by displacing complex goals by narrowly focused targets - rather than attacking the root causes of homelessness a simpler goal of ensuring a steady flow of housing land for new build can be more easily achieved. Central government is the lynchpin of the system, through its actions orchestrating policy direction and providing a redress of last resort on matters of procedural fairness. This explains the centralisation of the British planning system - its success and discretion depend on the firm steering hand of the Department of the Environment (DoE) and political masters. System controls are aided by central government appointees, criteria specification for access to additional funding and performance audits for participating executive agencies. Where central government power is exercised through controlling the political setting rather than specific policies, the discretion available to local level decision-makers gives the illusion of power, but this right is always exercised within the brief set out by central government which demands a major involvement by the private sector and conformity with central government policy. The emphasis on market performance and efficiency measures structures the way land is used and allocated in development plans and ensures that key local stakeholders are brought together to pursue their mutually intersecting goals and objectives.

3.   Planning for Housing

Within this broad policy context characterised by delegated decision-making sits two policy sectors dealing with the allocation of land for housing and land for mineral extraction, where the central state has sought to structure the decision-making process and the policy agenda through *more* detailed specification. The successful control of this policy implementation 'at a distance' can be witnessed as government priorities on quantities of land required and site identification are translated through the spatial tiers - from national to regional, sub-regional to local levels - where conflict between residents and developers is resolved. Government projections of national demand for new houses or aggregates over the plan period starts the process off and gives a privileged position to producers in the production of policy. Implanting commercial criteria into the policy equation and reliance on national projections has had three specific impacts. Firstly, the range of substantive issues addressed has narrowed; secondly, more emphasis has been placed on policy outputs in terms of providing a flow of sites for development; and thirdly, the resource costs for effective citizen participation have increased. These two sectors show the British planning system at its most centralised and inflexible, with the use of targets negotiated at regional level providing informal rights to develop in line with their translation into development plans. This paper focusses on the mechanisms for releasing housing land within the North East region using the documentary material prepared by the main participants in the reviews of both regional guidance and development plans. The analytical framework used is one that explores how these arenas are being used, in particular localities by different groups, to express and take control of the agendas framing the allocation of land for housing growth. Particular attention is given to the main participants and their conceptions of housing requirements and how these are expressed at the key stages of interaction in the decision chain. This involves appraising whose terms dominate the discussions, whose voice predominates as well as whose interests are not represented. A final normative objective is to assess the potential role that localities could play in cohering to produce local housing strategies which take a more holistic look at housing requirements than the narrow sectoral approach suggested by the centrally-determined decision criteria and procedures.

Policy making on housing land release can be considered as a cascading system of accretionary decisions from Ministerial statements and Planning Policy Guidance Notes (PPG) through Regional Planning Guidance Notes (RPG), Structure Plans and Local Plans where appropriate, to Unitary Development Plans, in which policy opportunities are foreclosed as accummulated decisions are made. Central Government guidance on housing is contained within PPG's issued by the Department of the Environment (DoE, 1992a; 1992b; 1992c; 1994) as well as the respective RPGs. These set the parameters to the coordinative regulation of the location of new housing at strategic level as well as the detailed, essentially restrictive controls over the phasing and capacity of individual sites (Goodchild, 1992; Monk, 1991).

Accommodating housing growth has been a key responsibility of local planning authorities since the 1947 Town and Country Planning Act in recognition of the political importance of housing as shelter and the impact of new housing developments on the demand for community facilities. Recent procedural changes to plan formulation have reduced the minimum period for consultation on the draft plan to six weeks. Yet this has been compensated by the intensity of the participation at the later Local Plan inquiry stage, with the draft consultation stage being used more as an expression of engagement in debate. The increased status of the development plan following s54a 1990 Act has alerted stakeholders to the issues to be contested at this agenda structuring stage. The approach to catering for housing growth during the post war boom period of the 1950s-1970s had been through planned neighbourhood extensions to the existing urban form either at the scale of a 'New Town' or a smaller development area. In each case the land use planning task was to cater for the comprehensive needs of the proposed neighbourhood(s) over a substantial time horizon aided by generous government subsidy. Since the 1980, and 1984 pivotal housing legislation and the shift to privatised forms of housing provision and management, there is a more prominent role for producer perceptions to frame the detailed decisions on quantity, quality and location of future housing. Procedural rules covering consultation and accountability in the formulation of land use development plans provide significant formal and informal arenas for stakeholders to both mobilise their support and influence policy direction. This enables the national Housebuilders Federation (HBF) and their regionally based organisations to play a diagnostic and presumptive role in assessing housing demand and in auditing housing land availability with each district local authority (DoE, 1980). Likewise the Council for the Protection of Rural England (CPRE) and certain developer interests have the resource capital to enter into a coherent and wider debate on housing issues.

4.   Case study: Strategic Issues

The North East region stretches from the Scottish border to the boundary with the County of North Yorkshire (Figure 1). The topography of the region is dominated by the Pennines mountain range which falls from a height of 800m OD in the west to the North Sea to the east. The main rivers of the Tyne, Wear and the Tees cut into this landform. Deposits of coal and iron and proximity to these rivers for transportation have fixed the high concentration of population along the coastal strip some 30km wide between the rivers Tyne and Tees. The remainder of the region is characterised by sparse population distributed among the agricultural and hill farming communities. Legislative powers in the 1960s relieved the population density along the banks of the Tyne and Wear through central and local government sponsored 'new' towns at Cramlington, Washington, Newton Aycliffe and Peterlee, shedding the population out of the riverine conurbations to locate on north-south road transportation axes. Substantial government-funded infrastructure was provided for these growth points through the 'new' towns and regional assistance programmes, including roads, social services and advance factories, to take advantage of new employment opportunities and diversify away from dependence on coal, steel and shipbuilding. These initiatives undoubtedly helped to stem the flows of outward migration the region has experienced since the 1960s.

Throughout the postwar period a three-tier system of land use regulation functioned to guide and regulate the use of land and buildings in the region. Broad strategic direction for investment decisions was provided by the central government's Department of the Environment based in London and the County Councils of Northumberland, Tyne and Wear, Durham and Cleveland, with the lower tier District Councils translating these strategic objectives to more detailed site policies for regulatory purposes. As part of their ideological crusade to remove Labour power bases in the metropolitan counties, the Thatcher government abolished the Tyne and Wear County Council in 1986. More recent 'populist' pretensions in the Major government seeking to bring power to the people has resulted in the abolition of the County of Cleveland in 1996. In each case, the separation of planning function at local level into strategic and and site-specific regulatory authorities has been collapsed into one unitary authority based on the existing District boundaries.

FIGURE 1. Map of the North East Region of England

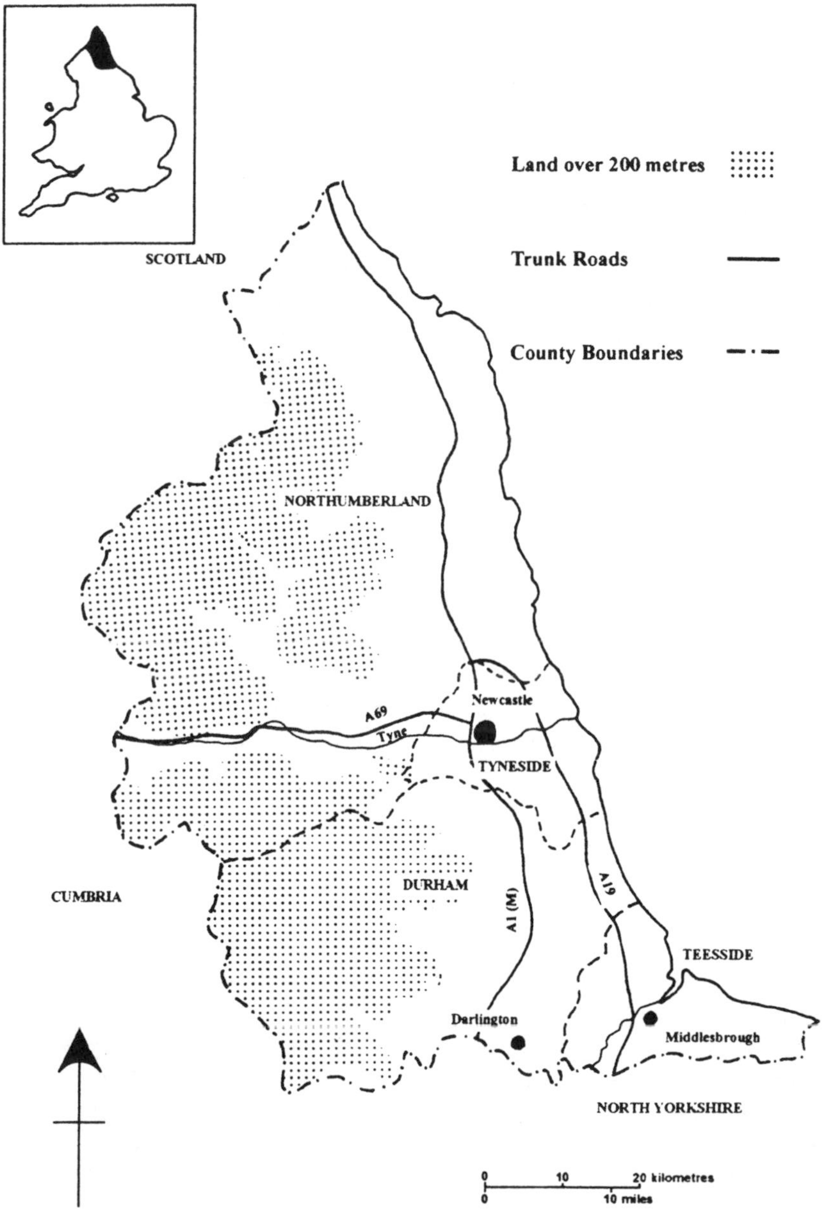

The removal of Tyne and Wear County Council in 1986 created a strategic impasse in the discussion and action on structuring issues wider than the individual unitary Districts. Around this time a new quasi-public agency, the Tyne and Wear Urban Development Corporation (TWUDC), had been installed in the region directly responsible to central government and bypassing the locally elected administrations. The TWUDC was given local planning responsibilities and local state landholdings in an area which stretched along the banks of both the Tyne and the Wear and crossed into the five Districts of Tyne and Wear (Newcastle, Gateshead, North Tyneside, South Tyneside and Sunderland). To address the wider strategic issues that this sub-region, with its 'cuckoo in the nest' institutional structure, would face the government initiated a series of regional planning guidance (RPG) notes. The procedural formulation of advice, to redress the removal of a strategic administrative and policy tier, required the local authorities to come together in planning conferences to share conceptions of how their regional space should be developed to inform the eventual guidance issued by the DoE.

The North East region currently has two RPG notes: RPG1 issued in 1989, and covering the period 1988-2001, for the metropolitan districts of Tyne and Wear (DoE, 1989) and RPG7 for the period 1992-2006, issued in 1993 for the remainder of the region (GONE, 1993). RPG1 advised local authorities that their plan strategies should be realistic about their chances of arresting future population decline, but should cater for both future household change and in-migration that might arise from better economic and housing prospects. The remainder of the region, outside Tyne and Wear, experienced population decline in the 1980s of around 12,000. RPG7 proposed a population strategy of stabilisation in Durham, reduction in population loss in Cleveland, and catering for a modest increase in Northumberland. Both RPGs specified the target number of dwellings to be built during the guidance period. Newcastle and North Tyneside Districts were combined for the purpose of allocating housing requirements because of their close links including the continuity of their built up areas, and given the perceived constraints on housing land availability in Newcastle.

RPG1 and RPG7 offer little advice on the appropriate locations for housing or other developments. The local planning authorities are expected to do a 'balancing' act; on the one hand, regenerating the region's existing urban areas in an effort to revitalise the economy of the region, while, pursuing policies to protect the environment through safeguarding the countryside, forests and the coastline. Local authorities were asked though to pay particular attention to:

- the scale and pace of urban regeneration, particularly efforts to stimulate economic development, new housing and to improve the environment; and
- trends in the factors affecting housing provision, in order to maintain an adequate supply and choice of housing land at all times; and

- transportation in terms of its performance and the functioning of the strategic road network and the adequacy of public transport and the success or otherwise of relating land use to transport and reducing the need to travel (RPG7 only).

4.1.   LOCAL PLAN POLICIES

The housing requirements specified in RPG notes are to be treated as guiding figures to inform the strategic decisions made at sub-regional level. They are based on projections of future households and therefore need to be tested through the development plan process, with variations being made on the basis of local circumstances, new evidence or specific policy proposals. All but five of the 22 Districts in the North East had placed their Local Plans on Deposit for the final stage of Local plan consultation by mid-1996.

Most local authorities were aiming to reverse the trends of decline over the last thirty years. Newcastle and North Tyneside had policies to cancel out in- and out-migration to zero over the plan period. Local Plan dwelling strategies have 'anticipated' future economic growth based on attracting inward investment to the region. Over the last decade, the Northern Development Company, a quasi-public body, has successfully spearheaded a public sector effort to improve the region's economic base through providing a one-stop shop for investors and the carrot of a substantial state-funded package from UK government and European Union regional programmes, derelict land grants from English Partnerships and Enterprise Zone assistance. Local Plans are catering for future household growth to 2006 through extensions on the edge of their built-up areas, with a small proportion of infill and recycling of inner urban sites. Environmental policy initiatives such as green belt extensions and community forests have been proposed to consolidate settlements and provide the tools to control future housebuilding pressures.

5.   The Review of Regional Planning Guidance

The government decided in 1995 to initiate the process of reviewing the regional guidance for the North East with the intention of producing a single guidance note for the entire region (Hull, 1996b). The North of England Assembly, a non-statutory body comprising representatives of all the elected local authorities in the region has been given the task to organise a regional view on what this guidance should contain. An Issues paper and background subject documents were released by the Assembly in May 1996 for a three month consultation period. The draft guidance to be submitted to the government has been delayed with the election of a Labour government in May 1997.

The consideration of regional growth requirements is being driven by the most recent government projections of population and household growth for the region (DoE,

1995). 60% of the growth to 2016 is expected in the urban unitary Districts in the Cleveland and Tyne and Wear sub-regions (Table 1). Since 1980, the housing concerns of the land use planning system have related only to the *amount, location* and *flow* of new dwellings and the infrastructure required to support substantial new developments (Hull, 1997). Land identified for housing in Local Plans caters primarily for private sector new-build, and if all the projected household growth was to be provided for by new build, the average annual requirement of 7113 dwellings is comparable to the recent yearly completions in the North East region.

TABLE 1. Household Projections for North East England by Sub-Region 1991-2016

Sub-Region	1991	1996	2006	2016
Nothumberland	122000	128000	140000	149000
Durham	243000	251000	269000	284000
Cleveland	210000	225000	238000	248000
Tyne & Wear	464000	481000	507000	532000

Source: Department of the Environment (1995)

There is a growing recognition that locations for growth should be sustainable. Recent government planning policy guidance notes on *Development Plans* (DoE, 1992) and *Transport* (DoE, 1994) have emphasised the restraint of car use for journeys to work, school and local facilities and this has added more clarity to the quest for 'sustainable' housing locations but there is still some doubt about the means of achieving sustainable development patterns. Increasingly, local authorities are having to demonstrate to vociferous environmental groups, such as the Council for the Protection of Rural England, that they have made maximum use of their existing urban sites before allocating greenfield sites, particularly those that are vacant and are accessible to a range of local facilities. Answerable to their local electorate, local authorities also wish to ensure that housing infill does not involve the loss of valued open space and that the cumulative effects of development do not damage the character and amenity of established residential areas (Joseph Rowntree Foundation, 1992).

Unless the regional conferences organised by the North of England Assembly to debate strategic spatial issues can break the simplified asssumption that every new household formed equates to an additional dwelling to be provided by the private sector, this vicious spiral of identifying housing land on the periphery of existing urban areas and fighting a rearguard action to protect the environmental quality of rural villages will continue. There are more complex housing goals which the land use planning system is choosing to ignore in its focus on providing a steady supply of sites for private sector housebuilders.

6.   Connecting with Housing Need

There is the view that the household projections should not be taken as an indicator of the demand for new houses in the next two decades but rather as an indicator of the requirement for future social housing. Table 2 shows that nationally 78 % of the additional housing demand will be from single person households mainly aged under 35. The increased demand from multi-person households is expected to be mainly from women over 45 due to divorce. The key housing debates *should* be how the region is going to meet the housing needs of these single, lone parent and multi-person households. This debate needs to consider issues of location and dwelling type. Disregarding unmet needs at 1991, Holmans (1995) has estimated that nationally 40% of the net increase in households will not be able to afford adequate accommodation unless it is partly or wholly subsidised. This equates to an annual national social housing requirement of 90,000 dwellings between 1991-2001 and rising to between 100,000 and 105,000 a year thereafter. This is in line with estimates by the Royal Institute of Chartered Surveyors (1995), Shelter (1995) and the Housing Corporation (1992). Current social housing provision is achieving 30,000-35,000 dwellings a year. There has been considerable concern expressed regarding the role of the planning system in securing affordable housing, with the powers available considered to be negligible (London Research Centre, 1991; Bramley, 1991; Feasey, 1992).

TABLE 2. Household Characteristics of the Projected Increase in Households in England 1991-2011

household type	Additional Households	% of total increase
single person	2760000	77.7
other multi-person	701000	19.7
lone parent	277000	7.8
all households	3554000	100

Source: Department of the Environment (1995) Tables 1 & 4

The extent of the social polarisation between tenure groups and present unfulfilled housing needs in the North East has been identified in recent research and in local authority housing strategies. In 1988, in the Newcastle Travel-to-Work area, owner occupiers received on average twice as much net income per week than housing association and local authority tenants (Cameron et al, 1991). State benefits made up nearly half of the average local authority tenant's net income, but only 8 % of owners' income. At that time, local authority rents were70% of market rents, housing association rents were 86% and controlled private sector rents were 83%. The gap between actual and market rents has narrowed over the last 8 years.

The Housing Corporation is the government's principal mechanism for addressing housing need through new forms of subsidised housing provision. Their development programme is drawn up each year in the light of local authority housing strategies. The strategies for the 1994/95 regional programme identified the housing needs of elderly households being a high priority in 13 of the 22 Districts in the region, single households in 14 Districts and homelessness as a major problem in 10 Districts. In these latter Districts, the number of homeless households has trebled over the last decade. For 5 Districts the key priority was targetting investment in the redevelopment of inner area cleared sites where, one District claimed, the maximum impact in terms of need and urban regeneration can be achieved (Housing Corporation, 1993).

Future household growth has implications for both the existing stock of dwellings and the regions's ability to add to that stock. The Government is relying on the private sector to provide for future housing needs, through subsidising owner-occupation and encouraging private renting. The rate of private sector starts has, in the recent past, revolved around the boom and busts of the economy, and has been targetted on a narrow range of housing needs. The low level of house prices in the North implies comparatively low levels of returns to housebuilders in certain housing markets which could inhibit private sector investment in new forms of housing provision. The switch by larger developers to prestigious inner city locations within the Tyne and Wear Urban Development Corporation area, subsidised by government funds may suggest a narrowing and declining nature of private sector housing developments. This and the local assessment of current housing needs supports the argument that the real requirement for additions to the housing stock is less than the household growth totals imply. So far regional and local debates are not engaging with these issues. In fact, housing association professionals either have not been invited or are not taking on a planning policy discourse role at either regional or local level. Debates on housing are dominated by housebuilder's perceptions of the marketability of land and the growth in additional households.

7.   Conclusions: Evaluating the Development Plan as a Regulatory Tool

Addressing first the substantive housing issues, the main outcome of this process is that local land use development plans concentrate on identifying a 10 years supply of housing land on sites deemed to be marketable and accessible to the housebuilders. Thus the principle of the quantity and location of development is now agreed at the plan stage. In terms of the mobilisation of institutional and political capital, the private actions of the HBF and CPRE are not seen by the Government Office in the North East as distorting regional priorities, rather their opposing interests of development versus conservation are seen as cancelling each other out. Yet because of the governmental desire not to intervene in policy outcomes, but to leave allocative decisions to the market, the debate revolves around how effective the local planning authorities are in releasing serviced sites for private sector new build to accommodate the expected household growth, with few voices questioning the need to assess either housing requirements or the spare capacity within the existing stock (Hull, 1996a).

The case study shows the narrow sectoral approach of the British planning system, in an issue area which is highly structured both procedurally and substantively, and failure to address and link in with the housing debates in other policy communities. Neither the efficacy of the housing benefit system for low income households in the social security policy community, nor the problems of negative equity discussed by the valuation and surveying societies, nor the resort to bed and breakfast accommodation discussed by housing managers have impacted on the town and country planning system. There is a tendency in British policy-making to treat individual issues in isolation, which allows as this case shows a constellation of interests (local and central regulatory institutions, environmental, housing suppliers, commercial) to participate in negotiation and persuasion in the forums established by central government. This has widened the debate away from the standards developed by the local regulatory authority to take on market perceptions. Market values and priorities, though, take the spotlight away from addressing issues of market failure particularly the undersupply of low cost accommodation and the underfunding of housing stock maintenance and revitalisation. Lack of resources and regulatory power has ensured that the British planning system is ineffectual in channelling where new investment takes place, of what kind, at what scale, and how it should be made to happen.

The case study also demonstrates how policy outcomes can be redefined through the restructuring of the political setting to encompass institutional and value change without the necessity to overhaul existing policy programmes. The presumption in favour of development per se with the encouragement of entrepreneurial flair is now well entrenched in British policy-making. This has been at the loss of a certain all-embracing integrated approach to infrastructure provision and societal growth. What coordination there is at the moment is through governmental financial initiatives and specified criteria for interaction. Policy coherence and implementation which relies on a collection of decentralised interests can only be achieved through the integration and coordination of the patchwork quilt of agencies involved and this will require a willingness to work together. Regional strategies, spatial and financial, need to address both social and economic issues and bring together the fragmented institutions that presently, through both positive and negative actions, decide how land and buildings will be utilised in the region.

The central state is regulating the policy process of housing land identification and release *at a distance and indirectly* through specific procedures and criteria. Control is being exercised through the participation of new organised interests who bring commercial information and new perspectives to the attention of policy makers. One can conclude that the British government lacks the foresight and the resources to invest in strategic intelligence so has co-opted organised private sector interests to provide the vision and ensure policy ownership. The leverage these new actors bring to debates coupled with the changing policy context of the 'centralised' British land use planning system is leading to a foreclosure of policy discussion and a lack of policy coherence.

References

Bramley, G. (1991) *Bridging the Affordability Gap in 1990*, The Housebuilders Federation and the Association of District Councils, London.

Cameron, S., Nicholson, M., and Willis, K. *(1991)* Housing Finance in Newcastle, *Joseph Rowntree Foundation, York.*

Department of the Environment (1980) Circular 9/80: Land for Private Housebuilding, *HMSO, London.*

Department of the Environment (1989) *Regional Planning Guidance Note 1: Strategic Guidance for Tyne and Wear*, HMSO, London.

Department of the Environment (1992a) *Planning Policy Guidance Note 12: Development Plans and Regional Planning Guidance*, HMSO, London.

Department of the Environment (1992b) *Planning Policy Guidance Note 7: The Countryside and the Rural Economy*, HMSO, London.

Department of the Environment (1992c) *Planning Policy Guidance Note 3: Housing*, HMSO, London.

Department of the Environment (1994) *Planning Policy Guidance Note 13: Transport*, HMSO, London.

Department of the Environment (1995) *Projections of Households in England to 2016*, HMSO, London.

Fainstein, S., and Fainstein, N.I. (1983) Regime strategies, communal resistance, and economic forces. in *Restructuring the City: The political economy of urban redevelopment*, (eds) by Fainstein, S., et al, Longman, New York, 245-282.

Feasey,D. (1992) Development Planning and Housing Provision, *Environmental Policy and Practice*, 2 (1), 70-82.

Francis, J.G. (1993) *The Politics of Regulation: A Comparative Perspective*, Blackwell, Oxford.

Goodchild, B. (1992) Land Allocation for Housing: A Review of Practice and Possibilities in England, *Housing Studies*, 7 (1), 45-55.

Government Office in the North East (1993) *RPG7 Regional Planning Guidance for the Northern Region*, HMSO, London.

Harding, A. (1996) *European Urban Regimes?* paper to the 1995/96 seminar series of the Centre for Urban and Regional Development Studies, University of Newcastle, 6 March 1996.

Harvey, D. (1982) *The Limits of Capital*, Blackwell, Oxford.

Holmans, A. (1995) *Housing Demand and Need in England 1991-2011*. Joseph Rowntree Foundation, York.

Housing Corporation (1993) *Policy Statement 1994/95*, Housing Corporation North East Regional Office, Leeds.

Hull, A.D. (1996a) Housing Need and Provision:The North East Region, in M. Breheny and P. Hall, (eds) *The People: Where Will They Go?* National Report of the TCPA Regional Inquiry into Housing Need and Provision in England, London.

Hull, A.D. (1996b) Rediscovering Strategic Planning: the role of regional guidance notes, *Local Government Policy Making*, 23 (1), 52-58.

Hull, A.D. (1997) Restructuring the Debate on Allocating Land for Housing Growth, *Housing Studies*, 12 *(3)*, 367-382.

Joseph Rowntree Foundation (1992), *Inquiry into Planning for Housing*, Joseph Rowntree Foundation, York.

London Research Centre (1991) *Much Ado About Nothing*, report by Mark London, Housing Studies and Social Surveys, London Research Centre, London.

Monk, S. (1991) *Planning, Land Supply and House Prices: The national and regional picture*, Land Economy Discussion Paper 33, Department of Land Economy, University of Cambridge, Cambridge.

Newman, P. and Thornley, A. (1996) *Urban Governance and Planning in London*, paper presented to European Urban and Regional Studies Conference: A Changing Europe in a Changing World: Urban and Regional Issues, Exeter, 11-14 April 1996.

Rittel, H.W.J. and Webber, M.M. (1973) Dilemmas in a general theory of planning, *Policy Sciences*, 4, 155-169.

Royal Institute of Chartered Surveyors (1992) *Housing the Nation: Choice, Access and Priorities*, RICS, London.

Shelter (1995) *1995 Housing White Paper - Shelter Briefing*, Shelter, London.

Thornley, A. (1991) *Urban Planning under Thatcherism: The challenge of the market*, Routledge, London and New York.

Whelan, R.K. Young, A.H. and Lauria, M. (1994) Urban Regimes and Racial Politics in New Orleans, *Journal of Urban Affairs*, 16 (1), 1-21.

PART II

Perspectives
In Evaluation
Methods

PERSPECTIVES IN EVALUATION METHODS

Introduction

A. BARBANENTE

The main stimuli for innovating evaluation methods from the notion of environmental sustainability

The contributions included in this section of the volume are united by the hints of particular interest they provide with reference to methodological innovation in evaluation. Following Altshuler and Zegans (1990), an innovation typically consists in a model integrating novel and familiar elements. The latter are typically more numerous, while the novel element may be a service concept, a technology, a way of putting the elements together, or some combination of these. Innovations in planning evaluation as emerge from these papers, seem to comply with such a model: thus, innovativeness must be appraised on the basis of a constantly evolving conceptual, technical and operational framework, adding to or modifying rather than radically substituting the core principles and components of methods and approaches.

Among the possible perspectives for reading the following papers, some seem able to highlight relevant common strands, without compelling differences into a coherent, but forced, framework. The first point concerns the *sources of innovation* in methods. A primary stimulus for innovation seems to be related to the environmental sustainability concept. There is no doubt that the explosion of the environmental problem arouses new and pressing questions for knowledge about urban and regional systems and justification for transformation actions. The shift from a notion of environmental protection as contrasting, and thus to be reconciled with, that of economic development, which was associated with the search of ways for compensating for and balancing between anthitetic demands, to a notion of environmental conservation as an essential precondition for local systems development and even for human survival, requires a radical change in the basic concepts and methods of planning evaluation.

Most of the contributions included in this section assume as a central notion that of sustainable development, essentially understood as a process of co-evolution of social, economic, physical and environmental systems meeting the needs of the present generation without subtracting future generations' access to resources. Despite its generic and ambiguous character, the concept of sustainability introduces a number of key *foci* in evaluation research as well as in practice: the relevance of equity in the

N. Lichfield et al. (eds.), Evaluation in Planning, 145–150.

distribution of, and accessibility to, resources; the acknowledgment of values of nature; the pursuit of economic goals compatible with the conservation of quality and quantity of cultural and natural resources.

The search for a method able to overcome mechanistic and reductionistic approaches to evaluation proposed by Lombardi arises from the conviction that planning for sustainable development at local level requires new holistic approaches suitable not only for integrating expert and non-expert knowledge, but also for ensuring the consideration of all relevant aspects of urban sustainability and quality of life in the decision at hand. A multimodal system of classification of sustainability indicators and criteria is then developed with the main purpose of guiding evaluation during the planning process.

Bizzarro and Nijkamp discuss the possibility of promoting a sustainable conservation, i.e. integrating the conservation of cultural built heritage (CBH) in a broader view on sustainable development by considering at the same time social, economic and environmental dimensions of urban revitalization. A portfolio of conservation policy options is assumed to be needed for achieving such a goal. A meta-analytic approach is thus applied in order to single out, within a number of case studies on sustainable revitalization policies, common and divergent components, success factors as well as impediments.

Assuming as a frame of reference the principle of weak sustainability, and looking to different research fields, namely economics and planning, the work by Clemente *et alii* focuses on the crucial issue of effectiveness, efficiency and equity of environmental policies at the regional level. (The same issue, with reference to conservation of cultural built heritage, is tackled in a similar perspective by Bizarro and Nijkamp in par. 4.) The search for a balance by the use of planning/economic instruments in order to make the need of profit of individuals and communities using the land compatible with the preservation of nature, is considered as an important factor of policy effectiveness, as it argues for the assessment of significant environmental resources existing in a regional context. Thus, the idea that the need for environmental protection strongly relates to equity in planning is affirmed together with a double operational purpose: that of making the implementation of planning instruments easier and even feasible, and that of assessing natural resources and natural risks in order to make the application of environmental policies more effective.

Although not so explicit as in the above papers, other works have in their methodological an awareness of the increasing relevance of environmental issues. It is well known that the evolution from Planning Balance Sheet to Community Impact Evaluation comes, on the one hand, from the increasingly more important role of Environmental Impact Assessment in planning practice, on the other hand, from the need to throw light on the "black box" of social, economic and environmental impact evaluation as had been conceived in PBS (Lichfield, 1985). A key methodological issue in the work by Dalia Lichfield is in the emphasized difference between "actual use" value and "passive use" or "altruistic" values which include values assigned to "products of the scheme by people who would not actually use them but nevertheless gain satisfaction from the knowledge that these products are available for future use to themselves, to others in this or future generations, or exist in their own right". Furthermore, Glasser argues for an integrated framework explicitly incorporating into the evaluation and courses of action selection process, detailed information about the interconnected, co-evolving causes of environmental problems.

This brings us to a further consideration: a search for *integration* is closely associated with the increasing importance of the question of environmental sustainability for evaluation. Here integration is considered both in terms of overcoming discipline 'barriers' and of rejoining different forms of knowledge within the evaluation process. Thus, integration is meant not only in the sense of *multi*-disciplinary and *multi*-dimensional approaches more traditionally experienced in the evaluation field (Clemente *et alii*) or of defining a combination of different tools able to promote urban revitalization (Bizzarro and Nijkamp), but also in a sense closer to the complexity paradigm, in this case attempting to comprehend the articulation, organization, complex unity, "relational circuit" of the environmental system, in its physical, biological, anthropo-sociological dimensions (Morin, 1977): the methodological contribution by Lombardi seems to be oriented in this direction, as it attempts to overcome approaches based on simple linear superimpositions of indicators/criteria, and strictly refers to the notion of environment as a complex system.

Evaluation in the face of new technologies

Another source of innovation in evaluation methods is linked to the use of new technology in planning research and practice: the paper by Batty and Densham indicates current and possible future developments in spatial and planning decision support systems. The most important innovative features for evaluation research and practice seems to be linked here, on the one hand, to the very powerful visualization facilities provided by GIS for display and manipulation, "giving immediate intuitive evaluation capabilities which a wide range of non-technical users and decision-makers can relate to", on the other hand, to the growth of WWW network, which will extend to non-technical users the possibility of being interactively involved with spatial and planning decision support systems in increasingly flexible, heterogeneous, dynamic processing environments. The hoary question of making technical steps and outcomes of evaluation more transparent and open to public debate may receive significant hints from the above technological improvements and thus should be of major interest for evaluators.

Procedural vs. substantive innovation

The second viewpoint concerns the *nature of innovation*, whether it is mainly substantive or procedural. My impression is that once again the focus is on how more than on why, and that this affects also the contributions including case studies. The main focus can be on the various steps of the evaluation process (D. Lichfield), on the indicators and criteria to be used (Lombardi, Bizzarro and Nijkamp), on making GIS and the related modeling technologies fit with elements of the generic problem solving process beginning with problem definition and description, and moving to prediction and prescription or design often involving the evaluation of alternative solutions to the problem (Batty), or on singling out descriptive and normative postulates intended as both guidelines for assessing frameworks for responding to complex, conflict-ridden policy analysis or project evaluation decision problems, and propositions for guiding the development of procedural reforms to better respond to such decision problems (Glasser).

Thus, the crucial importance for evaluation of particular cultural values embedded in local contexts especially as far as environmental problems are concerned seems to be neglected, as does the capability of setting up negotiable as well as non-negotiable community meanings/values in the course of real actions. The problem whether and when to use compensatory or non-compensatory criteria, which arose in some papers (e.g. Clemente *et al.* and Glasser), has probably to be faced in the light of "situated" community values, which should become an open, explicit aspect of evaluation.

Furthermore, embracing a participatory approach (Glasser) should lead to the recognition that the goal-oriented dominant paradigm of current evaluation could be radically questioned, as goals can not only change during the planning process, but can also be defined during or even after the action, subjected as they are to political pressures of stakeholders, to conflicting interests and values of involved groups, to the unavoidable ambiguities of the problem setting process in a decisional arena, to the different argumentative strength of the actors involved.

Nevertheless, it is worth noticing that current strands in evaluation methods stress the dynamic character of tools (D. Lichfield), more and more open to continuous adding of new information and refining of the factual and judgmental information as they emerge during the debate. It would be of great interest to discuss how and why changes are introduced and what is the role of conflict, negotiation, political as well as technical power (Benveniste, 1989) in such an "on-going evaluation" taking place during the decision process. Evaluation, in fact, is not only a way of structuring information but also a tool which decision-makers should use for interpretation and evaluation (D. Lichfield), and for consensus building on decisions to be taken.

Some perspectives for further explorations

So far I have discussed the main sources of innovation in methods and its nature, as they emerge from the contributions included in this section of the volume. Now I will briefly underline what, to my opinion, should be more deeply addressed by innovative efforts in evaluation.

While a number of implications of the sustainability notion have been fundamental sources of innovation in evaluation, as emerges from the above papers, these do not seem to consider enough to the risky, irreversible, uncertain, unknown and largely unpredictable character of phenomena and actions taking place within the environmental system, which closely underlies the complexity paradigm and the long-term perspectives and local/global intertwines inherent in the sustainability concept. Issues of *intergenerational fairness*, of *uncertainty* and *irreversibility* affecting *long-term environmental problems* (Arrow and Fisher, 1974), and the associated ethical principle of cautiousness in the evaluation of transformation options (Jonas, 1990; Perrings, 1991), if considered more thoroughly, would be able to give rise to major innovation in planning evaluation.

The *tension between the subject and the object* emerges as a crucial issue in some papers, namely in the impact chain proposed by D. Lichfield, as being the effect of the project objectively measured and being the impact subjectively valuated by those impacted (on this issue see Lichfield 1994, 67-68). The parallel *tension between the observer and the observed*, which recalls the absence of detached neutrality in knowing the world, of a physis isolated from human understanding, logic, culture, society, and the deep reciprocal embeddedness of the knower and the known (Morin, 1977), has been discussed by Lombardi. Following Dooyeweerd, who escapes both dangers of "realist" approaches, still dominant in evaluation, emphasizing the known and acted-upon object and denying the relevance of the subject, and of "nominalist" approaches, emerging also in evaluation (Guba and Lincoln, 1989), leaving only the knowing and acting subject, as they both seem inappropriate for dealing with sustainability issues, Lombardi upholds a "pluralist ontology" admitting the existence of an external reality independent from the acting and knowing subject which is affected by it but also affects it and has ideas and wishes about it. Abandoning neutral, detached approaches in evaluation poses serious challenges to the current research in the field, which have probably not been fully explored up to now.

These considerations recall the question of possible roles of evaluation in decision-making. While the traditional view of evaluation was mainly linked to a summative role, i.e. helping in final or continuative decisions, new perspectives in methods also emphasize a formative role for evaluation practice, i.e. contributing to structure information during the decision process, to highlight hidden values, interests and wishes of stakeholders and possible conflictual interactions among them, and to develop a mutual learning process between decision-makers and communities, building will-shaping (Voogd, 1997) and consensus around decisions. An effort to make methods congruent with decision contexts seems fruitful for evaluation to be utilized and its findings to be more useful (see Voogd in this volume). This probably implies not only building taxonomies which refer different strategies for evaluation to different types of decision context (e.g. Hill, 1985 on the basis of Friedmann, 1973), but also follow methodological approaches based on the observation of practices in concrete cases, especially if the main purpose of evaluation is to improve social action (Peatty, 1994).

In the first case, in fact, it is easy to remain anchored to the generic form of the rational model following which decision-makers, before they act, identify goals, specify alternative strategies for reaching them, and select an alternative on the basis of some satisfying criterion. In the other case, we would observe that only rarely are the "ideal" preconditions of evaluation met, and that a logic of consequentiality and rational calculation in which behaviors are driven by preferences and expectations about consequences is often contrasted by other logics (e.g. that of appropriateness, as stated by March and Olsen, 1989), and we would argue for more efforts to be made to revise current approaches.

References

Altshuler, A. and Zegans, M. (1990) Creativity and Innovation. Comparison between Public Management and Private Enterprise, *Cities* 7 (1), pp. 16-24.

Arrow K. J. and Fisher, A. C. (1974) Environmental Preservation, Uncertainty, and Irreversibility, *Quarterly Journal of Economics* 88, pp. 312-319.

Benveniste, G. (1989) *Mastering the Politics of Planning*, Jossey-Bass, San Francisco.

Faludi, A., Voogd, H. Evaluation of Complex Policy Problems: Some Introductory Remarks, in Id. (eds.) *Evaluation of Complex Policy Problems*, Delftsche Uitgevers Maatschappij, Delft.

Guba, E. G. and Lincoln, Y. S. (1989) *Fourth Generation in Evaluation*, Sage, London.

Hill, M. (1985) Decision-Making Context and Strategies for Evaluation, in Faludi, A. and Voogd, H. (eds.).

Jonas, H. (1990) Il Principio Responsabilità. Un'Etica per la Civiltà Tecnologica, Einaudi, Torino. (Orig. Title: Das Prinzip Verantwortung, 1979.)

Lichfield, N. (1985) From Impact Assessment to Impact Evaluation, in Faludi, A., Voogd, H. (eds.).

Lichfield, N. (1994) Community Impact Evaluation, *Planning Theory* 12 Winter, pp. 55-79.

March, J. G. and Olsen, J. P. (1989) *Rediscovering Institutions: The Organizational Basis of Politics*, The Free Press, New York.

Morin, E. (1977) *La Méthode I. La Nature de la Nature*, Editions du Seuil, Paris.

Peatty, L. (1990) An Approach to Urban Research in the Nineties, *Planning Theory 12*, Winter, 9-34.

Perrings, C. (1991) Reserved Rationality and the Precautionary Principle: Technological Change, Time and Uncertainty in Environmental Decision Making, in: Costanza, R. (ed.) *Ecological Economics: The Science and Management of Sustainability*, Columbia University Press, New York.

Voogd, H. (1997) On the Role of Will-Shaping in Planning Evaluation, in Borri, D., Khakee, A. and Lacirignola, C. (eds.) *Evaluating Theory-Practice and Urban-Rural Interplay in Planning*, Kluwer Academic Publishers, Dordrecht.

INTEGRATED PLANNING AND ENVIRONMENTAL ASSESSMENT

The Case of Torbay Ring Road.

D. LICHFIELD

The paper presents a case study of Integrated Planning and Environmental Assessment (IPEA). IPEA is a method which assists decision making authorities to consider Environmental Impact Statements within a wider planning framework (Lichfield, 1992). It provides a unified methodology for the analysis of environmental effects and of economic and social effects; it then translates the scientifically measured effects into impacts on people (Lichfield and Lichfield, 1992). As such it incorporates Community Impact Evaluation (Lichfield, 1996; see also paper in this volume). It has several advantages in terms of balanced decision making, accountability, public consultation and involvement of 'stakeholders'. It can also improve the design process. IPEA won an award from the British Royal Town Planning Institute in 1991 as a "contribution to the art and science of town planning in the public interest" (RTPI, 1991).

The case study presented here was commissioned by Devon County Council in the UK, to assist in the decision making over the Torbay ring road options.

1.   Background

The south-western coastline of England forms the bay of Torbay, surrounded by a densely populated area. A Ring Road around the built up area has been on the cards for some 20 years. Its aims were, and still are, to provide an alternative route to the coast road between Torquay and Paignton, thereby reducing congestion in the central areas of those towns, and to reduce traffic congestion and accidents on an existing peripheral road (Kings Ash Road). Devon County Council (DCC) as a Highway Authority have concluded that a ring-road is preferable to reliance on the existing road network, but a great debate developed between them and Torbay Borough Council as to the alignment of the ring road (Figure 1 -The Ring Road Options):

- A Plateau route option took it through the edge of a designated residential area on a plateau;

- A Valley route option filed its way through a secluded peaceful valley adjacent to the Torbay urban area.

N. Lichfield et al. (eds.), Evaluation in Planning, 151–175.

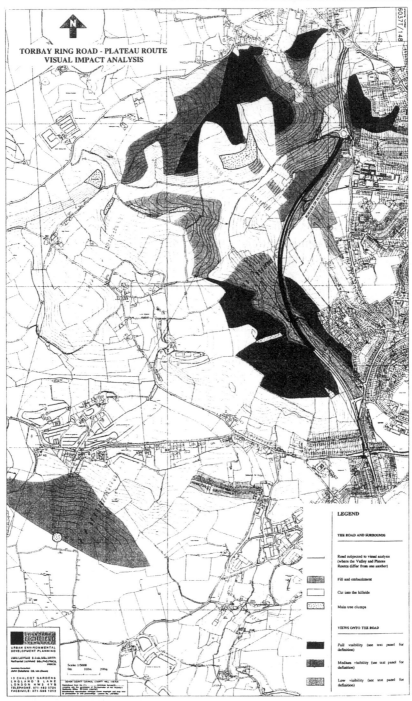

Figure 1 The Ring Road Options

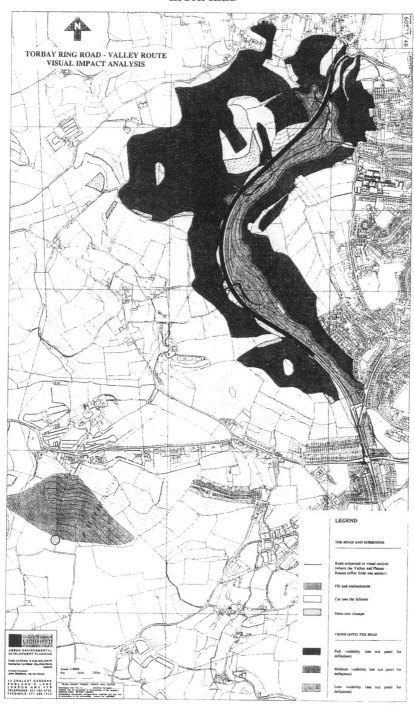

Figure 1 (Cont.) The Ring Road Options

In traffic terms there was little between them - both options started and ended at the same points and had similar capacity and travel time. In their earlier assessments of the options DCC addressed mainly four factors: engineering design; impact on housing land; impact on the environment; and costs (of land, construction and mitigating environmental impacts). They concluded that neither of the options showed a decided advantage on any of these counts, such as would override other considerations. There was also some difficulty in comparing costs and benefits amongst those four entirely diverse factors and generally comparing scientific Environmental Statement data with traditional Planning Policy criteria.

DCC nonetheless had to determine which of the options would best *serve the public interest,* which is the declared purpose of the British planning system (DoE, 1997, 7). They also had to be prepared to defend their choice against potential objections from Torbay Borough Council (the lower tier planning authority) and potential public criticism. Any major conflicts of view would be aired at a *public inquiry* - when an Inspector on behalf of the Secretary of State for the Environment (who is in charge of urban and regional planning in Britain) would hear the views of all concerned and determine the final choice of route. In fact their decision did conflict with the District Council's preference and the IPEA analysis was presented by the author as evidence at the public inquiry, supported by a written Proof of Evidence[i].

2.   Why IPEA Was Chosen

DCC decided to employ the method of IPEA to assist their evaluation of the options on the basis of impacts on different groups in the community. What attracted DCC to this method probably amounted to three sets of reasons:

(a) *The intrinsic qualities of the method* as a decision making tool:

- creating a *unified framework* and methodology which is sufficiently rigorous to satisfy the Environmental Effects Regulations, while also capable of being applied to other planning considerations, which are subject to less rigorous requirements;

- conducting a *rigorous analysis* of the changes which the proposed development would introduce to the physical (natural and man made), economic and social environment, expressed as *effects*; identifying all those groups of people who would experience each effect; then establishing the different *impacts* on each group.

- Providing a *comprehensive picture of groups or stakeholders affected* (including both residential and business groups, the local community as well as city wide or nation-wide interests) and thus reducing the risk of 'single issue' pressure groups dominating the decision making scene .

(b) *The ability to justify their decision to the public* (or to the Secretary of State at a public inquiry) as one which is in the public interest - at the most comprehensive and fair sense of the word.

- An IPEA format *enables stakeholders, including members of the public,* to identify themselves amongst the groups accounted for and make submissions if they find that their interests have not been accounted for correctly.

- The analysis tables can be *continually updated* with new data, including during the Public Inquiry.

- The findings about "who gains & who loses" can assist the road planners to mitigate adverse effects and increase benefits in advance.

Decisions made on that basis would thereby be well informed, accountable and justifiable.

(c) *The opportunity for capability building of the County staff,* with outside experts guiding the County's own staff, drawing on the County's detailed local knowledge and introducing new methods to them, rather than transferring the whole responsibility to outside consultants.

The analysis was accordingly conducted in collaboration between Dalia Lichfield of DNLA and a group of the County's own experts on engineering, noise, landscape, etc. In addition to the IPEA, Dalia Lichfield provided a visual impact analysis which identified the groups experiencing visual impacts and fed these findings into the IPEA.

3.   The Options : Valley and Plateau Routes

The Torbay urban area hugs the bay from the west and is densely built up. The western edge of the area includes a plateau on which a large site is identified in the Local Plan for housing development. Beyond it the land drops into the Westerland Valley which encircles the area. "The Valley is unusual in that it is so close to an urban area and yet so unaffected by it. For most of its length it is extraordinarily peaceful and quiet with a feeling of remoteness sometimes not found even in more rural areas where noise frequently infringes on the atmosphere." (Pickthorne, 1996) It is however little known and little used at the moment.

The two options are fairly similar in terms of traffic, pollution, and financial costs but pass through entirely different landscapes and have impacts on very different groups of people.

The Plateau route runs along the edge of the built up area and takes up some of the designated housing land on the plateau. It passes near a larger number of future houses, but as it goes for most part behind the edge of the plateau it is not visible from within the valley nor from a greater distance; it only affects the very end of the valley itself. The route in the Valley does not take up any housing land and is based on its supporters suggestion that by following the general curvature of the valley, half way up the slope, it would be contained visually and blend with the environment. This is contested by the opposition who demonstrate that the Valley route will introduce a dual carriageway and traffic which will greatly damage the tranquil rolling scenery of the valley itself and will be visible from some of the houses on the surrounding hills.

4.   IPEA Applied to Torbay Ring Road

4.1.   EVALUATING CHANGE: FROM DEVELOPMENT, THROUGH EFFECTS, TO IMPACTS

Road schemes have implications for example for traffic, land use and economic activities, the environment, as well as financial costs. These in turn will affect different groups in the community in different ways - be it as road users or as people exposed to traffic noise, as property owners or as tenants, as statutory authorities, as tax payers, etc. This is the result of a *process of change* introduced by the development into existing and evolving urban and environmental systems. We refer to this process with the following terminology:

- *Elements* of the development - these include *actions taken* to implement the development (e.g. land acquisition, construction and traffic,); the *physical products* of the development (e.g. roads, bridges) and its *operational products* (e.g. vehicular movements, employment, leisure);
- *Effects* of the development - each element causes some change in existing and evolving systems. The change, measured as far as possible in scientific terms, is the *effect* (e.g. reduction in farmland, change of scenery, increase in noise levels);
- *Impacts* of the development - each effect will be experienced by different people and will have different impacts on their life styles. The difference between *effect* and *impact* can be illustrated by a gun shot: the ballistic and noise effects can be described in scientific terms but, in the final account, what matters is *who* was hit and *what impact* it had on his/her way of life.
- *Groups* affected - people who *experience similar impacts* are identified as a 'group'. (e.g. a new structure may have different visual impacts on the occupiers of houses in different locations; additional jobs will be enjoyed by certain skill groups not others; changes in the countryside will be experienced differently by ramblers and by drivers, etc.). For convenience of analysis we divide these groups into two categories:

Producers/Operators are those involved in making the development happen and benefit or suffer in consequence (e.g. land-owners, builders, local authorities);

Consumers are those using the products or being passively subjected to the effects of the development, thus having a benefit or a loss in different ways.

- *Group preferences*: any group is considered to have the objective of increasing its benefits and reducing its costs (detrimental impacts upon it). A group that experiences different impacts as between the development options (or under 'do nothing') will therefore prefer the option that achieves the group's objectives.

- *Community Impact Analysis* - Presenting within one framework all the groups affected and their preferences between the various options.

- *Community Impact Evaluation* - striking the *balance* between impacts on the different groups, to indicate the overall public interest.

4.2.   PROCEDURE OF ANALYSIS

The Integrated Planning and Environmental Assessment follows a set procedure as shown in the *Framework Table* attached. The procedure governs the comparison of one option with an alternative, which may be another development option or a "do nothing". It comprises:

1. Description of the proposals - for each option a comprehensive description of its relevant elements.

2. Existing site and context - a description of the general baseline;

3. Scoping - identification of possible effects and selection of effects that are likely to be significant

4. Analysis of effects - conducted by various specialists for their respective areas of concern which reflect the matters raised in the Design Manual for Roads and Bridges and in schedule 3 of the Regulations as well as other planning considerations, e.g. employment, housing, etc. The predicted effects are normally summarised, following a uniform sequence including location and baseline for each effect; mitigation/enhancement measures. The analysis culminates with the changes that each option will bring into the existing and evolving urban and environmental systems, i.e. its predicted *effects* following mitigation/enhancement.

At this point starts the Community Impact Analysis:

5. Summary of predicted effects for each option and the groups affected - produced as a *working Table A;*

6. Impacts on groups- the groups which were identified initially within Table A alongside the effects are now arranged by category (producers/consumers; on site/off site) in *Table B.* For each of the groups the impacts of both options are compared;

7. The presumed group preferences as between the options, on account of *each* impact, are indicated. As far as possible the presumptions on the group's objectives and preferences are then confirmed in public consultation with the relevant groups.

8. Summary schedule of the differences in significant impacts and preferences - *Table 1* is a concise schedule, following deletion from the analysis of those items not showing impact differences that are significant for the decision. Examples are impacts which are identical in all options, those which cancel out one another and therefore would not affect the choice, as well as impacts which emerge as of relatively little significance. Additional quantitative information may be required at this stage, if the differences between the options are not clear.

9. Decision by elected members - The analysis described above provides a comprehensive picture of the considerations that should inform the decision. But it does not of itself produce the decision. The evaluation of the options requires that decision takers make value judgements on the weight of the various impacts and groups affected, in terms of the "public interest". Decision-makers with different values could conceivably reach different conclusions but with the aid of IPEA such judgements are exercised in a fully explicit manner.

TABLE 1 – Torbay Ring Road – Integrated Planning and Environmental Assessment. Summary of Community Preferences. Pre-completion
Dalia &Nathaniel LICHFIELD Associates

Code	1 GROUP/ INDIVIDUAL AFFECTED	2 IMPACT	3 MONTHLY EXPOSURE PER PERSON	4 SIZE OF GROUP	5 GROUP EXPOSURE UNDER PLATEAU	6 GROUP EXPOSURE UNDER VALLEY	7 GROUPS PREFER PLA-TEAU	8 GROUPS PREFER VALLEY	9 RATING (FROM DETAILED TABLE)	10 ADJUSTED WEIGHT PLATEAU PREF.	11 ADJUSTED WEIGHT VALLEY PREF.
PRODUCERS											
1	Devon County Council	public budget spending				additional cost £ 876,900	P		P		
7	Torbay labour force	Employment during construction only	150 hrs/mth	25 jobs		more jobs 3750 hrs/mth		V	VVV		
13	landowners (farmer)	land loss; compensated;	'one off'	1 net		un-quantified	P		P		
19	landowners (farmers)	land loss; compensated;	'one off'	1 net	un-quantified		P		PP		
41	landowners (farmers)	land loss; compensated;	'one off'	1 net	un-quantified			V	VVV		
35 39	landowners (financial interests)	land loss; compensated;	'one off'	2 net	un-quantified			V	VV		
49 51	landowners (financial interests)	land loss; compensated;	'one off'	2 net¹	un-quantified			V	V		

TABLE 1 – Pre-completion. Ctd.

Code	1 GROUP/ INDIVIDUAL AFFECTED	2 IMPACT	3 MONTHLY EXPOSURE PER PERSON	4 SIZE OF GROUP	5 GROUP EXPOSURE UNDER PLATEAU	6 GROUP EXPOSURE UNDER VALLEY	7 GROUPS PREFER PLA-TEAU	8 GROUPS PREFER VALLEY	9 RATING (FROM TABLE B)	10 ADJUSTED WEIGHT PLATEAU PREF.	11 ADJUSTED WEIGHT VALLEY PREF.
	CONSUMERS										
4 6 8	net 17+3+25= 45 houses[1] x 2.4 P per hh	Disturbance from works	average 4 hrs/d p x 26=104 hr/mth[2]	108 P	11,232 hrs/mth			V	VV		
14 16	2 farmers	Disturbance from works	104 hrs/mth	2 P		208 hrs/mth	P		P		
24	Pontins labour force	Disturbance from works when outdoors;	average 4 hrs p.d. x 26= 104 hr/mth[3]	113 P[a]		11,752 hrs/mth	P		PP		
44	Nursery labour force	Disturbance from works when outdoors;	average 4 hrs p.d. x 26= 104 hr/mth[3]	5	520 hrs/mth			V	VVV		
52	Walkers in the Valley	Disturbance from works	30 hrs/mth	41 P[a]		123 hrs/mth	P		PPP		
54	Wild life enthusiasts	Disturbance from works	30 hrs/mth	22 P		66 hrs/mth	P		PPP		
56	Nature in its own right	disturbed habitats	un-quantified	?		un-quantified	P		P		

Table 1 – Post Completion

Code	1 GROUP/ INDIVIDUAL AFFECTED	2 IMPACT	3 MONTHLY EXPOSURE PER PERSON	4 SIZE OF GROUP	5 GROUP EXPOSURE UNDER PLATEAU	6 GROUP EXPOSURE UNDER VALLEY	7 GROUPS PREFER PLA-TEAU	8 GROUPS PREFER VALLEY	9 RATING (FROM DETAILED TABLE)	10 ADJUSTED WEIGHT PLATEAU PREF.	11 ADJUSTED WEIGHT VALLEY PREF.
	PRODUCERS										
	No significant difference										
	CONSUMERS										
2	Accident injuries	slight only	2P per year	? P	un-quantified			V	VVV		
8 10 12	net 20+5+24= 49 new and existing houses x 2.4 P/hh	Visual and some noise intrusion by Road	120 hrs/mth[b] (4hr/d x 30)	117 P	14,112 hrs/month			V	VV	*[6]	
18 20	2 farmers	Visual and some noise intrusion by Road	average 4 hrs/d p x 26= 104 hr/mth[2]	2 P		208 hrs/mth	P		P		
28	Pontins labour force	disturbance from works when outdoors;	average 2 hrs p.d. x 26= 52 hr/mth[3]	113 P[3]		5,876 hrs/mt + say 4 jobs	P		PP		
42	Hilltop Nursery labour force	disturbance from works when out-doors; jobs	average 4 hrs p.d. x 26= 104 hr/mth[3]	5		520 hrs/mth + say 1 job		V	VVV		

Table 1 – Post Completion Ctd.

44	Walkers in the Valley[8]	loss of amenity	30 hrs/mth (average 1 hr/d[4] x30)	41[4]	1,230 hrs/month	P	PPP
48	Wildlife enthusiasts[8]	loss of amenity	30 hrs/mth (average 1 hr/d[4] x30)	22[4]	660 hrs/month	P	PPP
50	Surrounding residents with option value[8]	loss of option to enjoy Valley	diffused awareness	at least 1850	un-quantified	P	PP
52	General public[8]	altruistic values dented	diffused awareness	Torbay Devon, or UK?	un-quantified	P	P
54	Nature in its own right	severance of habitats	un-quantified	?	un-quantified	P	?
58	The Nation	loss of agricultural land	un-quantified		un-quantified	P	P

NOTES TO TABLE 1:

Abbreviations: P = 'people';   hh = 'households';   p = 'per';   hr = 'hour';   d = 'day';   mth = 'month';
P = Plateau Preference   V = Valley preference

1. 'Net' is the balance after similar interests of opposing preferences have been cancelled out.

2. It was assumed that not all members of a household will be at home throughout the construction hours; a 6 days construction week comprises 26 construction days per month.

3. We have assumed 50 FTE (full time equivalent) staff experiencing the effects for an average of 4 hours per day.

4. The survey revealed an average of 63 visitors per day, of which 22 were wild life enthusiasts and 41 simply enjoy walking. The average stay per visit was approximately 1 hour.

5. We have assumed that all additional local plan-making work and all land sale or CPO transactions will have been settled before completion of the works. To the extent that the same person as sold some land will continue to live or work in the same property, their experience following completion is regarded as a 'consumer' interest.

6. We have assumed an average of 4 hours per person per day during which the person will be at the scene and aware of the visual and noise impacts, for 30 days per month.

7. Accidents may cause as little as a couple of hours check-up which is a very low average exposure per month over a lifetime, or, more rarely, they could result in permanent impairment or even death. Predictions are unreliable for such a small sample.

8. The concern for the Valley comes in three spheres : those who make actual use of it; those who live within reach and would like to keep the option of using it; and those of the general public who may never use it personally but wish to keep it for future generations under a general strategy of sustainability, particularly since it is an important 'green lung' for the expanding urban area of Torbay.

- Figures on actual use were obtained through a users survey;

- Figures on surrounding residents who expressed an interest were derived from responses to the exhibition questionnaire but the response level was only 11%. Therefore 1850 is likely to be an under estimate

- Figures on the general public are difficult to obtain and would depend also on the geographical boundary adopted - Torbay, Devon, or the UK..

TABLE 1-b Torbay Ring Road- Integrated Planning and Environmental Assessment. Summary of Community Preferences. Post Completion

Dalia &Nathaniel LICHFIELD Associates

Note: underlined figures provide my personal judgement and weighting

Code No	1 GROUP/ INDIVIDUAL AFFECTED	2 IMPACT	3 MONTHLY EXPOSURE PER PERSON	4 SIZE OF GROUP	5 GROUP EXPOSURE UNDER PLATEAU	6 GROUP EXPOSURE UNDER VALLEY	7 GROUPS PREFER PLAT'U	8 GROUPS PREFER VALLEY	9 RATING (FROM DETAILED TABLE)	10 ADJUSTED WEIGHT PLATEAU PREF.	11 ADJUSTED WEIGHT VALLEY PREF.
PRODUCERS											
	No significant difference										
CONSUMERS											
2	Accident injuries	slight only		2P per year	average 5 hrs/month			V	VVV		$\underline{30x2=60}$
8 10 12	net 20+5+24= 49 new and existing houses x 2.4 P/hh	Visual and some noise intrusion by Road	120 hrs/mth[6] (4hr/d x 30)	117 P	14,112 hrs/month			V	VV	*[6] $\underline{1}$	$\underline{28,220}$
18 20	2 farmers	Visual and some noise intrusion by Road	average 4 hrs/d p x 26= 104 hr/mth[2]	2 P		208 hrs/mth	P		P	$\underline{210}$	
28	Pontins labour force	disturbance from works when outdoors; say 4 jobs lost	average 2 hrs p.d. x 26= 52 hr/mth[3]	113 P[3]		5,876 hrs/mt + 4 jobs	P		PP	$\underline{11,750}$	

TABLE 1-b - POST COMPLETION Ctd

#								V	VVV		
42	Hilltop Nursery labour force	disturbance from works when out-doors; say 1 job lost	average 4 hrs p.d. x 26= 104 hr/mth[3]	5	520 hrs/mth + 1 job				VVV		1560
44	Walkers in the Valley[8]	loss of amenity	30 hrs/mth (average 1 hr/d[4] x30)	41[4]		1,230 hrs/month	P		PPP	3690	
48	Wildlife enthusiasts[8]	loss of amenity	30 hrs/mth (average 1 hr/d[4] x30)	22[4]		660 hrs/month	P		PPP	1980	
50	Surrounding respondents with option value[8]	loss of option to enjoy Valley	diffused awareness	1850		say 10 hrs/month	P		PP	18,500	
52	General public[8]	Passive use values dented	diffused awareness	8000		say 2 hrs/month	P		P	16,000	
54	Nature in its own right	severance of habitats	un-quantified	?		un-quantified	P		?	10,000	
58	The Nation	loss of agricultural land	un-quantified			un-quantified	P		P	5,000	
TOTALS										67,130	29,820

NOTES TO TABLE 1-b:

Abbreviations: P = 'people';   hh = 'households';   p = 'per';   hr = 'hour';   d = 'day';   mth = 'month';

P = Plateau Preference   V = Valley preference

1. 'Net' is the balance after similar interests of opposing preferences have been cancelled out.

2. It was assumed that not all members of a household will be at home throughout the construction hours; a 6 days construction week comprises 26 construction days per month.

3. We have assumed 50 FTE (full time equivalent) staff experiencing the effects for an average of 4 hours per day.

4. The survey revealed an average of 63 visitors per day, of which 22 were wild life enthusiasts and 41 simply enjoy walking. The average stay per visit was approximately 1 hour.

5. We have assumed that all additional local plan-making work and all land sale or CPO transactions will have been settled before completion of the works. To the extent that the same person as sold some land will continue to live or work in the same property, their experience following completion is regarded as a 'consumer' interest.

6. We have assumed an average of 4 hours per person per day during which the person will be at the scene and aware of the visual and noise impacts, for 30 days per month.

7. Accidents may cause as little as a couple of hours check-up which is a very low average exposure per month over a lifetime, or, more rarely, they could result in permanent impairment or even death. Predictions are unreliable for such a small sample.

8. The concern for the Valley comes in three spheres : those who make actual use of it; those who live within reach and would like to keep the option of using it; and those of the general public who may never use it personally but wish to keep it for future generations under a general strategy of sustainability; particularly since it is an important 'green lung' for the expanding urban area of Torbay.

• Figures on actual use were obtained through a users survey;

• Figures on surrounding residents who expressed an interest were derived from responses to the exhibition questionnaire but the response level was only 11%. Therefore 1850 is likely to be an under estimate

• Figures on the general public are difficult to obtain and would depend also on the geographical boundary adopted - Torbay, Devon, or the UK..

11 ´ PUBLIC PARTICIPATION

The IPEA inherently seeks to identify the interests of different groups amongst the relevant community. Projects presented for public consultation in an IPEA format enable members of the public to identify themselves amongst the groups accounted for. If they find that their interests have not been accounted for correctly and make submissions to that effect, the analysis is amended following consultation. If they disagree with the values placed by their Councillors on any of the groups, they can express their disagreement during the public participation debate or the public inquiry, or at elections.

The IPEA is open to continued amendments, as new groups surface or retract and as information about effects and impacts is being fine-tuned. This arises both before and during public inquiries. Studies, consultations and Proofs of Evidence may reveal new facts or lead to re-assessment of earlier information. Therefore the evaluation could require updating even at this stage. Such amendments do not invalidate the analysis itself but, if arising, I may have to reconsider and re-confirm the earlier balance between the options. Often the addition or subtraction of a few groups, while slightly changing the relative weights, does not tip the balance between the options. A major change or a cumulative change may however do so.

The Integrated method, if properly used, provides for a fully informed and balanced decision, accountability and public participation.

I now go on to show in greater detail the application of this method to the Torbay Ring Road options, in Sections 5-7below.

2.   The Analysis

When conducting the analysis in the Torbay case it quickly became apparent that to interpret its results some key concepts of environmental economics (Pearce, …) needed to be introduced, emphasising the difference between *actual use* value, expressing benefits/costs derived from actual use of the urban and natural environment, and *passive use* or *altruistic* values (including 'option value' 'bequest value' and 'existence value') i.e. values attached to products of the scheme by people who would not actually use them but nevertheless gain satisfaction from the knowledge that these products are available for future use to themselves, to others in this or in future generations, or exist in their own right.

The distinction arose from the initial finding that, by and large, those experiencing impact on 'actual use' values (e.g. development land owners, residents, and the Torbay District planning authority) will fare better under the Valley route option which does not disrupt the Plateau, while the much larger number of people in 'passive use' would prefer the Plateau route which does much less damage to the Valley.

The results of the analysis were first presented in a detailed Table which recorded for the Plateau route option all impacts on all groups affected, and compared them line by line with the respective impacts of the Valley route.

In comparing the two options we took the Plateau Route as a 'datum' and analysed the differences of impacts between it and the Valley Route. In this way we avoided the need to determine the absolute magnitude and value of the various effects, and so handled the simpler task of assessing the extent to which the one option performed better or less well than the other.

Although the detailed Table was itself a summary of earlier analysis, and a necessary record of all the groups affected, it was too extensive to readily point to a conclusion. It was therefore further simplified by 'netting out' - i.e. eliminating impacts of similar nature and size but of opposite preferences. The outcome, a more manageable "net" balance of groups with significant impacts, was presented in a final Table (Table 1).

The concise Table 1 presented the preferences at each of two phases: *pre-completion* in which some disturbance through construction is normally taken for granted, and *post completion* in which impacts last beyond the completion of the project.

To facilitate comparisons between groups a finer grain of analysis had then to be applied. Measurement of impacts through monetary values presents difficulties and would have been too time consuming in this case. We have instead indicated, where possible, three other relevant dimensions which could help in the balancing of impacts. Of these the first two are estimates of fact and the third a judgement made by DCC officers and myself. The dimensions are, for each group:

(a)   the number of people affected;
(b)   the length of monthly time exposure to the impact; and
(c)   severity or intensity of impact.

Columns 1-8 in Table 1 provide factual information (subject to correct estimates of the number of people and duration of exposure); while columns 9-11 provide for judgements and weightings. Column 9 presents the Officers' and my judgement on severity of impacts and columns 10-11 allow decision makers to summarise their weighting of each item in accord with their own judgement.

This information thus produces a picture which indicates a net balance of preferences between the options, but still allows different weighting to be applied to particular groups. When doing so one ought to consider the following:

• The group exposure figures do not in themselves provide the full answer on the relative weight of groups, and the severity or intensity of impact has to be taken into account. A given decision taker may personally attach greater or lesser weight to particular impacts (e.g. to environmental conservation; to employment) or to groups, (e.g. the farmers; the voters, etc.).

- The group exposure figures do not apply to all groups and where absent a judgement must be made on the relative weight of the impact.

Consequently, once the data in columns 1-9 has been analysed, further judgement and weighting have to be introduced. I have displayed my own judgement and weighting in Table 1-b. The *decision makers* were free to accept the result or apply their own values. DCC's decision as highways Authority to adopt the Plateau route reflects similar judgement and values. Torbay Borough Council (the local planning authority) came to a different conclusion, reflecting different values. It was then left to the Inspector to accept the balance or adjust it if he so chose by applying his personal weighting.

The method allows to test the sensitivity of the conlusions to variations in the weighted impact on particular groups. This indicates that efforts need not be spent on the finer details of those impacts which cannot reverse the final conclusion. The evaluators can thus focus on impacts which merit more detailed analysis since a variation in their weighted impact could reverse the conclusion. Indeed some of the discussions at the inquiry revealed that a heated debate on the difference in weighting between close views and medium views of the road would be wasteful whereas the weight attached to Passive or Altruistic use value, as against the weight of Actual Use value, was of utmost importance.

3.   Conclusions from the Analysis

3.1.   ASSESSMENT OF FINDINGS IN TABLE 1

Bearing in mind the distinction between 'actual use values' and 'passive use values' I found that in the final account the most significant factors in this evaluation are:

(a)   the inclusion or exclusion and the weight attached to the *passive use* value of conservation of the natural environment for potential use, and

(b)   which criterion is used to weigh the benefits in *actual use* of the built and natural environment - 'head count' or the 'exposure time'.

The evaluation of the net impacts in Table 1 indicates the following:

Pre-completion
The majority preference of the *producers* is on balance for the *Valley* route and is based entirely on 'actual use' value. Those favouring the Valley route include: labour

force (25 man/years or approx. 40% of one working life), 1 farmer and 4 commercial land owners whose land is compulsorily purchased. Against them stand the DCC treasury (cost of £876,900) and 2 farmers.

Amongst *Consumers* the balance of preferences is reversed:

The Valley route is preferred by a net of 113 people including 108 residents and farmers who would experience greater disturbance from works on the Plateau route and 5 at the Hilltop Nursery who would suffer environmental and operational disturbance and loss of say 1 job during construction. Their total exposure time to impacts from the Plateau route, as assessed by us, is *11,752 hours* per month.

Against them stand 178 people who prefer the Plateau route, including 2 farmers and 113 staff and management at Pontins who would suffer visual and noise intrusion and loss of say 4 jobs during construction, as well as some 41 walkers and 22 wildlife enthusiasts for whom the pleasure of actual use of the valley would be disrupted during construction. Wildlife in its own right will also suffer more disturbance during construction of the Valley route. Their total exposure time to impacts from the Valley route, as assessed by us, is 12,249 hours per month, higher than the exposure of the Valley route supporters.

All these groups will place an *actual use* value on their preference. Longer term *passive use* values are of little significance in the Pre-Completion phase which is inherently of short duration. The longer term effects are accounted for in the analysis of Post Completion impacts.

Irrespective of this outcome, pre completion disturbance is a transitory stage and it is generally accepted as unavoidable if development is to take place since it is short term by comparison with the life long benefits of the ensuing product.

Post Completion
Producers Post completion include only the maintenance functions of the DCC and they do not see a significant difference between the options.
Consumers are presenting a more complex picture.
The preference for the *Valley route* is supported by:

- a net number of 117 people in houses along the Plateau route who will suffer greater visual and noise disturbance from the Plateau route and whose preference is related to actual use.

- The 5 labour force of the Hilltop Nursery, who will suffer exposure to the new road view and noise, operational difficulties, and poor customer access. As a result say 1 job is lost. These too are actual use impacts;

- An average of 2 people injured per year, none of them severely, as a result of accidents at the additional roundabout on the Plateau route;

The total number of Valley route supporters is 124 and their total exposure time is 14,632 hours per month, of which the 117 residents suffer 14,112 hours and the 5 Nursery staff suffer 520 hours and 1 job loss. Another 2 people are lightly injured per year. All these come under the category of impacts on actual use.

Against them, with preference for the Plateau route stand:

- Pontins 113 staff and management who will suffer from visual and noise intrusion and whose holiday village may become less attractive to customers with the result of say 4 job losses - all these are impacts on actual use;

- 2 farmers suffering view and noise impacts in actual use of their fields;

- 41 walkers in the valley, who will also suffer in actual use;

- 22 wildlife enthusiasts in actual use;

- Approximately 1840 surrounding residents appreciate the Valley and wish to retain it, probably as an option for future use. This number is derived from responses to the postal questionnaire that accompanied the exhibition in February 1994. Out of 3000 local respondents some 1840 favoured the Plateau route, mostly on grounds of preserving more of the Valley. The 1160 that favoured the Valley route did so mostly on grounds of protecting residential and land interests (the ones directly affected are already accounted for in our analysis) or on the mistaken assumption that the Valley route could be built much sooner than the Plateau route.

- Besides those who took the trouble to reply to the questionnaire, an unknown number amongst the general public may never use the valley personally. I am nevertheless making the assumption that they would gain satisfaction from it being preserved in its own right and from reserving the option for others, or future generations, to use the valley. They can be regarded as applying the principle of 'environmental sustainability' particularly since the Valley is an important 'green lung' for the expanding urban area of Torbay and future generations should not be denied its enjoyment. These views may be held by members of the public in Torbay, in Devon, and at large in the UK who value the principles of 'environmental sustainability'. Some indication of the presence of the wider public interest is evident from the public consultation replies by various interest groups[2].

The total number of Plateau route supporters including both in *actual use* and in *passive use* is *at least 2,018* (178 + 1840 + unknown public). Of these, the number in actual use is 178 and their total exposure time is 7,974 hours per month + 4 job losses. In passive use there are at least 1840 but their exposure time is difficult to estimate, it being a matter of sentiment and of potential, rather than actual, exposure. To these should be added *agricultural* land reserves and *Nature* in its own right.

These figures are set out in brief in Table 2 (Summary Exposure To Route Options), indicating the preferences resulting from the exposure:

- Producers Post completion are indifferent between the options.
- Consumers Post Completion show a majority preference for the Plateau route (Table 2)

Exposure time is the only criterion by which the Valley route scores higher than the Plateau route. However 'exposure time' only applies to clear 'actual use', yet people with an unknown exposure time do have to be taken into account. These include:

- For the Valley route - 2 people injured
- For the Plateau route - a probably far larger number who would gain a *passive use* value from maintaining the principles of environmental sustainability and from retaining their options to use it, but have not come forward during public consultation.

Table 2   Summary Exposure To Route Options

QUANTITATIVE CRITERIA	PREFER PLATEAU ROUTE	PREFER VALLEY ROUTE
Survey - head count	1840	1160
Actual use - head count	178	124
Exposure time (of actual use only) hours/month	7,974	14,632
"Passive use" of the environment - head count estimate	8,000 (see below)	0

While data on actual users (whether in houses, businesses or walkers in the valley) was readily available, data on passive users was not. The study was conducted within strict time and financial constraints and there was not the opportunity to carry out tailor made surveys - whether contingent valuation, hedonic or other - to establish the distribution and value of the benefits of conservation. In the absence of more reliable information I proposed an approach to estimation of the 'passive users' numbers and concluded with the following conservative assumptions:

a) Only 33,000 (or half the population of TQ3, TQ4, TQ5) is within the area of priority access to the Westerland Valley; Of these 5,000 would attach 'passive use value' to the valley;

b) Of the remaining 33,000 only 2000 would attach 'passive use value' to the valley;

c) Amongst the general public in Devon another 1000 would attach value to the valley.

d) The intensity of sentiment will be greater in the first and third of these groups and less in the second, but I have regarded all under a P rating.

In consequence I have assigned 8,000 people to the category of 'General Public' who would consider the preservation of the valley as a benefit, rated P. This indicates a higher still level of support for the valley.

4.   Drawing the Balance - Decision Criteria for Evaluation

In my Proof of Evidence I discussed in some detail key questions related to decision criteria which the Inspector (and decision makers generally) may be concerned with. In essence these were:

1. Should the decision criteria include both 'passive use' and 'actual use' values, or actual use values alone?

Current thinking recognises these 'passive uses' as part of a total *economic value* of the environment and leaving them out will not be acceptable. I do therefore include the 'passive use values' in my analysis.

2. Which of the 'actual use' criteria ('number of people', 'exposure time', 'severity rating') ought one to use?

These criteria serve different purposes:
- The number of people or *head count* gives a simple indication of public support.
- The *exposure time* gives some indication of the extent of the impact on individuals and thus is a step towards an equitable solution.
- The *severity rating* is a further step towards an equitable balancing of impacts. Although subjective, being explicitly presented it is preferable to the intuitive subjective judgements implicit in all decision making.

3. Could the three criteria be summarised in a single figure for each option?

It would have been all too easy in this analysis to produce just the results of an arithmetical calculation that would ease the reader into an effortless conclusion (with imputed figures where these cannot be calculated). We have refrained from relying on such a single figure because it is important to expose the assumptions and judgements underlying each step of the way. Moreover, decision makers should attach their own values/weightings to the arithmetical figures. I have demonstrated my own composite evaluation and weighting in Table 1-b, columns 10 and 11.

4. How can personal values be brought to bear on the conclusions?

The Inspector (or other readers of this Proof) may disagree with my judgement on the relative values of various groups in Table 1-b. Table 1 therefore leaves open Columns 10 and 11 to accommodate an adjusted set of figures.

Having considered these questions I explicitly applied my own views weightings in Table 1-b to Post Completion Consumers. I concluded that there was clear preference for the Plateau route. This was based on the belief that impacts on both *actual use* and *passive use* of our natural and built environment should be taken into account and that both 'head count' and 'exposure' have significance.

DCC in their function as highways authority have after full consideration taken a decision to favour the Plateau route and have thus reflected in practice a similar set of values.

5.   Conclusions On the Use of IPEA

The tables have been through many drafts, continually adding new information and refining the factual and judgmental information. Such amendments are fully recorded and explicit and several drafts have been exposed to committees and to public consultation.

At the time of writing the Public Inquiry was still in progress. More detailed information was coming forward through evidence of both supporters and objectors. Torbay District Council, who favoured the Valley Route, have helpfully presented evidence in an identical format to IPEA, thus presenting a clear basis for argument. The tables were being adjusted at the Inquiry. Factual amendments were fully incorporated while judgmental ones were debated.

The Inspector and all those present appeared to agree that the analysis should be used as a dynamic tool, rather than as a finite pronouncement which should be shot down when a detail needed amending.

Used in this way, IPEA serves for greater professional responsibility and it focuses the mind on benefits/costs to all groups of the community - the public - which, after all, is the main purpose of our planning system.

However it would be difficult for an individual Inspector to greatly vary the standard format of inquiries and reports which the Government and the parties to the debate expect to see. There has to be public and Government recognition of the value of

IPEA to encourage its wider use. In this publication we hope to make a small step towards such recognition.

References

Lichfield, D. (1992) Making the Assessment Link, *Planning*, 3 July, pp.4-5.

Lichfield, N. and Lichfield, D. (1992) The Integration of Environmental Assessment and Development Planning, *Project Appraisal*, Sept. 1992, pp. 175-185.

Lichfield, N. (1996) *Community Impact Evaluation*, UCL Press, London.

Royal Town Planning Institute (RTPI) (1992), *Award for Planning Achievement 1991*, London.

Department of the Environment & Welsh Office (February 1997) *PPG1 (Revised) - Planning Policy Guidance: General Policy And Principles*, HMSO, London, page 7

Pickthorne, M. (1996) *Proof of Evidence on Landscape,* Devon County Council (DCC), Torbay Ring Road Public Inquiry.

Pearce, D.

---

[1] The essence of The IPEA, its methodology and terminology are described in the Proof of Evidence presented to the Public Inquiry in April 1996. A summary of it is available from the author.

[2] Respondents to public consultation, who represented wider interests supporting the Plateau route, included: Devon Wildlife Trust: the R.S.P.P.; Dartmoor Badgers Protection League; Devon Bird Watching and Preservation Society; Marldon Parish Council; South Hams District Council. The number of people represented by these groups is not known, but it would appear that some overlap with those who responded to the questionnaire and other have not responded or come from a wider area of Devon.

SUSTAINABILITY INDICATORS IN URBAN PLANNING EVALUATION

A new classification system based on multimodal thinking

P.L. LOMBARDI

Abstract

Planning evaluation in the age of sustainability involves a change of emphasis and change in the criteria by which development is judged towards environmental protection and socio-economical objectives. It requires to build social consensus as well as to improve technical demonstration. At different levels of government - international and national - a number of sustainability indicators have been suggested in order to audit environmental performance of an urban system and to guide decision making toward the achievement of a sustainable urban development. All these indicators are quantitative in natura and are usually classified on the base of a reductionistic view of reality. This is unable to handle all the elements and components of the system, both deterministic and nominalistic ones. Utility theory and other mechanistic conception of life are not satisfying since they approach reality and decision-making problems from a very narrow and technical perspective. On the other side, nominalistic and post-modern approaches do not always leave space for objective or rational solutions. Decision-making for sustainability planning requires new approaches which can be able to integrate and synthesise all dimensions of an urban systems and different point of views, in a holistic manner.

This paper postulates a new scientific paradigm, named multi modal thinking, moslty because it offers a very useful checklist to guide planning evaluation process, ensuring that all aspect of human life be present in the design. In particular, this is used for establishing a more holistic and integrated classification system for sustainability indicators in the evaluation of the built environment at local planning level.

1. Introduction

The notion that urban development should be sustainable has received recognition by the European and global community through a number of international forums and reports, including the World Commission on Environment and Development (the Brundtland Report: WCED, 1987), the United Nations Conference on Environment and Development (Agenda 21: United Nations, 1992), the Expert Group on the Urban

N. Lichfield et al. (eds.), Evaluation in Planning, 177–192.

Environment (the report "European Sustainable Cities": EC 1994) and the European Environment Agency Task Force (the Dobris Assessment, 1995).

One of the man contribution of the debate on sustainability to planning and construction is the particular concern for all the following principles (Mitchell et al., 1995): integrity of eco-systems (environment), equality of access to the world's global resources (equity), future generation needs (futurity), and possibility to participate in decision-making processes (public participation). However, there is still a need to incorporate sustainability principles and criteria in current practice (Lombardi and Brandon, 1997).

An obstacle to this goal is the multi-aspectual meaning of sustainability. This concept is (and always be) above definitions. An example of the variety of current approaches to sustainability is illustrated in Brandon et al. (1997). For instance, it has been recognised that developers of assessment model for sustainability at urban scale, such as May et al. (1997), Jones and Vaughan (1997), take into account most economic-social and physical aspects of a sustainable development, while environmental assessment methods at building scale, such as the BREEAM in the UK (Prior, 1993) and the BEPAC in Canada (Cole et al., 1993), emphasise mainly the environmental and ecological issues related to sustainability and quality of life.

Camagni (1996) has classified different approaches to sustainability as follows:

a) 'input' or 'output' oriented approaches (i.e. 'strong' and 'weak' sustainability), according to their emphasis on, respectively, limitation in the use of not-renewal resources and guarantee of well-being in long term; and

b) approaches based on 'boundered' or 'procedural' rationality (Simon, 1975), in relation to the scientific theory they are referring, respectively, neo-classical economy (i.e. rational behaviour and full information of decision-makers) and decision theory developed in psychology (i.e. process by steps without full information).

According to him and other authors, a sustainable urban development is a process of balancing and of synergetic integration (or co-evolution) between sub-systems, i.e. economic system, social system, physical system (including the built and cultural heritage) and environmental system (Costanza, 1993). This process should be able to guarantee both a non-decreasing level of well-being to local community in long term and the reduction of negative effect in the biosphere.

However, an evaluation of sustainability in planning faces many problems and requires adequate ex-ante evaluation approaches (Voogd, 1997).

These should be able to consider the following main criteria introduced with the notion of sustainability in local planning agenda of many E.U. countries: the integration of different dimensions in policy analysis, the consideration of a long-term view in forecasting and the improvement of information to all citizens for helping public participation in decision-making.

This paper will firstly provide an overview and a critique of the principal classification systems for sustainability indicators as suggested by international organisations at urban level. Secondly it will illustrate and discuss the main benefits of the multimodal systems thinking (de Raadt, 1994), for tackling the above issues. This not only provides integration between aspects and disciplines, but also may be able to offer specific guidance to the evaluation in planning and the built environment.

In previous studies, this approach has been postulated for structuring a framework for evaluation of the built environment quality by integrating different work-methodologies (Lombardi and Brandon, 1997); for understanding the multi-aspectual nature of sustainability in relation to system development (Lombardi and Basden, 1997) and, finally, for assisting the resolution of the conflicts between actors and parties in re-development processes at local planning level (Lombardi and Marella, 1997). In this study, it will be used for developing a new classification system for different sustainability criteria and indicators in a systematic and holistic manner.

2.   Sustainability indicators and classification systems

International organisations such as United Nations, World Bank, and European Union have recently developed a number of sustainability indicators which illustrate current environmental problems, identifying their causes and effects, in order to improve decision making processes at all levels, local, national and international. All the selected indicators are quantitative and statistical in nature (Istituto Ambiente Italia 1995).

At urban level, the development of sustainability indicators specifically aims to: audit the urban development, evaluate policies performance and decision-making processes and assist local administration in finding possible solutions and correct strategies for environmental and social problems. These indicators should represent both a fundamental vehicle for improving communication with local community and an efficient technical tool for supporting decision making process. In accordance to Agenda 21, the selection of indicators should be operated through a bottom-up approach where citizens play a crucial role in identifying the more appropriate ones. However, many European countries has adopted a top-down approach, leaving to experts the responsibility of this choice.

At international level, this selection has been linked to a possibility of comparison between regions and countries, by harmonising the list of selected indicators. In addition, different international organisations have tried to establish a classification system for sustainability indicators with the aim to find integration between them, improving the evaluation of urban developments and policy decisions. For example, the Organisation for Co-operation and Economic Development (OECD, 1994) have classified the selected indicators in three main groups:

a) *State,* describing environmental resources quality or depletion;
b) *Pressure,* describing the carrying capacity of environmental resources;

c) *Response,* describing public policies-programmes and private behaviour.

A similar classification system for sustainability indicators has been developed by United Nations (1995) as *"Driving force-Sate-Response"* where more emphasis is placed on the effects that human actions and processes have on the natural environment. These indicators are grouped in four main sustainability issues, i.e. social, economical, environmental and qualitative aspects. However, there is not a specific integration between them in practice.

The list of indicators which has been selected among 72 European cities by the Task Force (EU,1995) is specifically concerned to the built environment. This includes 55 quantitative indicators classified in relation to: i) urban structure (population, land occupation, number of journeys for mode, etc.); ii) urban performance (water consumption, energy consumption, number of waste disposals, etc.) and iii) urban quality (e.g. number of days in which there is an exceeding of environmental standards).

A different method for harmonising the selection process of sustainability indicators, named ABC-indicator-model, is illustrated in "The European Sustainability Index Project" (International Institute for Urban Environment). This is based on a classification of indicators in three main groups:

a)   *area specific indicators,* developed by local organisations or administrations, and related to specific problems or characteristics of the area;
b)   *basic indicators,* which support the following indicators (c), clarifying the context and specifying the results;
c)   *core indicators,* represents the principal ones, providing the more essential and fundamental information for measuring local sustainability.

This project has included 12 European cities; it represents a way forward in the development of a common understanding of sustainability and toward an homogeneous method for the development of sustainability indicators (AA.VV., 1997).

2.1.   A CRITIQUE TO CURRENT CLASSIFICATION SYSTEMS

A major limitation of all the above classification systems is the focus on the environment rather than sustainability. Mostly of the selected indicators are related to a description of the environment as such, without an identification of the multiple effects this state has on human and natural resources. This often leads to immediate and short term solutions rather than to a prevention of negative effects.

A second problem with the above sets of indicators is the unique utilisation of quantitative measures for describing sustainability in the built environment. This is a very narrow way to represent the problem which leave uncovered a large number of fundamental aspects, such as the spatial (morphological indicators for describing urban form), the analytical (related to teaching and learning, such as educational programmes and incentives), the lingual (the role that communication and media may play in development processes), the ethical (ways to go beyond the traditional NIMBY - not in my back yard - defensiveness) just to mention few.

In addition, the very seeming comprehensiveness of some of these lists can be misleading, and they are likely to be unbalanced, putting more emphasis on certain issues than others.

Elevation of aspects, and consequent imbalance and ignoring of others, is a particular danger in sustainability planning, particularly for urban environments, because it places heavy reliance on specialist knowledge. Planning should be seen, not as a branch of higher abstraction, but as everyday thinking and acting (Kuhn, 1962; Ravetz, 1971).

In any planning situation, there are inter-dependencies which need to be understood, and in particular the effect that action in one aspect might have in others. Consequently, it is important to keep a systemic and pluralistic approach to the problem (Costanza, 1993).

Finally, all the above classifications fail in their aim to integrate all the dimensions of urban sustainability, i.e. environmental, social and economical. One reason for this problem is that rigid and definite classifications usually lead to difficulties in defining interactions and relationships between sub-systems. A second reason is related to a difficulty in defining sustainability in substantive terms. This assumes different meanings in relation to different contexts or to different approaches, a bottom-up approach, improving participation or top-down approach, based on expert opinions.

Considering the multi-aspectual nature of sustainability at local level, all aspects of reality are important for the true long term sustainability of any built environment and its community (Lombardi and Basden, 1997). These aspects are a reality that is independent of us, and they pertain to all our functioning; even though we are usually not aware of them. In higher abstraction we isolate and deeply analyse each aspect, making an explicit knowledge of its laws available to the human community. Scientific knowledge can then be used in our everyday living, but usually requires the services of specialists. Unfortunately, the availability of specialist knowledge of an aspect tends to elevate that aspect, and thus result in less then "healthy" living (Dooyeweered, 1968).

We need a framework and mechanism that brings unity within diversity. If unity in diversity is important for urban sustainability, the different experts, the stakeholders and all concerned citizens need to be able to communicate at a sufficiently deep level and the earning of sustainability must be clear and agreed.

A useful classification system for sustainability indicators at a local planning level is the one able to:

- recognise specific and definite relationships between the components of an urban system;

- be comprehensive of all the technical-deterministic and non-technical-nominalistic variables such as those in the social, cultural and political realms;

- handle all the above variables which are essential to the progress and management of a sustainable society.

These issues are tackled by the multimodal system thinking which is based on the philosophical approach of Herman Dooyeweerd (1953-55). This provides a number of benefits and particularly the followings. It is explicitly trans-disciplinary yet provides integration between disciplines and an holistic approach to the problem. It considers different levels of information nesting all aspects of reality in an ordered manner yet provides structure and continuity. And it recognises the importance of multi-people action across different time scales keeping a western view.

3.   Development of a multimodal evaluation system in planning

3.1.   PHILOSOPHICAL ASSUMPTIONS

The philosophy of Herman Dooyeweerd is not only comprehensive and wide-ranging, but it is also orthogonal to most philosophical and theoretical debate during the last few centuries. This is why it might throw useful light on issues that have emerged from Western activity over the same period to haunt our thinking in recent decades. It is most fully explained in his four volume magnum opus, 'A New Critique of Theoretical Thought' (1953) and since then expanded and developed in his other writings on Western thought and decline. It is still in the process of development.

It takes far more than one paper even to introduce it and there are several ways of approaching it. While Dooyeweerd (1953) himself starts with the flows of thought over the last 3000 years, Kalsbeek (1975) starts by introducing Dooyeweerd's irreducible aspects of reality, de Raadt (1994) starts from systems theory and Clouser (1991) starts from the relationship between theory and practice, we will start from the problem of the subject-object relationship.

We can see many of the problems of sustainability as emanating from the artificial (but time-honoured) separation of subject and object. Two main streams of philosophical and theoretical thought over the last 500 years - realism and nominalism - have emphasised one or the other, and these two are part of Dooyeweerd's longer term (3000 year) analysis of theoretical thinking.

Traditional realism emphasises the   known and acted-upon object and de-emphasises the knowing, acting subject. In extreme versions (such as positivist science) the relevance of the subject (e.g. the experimenter as person) is denied altogether, leaving only the object. Laws must then be deterministic rather than normative.

Approaches to planning based on realist philosophies have the danger of reductionism: of the above problems of ignoring and imbalance (Lindblom, 1965; Simon, 1972). It may not be absolute but shows itself in an imbalance, in which one aspect is given undue emphasis to the detriment of others. Doing so threatens sustainability of the built environment. Another problem with planning based on realist philosophy is that because the subject is de-emphasised the effects of the action and knowledge of the subjects are often ignored (Kuhn, 1962).

The opposite approach to the subject-object theme is based on nominalist philosophies, of which existentialism is an extreme form. In these approaches the object is denied, leaving only the knowing and acting subject. This approach pervades post-modernity (Lyon, 1995) and, in scientific circles, constructionist and interpretivist paradigms (Guba and Lincoln, 1989). It claims to avoid the dangers of reductionism by acknowledging the views and wishes of all and sundry (Fisher, 1990). While it has some success in this, there are three problems. No external reference point is acknowledged or even allowed, so there is no certainty that planning according to these wishes will in fact lead to sustainability in the longer term. Second, when wishes and views of different people or groups appear inconsistent there is no standard by which to arrive at consensus. Third, there is the danger in practice that those who shout loudest get heard, while less articulate groups and those who cannot represent their right, such as animals or young children, tend to get ignored unless their cause is championed by others. Therefore while less reductionist than approaches based on realist philosophies, there is still no guarantee of sustainability. There is not even any guarantee that sustainability will be greater than when adopting approaches based on realist philosophies. Approaches based on nominalism find integration and inter-communication, mentioned above, difficult.

Unlike nominalist approaches Dooyeweerd acknowledged an external reality that is independent of the acting and knowing subject. We are affected by it but also affect it and have views and desires concerning it.

Realist philosophy drives its adherents to reducing all types of laws to one, such as the laws of physics, of logic, or evolutionary biology, etc. Nominalist philosophy drives its adherents to a denial of all laws. Dooyeweerd wished to escape both dangers, proposing a pluralist ontology, in which temporal reality has fifteen aspects or dimensions, each of which has a kernel meaning, as it is illustrated in Table1.

Each aspect provides a set of laws - e.g. laws of arithmetic, laws of physics, laws of aesthetics, laws of ethics, etc.- which not only guide but enable entities (people, animals, etc.) to function in a variety of ways. The laws of the earlier aspects are more determinative while those of later aspects are more normative. Normativity gives a degree of latitude in choosing between valid courses of action. But this latitude also extends to non-valid, non-*healthy*, options such that when acting as subject of normative laws we can act against them. In this latter case, there will be consequences.

TABLE 1. The pluralistic ontology of Dooyeweered

Levels of Information

Quantitative	Awareness of 'how much' of things
Spatial	Continuous extension
Kinematics	Movement
Physical	Energy, mass
Biological	Life function
Sensitive	Senses, feeling
Analytic	Discerning of entities, logic
Historical	Formative power
Lingual	Informatory, symbolic representation
Social	Social intercourse, social exchange
Economic	Frugality, handling limited resources
Aesthetic	Harmony, beauty
Juridical	Retribution, fairness, rights
Ethical	Love, moral
Credal	Faith, commitment, trustworthiness

These aspects are irreducible to each other ('sphere sovereignty'). However, there are definite relationships between them ('sphere universality'), which allows an entity to function in a coherent rather than fragmented manner.

These relationships are of three kinds:

a) Dependency. The laws of later aspects depend on and require those of earlier ones. Thus, biotic laws require those of physics, which require those of movement, etc.

b) Analogy. Components of each aspect are mirrored, echoed in others. Thus, for instance, we have a feeling (Sensitive) for justice (Juridical) which is different from a feeling for harmony (Aesthetic) or logical clarity (Analytic). Causality is rooted in the physical aspect but is echoed in others (e.g. logical entailment). Such analogy is the basis for symbolic representation of knowledge on computer (de Raadt, 1994). So, when functioning in each aspect we encounter a rich interweaving of analogies.

c) Functioning. What Dooyeweerd calls individuality structures (and we call entities and systems) function in all aspects. An individuality structure functions in each aspect either as subject or object. This functioning individuality structure serves as an integration point for the aspects; there is no direct causal link between aspects (e.g. better lingual communication does not automatically bring better social relations).

What this means is that no amount of planning can actually guarantee sustainability. All we can do is maximise the balance between aspects and attend to these types of relationships between them, and then we will lay the path towards the necessary integration.

3.2.   ILLUSTRATION OF A NEW CLASSIFICATION SYSTEM

Based on the above assumptions, a new classification system has been developed, named BEQUEST (Built Environment Quality Evaluation for Sustainability through Time). This includes the fifteen modal aspects, as levels of information and their relationships for structuring a multimodal check-list of sustainability indicators and evaluation criteria in planning and the built environment.

A synthetical illustration of this list is given in Table 2 (see also: Lombardi, 1995; Lombardi and Brandon, 1997 and Lombardi and Marella, 1997). However, a better explanation of the relationships between the variables included in it is provided by the "98-crossings matrix" illustrated in Table 3. This matrix particularly illuminates the existing interrelation between the aspects which defines their position (dependency relationship). This corrispondence between the orders of different aspects allows one aspect (named *source*) to be used as metaphoric representation of another or several aspects (named *idioms*) (analogy relationship). However though every aspect can be an idiom for another its effectiveness as an idiom varies and the degree of correspondence declines as the distance between one aspect and another increases. For example the numeric aspect is not a very suitable idiom for the juridical and very few quantitative indicators may be found in this area. Closer aspects such as the ethical are more useful for explain juridical aspect.

The form of this new matrix is triangular with a number of columns and rows corresponding to the number of aspects, i.e. fifteen, and a number of boxes corresponding to all possible crossings (or interactions) between them, i.e. $[(15 - 1)$ x $(15 - 1)] : 2 = 98$. Each box includes a set of sustainability indicators and performance criteria for planning evaluation at local level. Thus for example an indicator such as So/Ph = resources consumption/depletion is included in the crossing beteen the social and the phisical because it aims to explains a social behaviour or activity by using a physical representation i.e. the availability of natural resources is a direct consequence of such behaviour. In this representation, the social dimension provides the source and the physical the idiom. A more detailed example of this new classification system for sustainability indicators with multiple measurement scales, is illustrated in Table 4 and 5 for the 'economical' dimension.

TABLE 2. Checklist for sustainability indicators

Aspects	Key-factors for urban sustainability	Description
QU	Population, Amount of various resources, Number of species, Accumulation	*It deals with numerical data, statistics and mathematics.*
SP	Layout, Shape, Density, Location, Proximity, Terrain shape	*It deals with land occupation, spatial differentiation, shape and layout of buildings, geographical position, areas and form*
KIN	Viability, Transportation, Mobility, Accessibility, Wildlife movement	*It deals with transport system, road network, accessibility to services and parking, drainage system*
PHY	Prosperity, Energy for human and for biotic activity, Structure of ground on which to build.	*It deals with energy, water, air, soil, natural materials, construction and agriculture*
BIO	Food, Air, Water, Shelter, Health and health services, Pollution, Soil quality, Biodiversity, Habitat diversity	*It deals with the vital functions (to live, grow or develop, the protection of the biosphere and biodiversity, the presence of green areas*
SE	Feeling of well-being, Feelings engendered by living there, Comfort, Security, Noise	*It deals with housing and hospitals, comfort, noise, visual impacts, pollution, physical and mental health*
AN	Diversity, functional mix, Quality of analysis for planning, Education services, research	*It deals with no-uniformity of the built environment and with education and teaching*
HIS	Built Heritage, Technology employed, Creativity	*It deals with cultural values, built heritage, historical development, maintenance and/or transformation*
LIN	Communication and Networks Advertising, The Media	*It deals with communication aspects such as information facilities, media, etc.*
SO	Social relationships, social climate, Clubs and societies, competitiveness, collaboration	*It deals with the sum of elements that enhance and sustain social interaction (Sociophilia)*
EC	Efficiency, Attitude to finance, Use of land and of resources	*Minimisation of effort, cost and waste in acting and producing*
AE	Beauty, Harmony, Architectonic Style	*Aesthetically satisfying* *Artistic character and significance*
JU	Democracy, Rights, Political structure, Legal institutions, Regulations, Laws	*It deals also with property and planning laws, land titles, regulations.*
ET	Equity, Solidarity, Sharing	*Reduce social inequality* *Some opportunity to people*
CR	Shared vision of what we are and of the way to go; Futurity; Equilibria	*A concern for future generations, No-conflict*

TABLE 3. The crossings matrix for integration of aspects

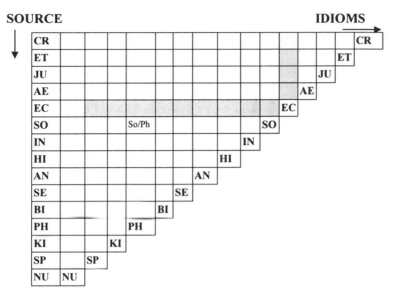

Each aspect can be both a source or an idiom for different indicators. Reading the matrix by row, the indicators belong to a 'source' dimension. Reading it by column, they belong to an 'idiom' dimension. For instance So/Ph=resources consumption/depletion. Here, the social dimension provides the source and the physical the idiom.

TABLE 4 Example of sustainability indicators in the economical dimension (source)

ECONOMIC - source	
Ec/ Numerical	Number of economic sectors and financial institutions involved in this development / Financial accounting of the project
Ec/Spatial	Location and land occupation by economic activities
Ec/ Kinematics	Min and max time and costs for accessibility to city center, to urban facilities, to public services and job places
Ec/Physical	Provision/self-sufficiency in goods and services
	Energy consumption by economic sector per year
	Resources consumption/depletion by economic sector
Ec/ Biological	Total production of domestic non-domestic waste
	Efficiency of sanitary services
Ec/Sensitive	Effect of noise and thermal dispersion on price of buildings
Ec/Analytical	Analysis of costs for construction and maintenance
	Economic sectors mix, by sector
Ec/Historical	Positive/negative growth rate / Effect of Heritage on the use of land
Ec/Lingual	Influence of communication and networks on the use of resources
Ec/Social	Employment mix, by sector, public/private split
	Influence of urban-rural immigration/emigration on economic activities
	Employment of local labour force in construction activities

TABLE 5 Example of sustainability indicators in the economical dimension (idiom)

ECONOMIC - idiom	
Aesthetic/ Ec	Willing to pay for aesthetic benefits and improvements on buildings and urban services
Juridical/Ec	Monetary implication for juridical responsibilities of each sectors participating to the project
Ethical/Ec	Redistribution of resources among community and sectors

At a conceptual level, this model may be effectively represented in a two-axes Cartesian diagram as it is shown in Figure1: where the x-axe corresponds to deterministic measurements (it is 0 in the credal) and the y-axe correspond to normative measurements (it is 0 in the numerical). A third z-axe may be included, representing the time which separates out the aspects and gives direction to the progress.

The numerical and the credal aspects find their fully explanation among the two respectively x-y axes. In particular, the x-axe contains all the quantitative indicators and measurements for sustainability, while the y-axe contains local community targets and goals to be achieved for sustainability. All the others boxes include a variety of measurement units, considering different degree of variation between the above two aspects: numerical and credal (i.e. quantity and quality). In other words, the evaluation within the 98 crossing areas of the diagram requires both, deterministic-quantitative and normativistic-qualitative assessments, technical expertise and communication between concerbed citizens.

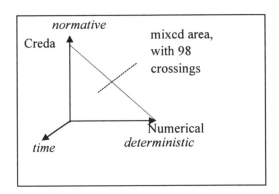

Figure 1. Synthetical representation of the model

At the moment, the BEQUEST model it is mainly a framework for guiding the evaluation among conflicting parties rather than a formalised methodology. It does not include metrics for calculation or assessment techniques, leaving to each scientific area the development of specific methods and tools which can cover the evaluation in different aspects. However, it recognises the importance of soft multicriteria analysis and specific interactive tools for improving participation in decision-making process.

4.   Conclusion and further research

The Bequest model is based on the theory of Dooyeweered, a Dutch philosopher of the middle of this century, who is the ideological father of a multi-modal system thinking approach developed in cybernetic by de Raadt in Sweden.

Both these sources have given the theoretical foundation to this study for:

- understanding the complex trade-offs between man and nature within the built environment and trying to synthesis them into an environmental and human oriented context;
- establishing an integrated, meaningful and holistic classification system for the evaluation in planning and the built environment.

The main benefits of this classification system are the followings:

- It links all relevant evaluation criteria and indicators in an holistic manner
- It nests them in an ordered manner
- It considers different levels of information, and different disciplines
- It avoids ignoring and imbalance among them
- It provides structure and continuity for the evaluation
- It helps discussion among the parties concerned

For developing the Bequest model, the methodology has undertaken a number of steps as follows:

1. a collection of aspects which qualify the built environment as being sustainable, through literature review, international reporting and Delphi exercise;
2. a deep stated study of Dooyeweered's theory and multimodal system thinking;
3. an understanding of sustainability in planning and the built environment and a definition of it on the base of the above theory;
4. a meaningful correlation of the above sustainability indicators and criteria among the multi-modal aspects;
5. a test of the model through two synthetical applications in different planning contexts.

In addition, the process of developing this holistic framework on a new way of thinking has involved an understanding of difficult issues, such the meaning of sustainability in relation to urban development and the built environment quality, the advantages and limitations of current planning evaluation methods and interpretative theories, etc.

The results obtained from this research at the present time are successful and give hope for further developments. The applicability of Dooyeweerd's aspects to urban planning and decision-making has been illustrated in previous studies, by using an example of regeneration process with conflicting interests (Lombardi and Marella, 1997) and of designing a car parking (Lombardi and Basden, 1997). Research is underway to explore the possibility to develop an information artefact on the base of this new classification system.

Acknowledgements

This study is part of the PhD research of the author at the University of Salford (UK), supervised by Professor Peter Brandon. The BEQUEST project has provided both the name and the background for developing a new European network between University and Research Institutions on this subject which has been recently supported by the financial contribution of the European Community within the Framework 4 (1996-99). Section 3.1 has been developed with the contribution of Andrew Basden of the I.T. Institute of Salford University.

References

AA.VV. (1997) *Progetto Venezia 21*, Fondazione Eni Enrico Mattei.

Bocchi, M. and Ceruti, M. (eds.) (1994) *La sfida della complessità*, Feltrinelli, Milano.

Brandon, P.S., Bentivegna, V. and Lombardi, P. (eds) (1997) *Evaluation of Sustainability in the Built Environment*, Chapman&Hall, London.

Brchcny, M.(ed.) (1992) *Sustainable development and urban form*, Pion, London.

Camagni, R. (ed) (1996) *Economia e pianificazione della città sostenibile*, Il Mulino, Milano.

Clouser, R.A. (1991) *The Myth of Religions Neutrality*, University of Notre Dame, London.

Cole, R.J., Rousseau, D. and Theaker, I.T. (1993) *Building Environmental Performance Assessment criteria, Version 1, Office Buildings,* December 1993, BEPAC Foundation, Vancouver.

Costanza, R. (1993) Ecological economic systems analysis: order and chaos, in Barbier E.B. (ed), *Economics and Ecology*, Chapman & Hall, London, pp. 29-45.

de Raadt J.D.R., 1991, Cybernetic Approach to Information Systems and Organization Learning, *Kybernetes,* **20**, pp. 29-48.

de Raadt, J.D.R. (1994) *Enhancing the horizon of information systems design: information technology and cultural ecology*, unpublished.

Dooyeweerd H., (1968) *In the twilight of Western thought*, Nutley, New Jersey, USA.

Dooyeweerd, H. (1955) *A New Critique of Theoretical Thought,* 4 volumes, Philadelphia, Presbyterian and Reformed Publisher Company, Pennsylvania.

EU Expert Group on the Urban Environment (1994) *European Sustainable Cities. Consultation Draft for the European Conference on Sustainable Cities and Towns,* first annual report, Aalbourg, Denmark, 24-27, Commission of the European Communities, Directorate XI, XI/307/94-EN.

European Commission (1990) *Green paper on the urban environment,* Commission of the European Communities, COM (90) 218 final, Brussels 27/6/1990.

European Environment Agency Task Force (1995) *Europe's Environment: The Dobris Assessment,* Earthscan, London.

Fisher, F. (1990) *Technocracy and the Politics of Expertise,* Nebury Park CA., Sage.

Guba, E.G. and Lincoln, Y.S. (1989) *Fourth Generation Evaluation,* London, Sage.

International Institute for Urban Environment, *The European Sustainability Index Project*

Istituto Ambiente Italia (ed) (1995) *Ambiente Italia 1995. Rapportoannuale di Legambiente,* Edizione Ambiente, Milano.

Jones, P. and Vaughan, N. (1997) Development of an environmental and energy performance model for sustainable cities, in Brandon, P.S., Bentivegna, V. and Lombardi, P. (eds) *Evaluation of Sustainability in the Built Environment,* Chapman & Hall, London.

Kalsbeek, L. (1975) *Contours of a Christian Philsophy,* Wedge Publishing Company, Toronto.

Kuhn, T.S. (1962) *The Structure of Scientific Revolutions,* University of Chicago Press, Chicago.

Lindblom, C. (1965) The Intelligence of Democracy, New York, Free Press.

Lombardi, P. and Basden A. (1997) Environmental Sustainability and Information Systems, *Systems Practice,* Hull (forthcoming).

Lombardi, P.and Marella, G., 1997, A multi-modal evaluation of sustainable urban regeneration. A Case-Study related to ex-Industrial Areas, in *Second International Conference: Buildings and the Environment,* CIB-CSTB, vol.2, Paris, 9-12 1997.

Lombardi, P. (1995) Evaluating Sustainability at a local planning level, in *Proceedings of the International research Workshop* CIB TG-8 on *Linking & prioritising environmental criteria,* Toronto, pp. 89-100.

Lombardi, P. and Brandon, P. (1997). Toward a multi-modal framework for evaluating the built environment quality, in Brandon, P.S., Bentivegna, V. and Lombardi, P. (eds), *Evaluation of Sustainability in the Built Environment,* Chapman & Hall, London.

Lynch, K. (1960) *The image of the City,* The Technology Press & Harvard University Press, Cambridge, Mass.

Lyon, D. (1995) *Postmodernity,* Open University Press, Oxford.

May, A.D., Mitchell, G. and Kupiszewska, D. (1997) The Development of the Leeds Quantifiable City Model, in Brandon, P.S., Bentivegna, V., Lombardi, P. (eds.), *Evaluation of Sustainability in the Built Environment,* Chapman&Hall, London.

Merret, S. (1995) Planning in the age of sustainability, *Scandinavia Housing & Planning Research,* 12, 5-16.

Mitchell, G., May, A.D. and McDonald, A. (1995) PICABUE: a methodological framework for the development of indicators of sustainable development, *International Journal Sustainable Development World Ecology,* 2, 104-123.

OECD (1994) Report on *Environmental Indicators,* OECD, Paris.

Parelman, C. (1982) *The Realm of Rhtorics,* University of Notre Dame, London.

Pearce, D., Markandya, A. and Barbier, E.B. (1989) *Blueprint for a green economy,* Earthscan Publications Ltd, London.

Polanyi, M. (1967) *The Tacit Dimension,* Routledge and Kogan Paul, London.

Prior, J. (ed.) (1993), *Building Research Establishment Environment Assessment Method, BREEAM, Version 1/93, New Offices,* Building Research Establishment Report, Second Edition.

Ravetz, J.R. (1971) *Scientific Knowledge and its Social Problems,* Oxford, Oxford University Press.

Simon, H. (1975). Theories of Bounded Rationality, in McGuire, C.B. and Radmer, R. (eds.) *Decision and Organisation,* Amsterdam, North-Holland.

United Nations Conference on Environment and Development, (1992). *Earth Summit '92 (Agenda 21),* Regency Press, London.

United Nations Conference on Human settlements (1996) *The indicators programme: monitoring human settlements for the global plan of action*, paper - United Nations Conference on Human Settlement (Habitat II), Instanbul.

Voogd, H. (1997). The communicative ideology and x ant planning evaluation, Third International Workshop on evaluation in Theory and Practice, London, September.

Voogd, H. (1995). Environmental management of social dilemmas. *European Spatial Research and Policy,* **2,** 5-16.

WCED (World Commission on Environment and Development) (1987). *Our Common Future (The Brundtland Report)*, Oxford, Oxford University Press.

Zeppetella, A. (1996) *Retorica per l'ambiente*, Franco Angeli, Milano.

CULTURAL HERITAGE AND URBAN REVITALIZATION: A META-ANALYTIC APPROACH TO URBAN SUSTAINABILITY

F. BIZZARRO & P. NIJKAMP

Abstract

This paper will address the issue of cultural heritage and urban revitalization based on a comparative meta-analytic approach regarding European urban policy experiences.

Cultural heritage of historical urban centres assumes a central role in sustainability strategies, focussing on a long-term (past and future) perspective in order to recognize the variety of values, objectives, interest groups and aiming at socio-economic and cultural development, conflict resolution and inter-generational equity goals.

From a policy-analytical perspective, it is important to focus on several issues of strategic importance, in particular: (i) linkages between conservation and socio-economic opportunities from a sustainability perspective, (ii) strategies for building a multi-faceted portfolio of policy options for cultural built heritage (CBH), (iii) financial and managerial mechanisms stimulating incentives for public-private partnership (e.g., a leverage perspective), and (iv) conflict management via the application of integrative bargaining principles.

Following a meta-analytic sustainability approach, we will in this paper try to identify common elements, success factors and impediments in prevailing policies, practical strategies and policy measures regarding urban conservation plans. This will be done by comparing, through the application of rough set method, interesting European CBH policy cases in order to identify political, social, physical, cultural and economic conditions that generated the success of a sustainable revitalization policy.

1. Integrated Conservation of Cultural Built Heritage in the Context of Urban Sustainability

The Cultural Built Environment is an asset of both historical and socio-economic importance in any society. It serves as an identification and orientation point for all individuals and communities. Conservation of the Cultural Built Heritage (CBH) should therefore be based on resource-efficient principles in order to properly respond to the contemporary demands on scarce historical resources. Heritage Conservation is thus a

N. Lichfield et al. (eds.), Evaluation in Planning, 193–211.

fundamental strategy through which groups of different socio-cultural backgrounds are able to preserve their socio-cultural identity and to identify the built environment as the physical manifestation of the meaning of the past. CBH, therefore, should be preserved for its externality value for the wider community (Hubbard, 1993). Moreover, urban conservation has to be seen as a valuable policy instrument in helping to fulfil economic objectives and, at the same time, to meet social and cultural needs.

In recent years, the contribution which urban conservation can make to urban regeneration has increasingly been recognised. Conservation is a special case of urban renewal. The notion of urban renewal includes essentially both the awareness of urban decay (in the local economy, in the use of land and the built environment, in the quality of the environment and of social life) and the search for a new basis in order to contribute to the socio-economic welfare of the community at large (Lichfield, 1996). It is, therefore, necessary to emphasize the strict conditioning relationship that exists between the revitalization of the urban CBH and the role of economics.

Economics can certainly provide significant arguments in favour of the conservation of historical resources, especially in the case of limited resource availability. It concerns also the management of scarce resources in order to satisfy contemporary and future human needs, while recognizing that we are often faced with resources that are unique and cannot be reproduced (because of irreversibilities).

Urban revitalization, therefore, has to be seen as an economic process with physical, economic, environmental and cultural effects (direct, indirect, induced) at a local, regional and national level. Conservation of CBH can hence yield monetary and non-monetary returns that are strictly related to urban economic development.

It is, moreover, possible to distinguish "upstream" and "downstream" socio-economic effects of urban CBH revitalization (for more details see Lichfield, 1996): different types of employment and income multiplier effects, tourism impacts, a stimulus to accelerated development (e.g., support for the commercial business sector) and regional growth, sustainable urban evolution, private sector stimulation, resource efficiency, available financing, etc.

From the above general considerations, we may infer that urban revitalization can be achieved only if it is envisaged in a wider policy context. We are referring here in particular to the concept of Integrated Conservation of CBH, including plan-making, financial planning, plan implementation , monitoring and evaluation processes. In this perspective, it is important to focus on several issues of strategic importance, notably: (i) linkages between conservation and socio-economic opportunities from a sustainability perspective; (ii) strategies for building a multi-faceted portfolio of policy options for CBH; (iii) financial and managerial mechanisms generating incentives for public-private partnership (e.g., a leverage perspective), and (iv) conflict management through the application of a proper bargaining theory. These issues will successivelly be discussed in subsequent sections.

2.   Conservation and Socio-Economic Opportunities of Cultural Built Heritage in a Sustainability Context

It has been argued above that urban CBH revitalization may provide a significant contribution to socio-economic welfare of the community at both a local, regional and national level. This implies also that the most successful revitalisation policy will be the one that best balances efficiency (maximization of total net benefits) and equity (acceptable distribution of costs and benefits) for the community.

From this point of view, it is possible to integrate the conservation of urban CBH in a broader view on sustainable development, by considering at the same time social, economic and environmental dimensions of urban revitalization. A *sustainable conservation* is essentially able to meet the needs of the present generation by generating both sufficient support for the local business and service sector and greater social equity, without compromising the future generations' access to such historico-cultural resources.

Some authors argue that sustainable development and conservation are necessary and mutually reinforcing goals (Tisdell, 1991). In fact, support for urban conservation can be pursued only in the wider policy context of sustainable development, by focusing on the maintenance of intergenerational economic welfare and on the sustainability of both production and economic systems and of the community at large. We aim to show in this paper that historical urban centres can be seen as strategic areas and starting points for development strategies, from both an environmental and a socio-economic perspective.

Urban centres represent essentially driving forces of economic, cultural and social development. From this point of view, it is plausible that revitalization of urban centres, in the context of a sustainable urban development, is closely linked to regional development. This implies the need for integrative strategies that are able to ensure the efficiency of micro-economic urban management, to reduce negative externalities on the environment ("precautionary principle") and to provide for an equal redistribution of resources and outcomes that cannot be supplied by private markets. Of course, these strategies have to be related to macro-economic policies due the fact that a sustainable urban conservation strategy requires a strong coordination and control at the national level, in particular if an active policy implementation can only be pursued at the local (urban or regional) level.

Seen from this perspective, the need for a coordination of efforts in different sectors and at different institutional levels has to be stressed. Faced with the actual market incapability in ensuring sustainability, the government may need to intervene in the context of integrated conservation of the urban CBH (for more details see Bizzarro and Nijkamp, 1996). For example, economic gains that can be achieved via a tourism policy based on historical urban resources represent a powerful force favouring conservation, but in the absence of a co-ordination-oriented planning by governments, it is not possible to promote an effective implementation of the conservation of the CBH.

One of the main reasons for urban system's degradation is the fact that the importance of historico-cultural resources is not yet fully reflected in economic planning and decision-making. Local planning processes should, in effect, be arranged in ways that stimulate a global awareness on the *social complex value* (Fusco Girard, 1993) of our historico-cultural resources and on implications of a sustainable conservation for the present and future generations.

Moreover, in order to pursue successful revitalization programmes, it is essential to look at financing (strategies for public-private partnerships, based e.g. on a leverage perspective) and co-ordination (based e.g. on a bargaining theory for conflict management) criteria in the framework of a broader urban planning process.

3.   A Portfolio of Policy Options for Integrated Conservation of Cultural Built Heritage

In many cases a combination of historico-cultural resources and efficient urban plans, including both all actors involved and the underlying economic driving forces, may create favourable conditions to achieve conservation objectives aiming at both efficiency and equity goals. In principle, it is possible to compose a *portfolio of conservation policy options* to develop strategic linkages between different kinds of local plans whose impacts and spin-offs can have a much broader coverage than a single conservation project. Especially the synergy and value added of a portfolio can positively influence the availability and effective willingness of the private sector to be actively involved in such conservation and revitalization processes. This idea can also be derived from the statement that "conservation objectives developed for specific areas may have important consequences for the management of change at the wider strategic level" (Pearce, 1994).

Following a portfolio perspective, we need to look at a wide range of complementary urban policy initiatives, by considering all financial and institutional implications involved. Thus, facing the limited budget of public funds in conservation policies as well as the conflicts existing between all agents concerned, it becomes necessary that governments develop supportive actions that are able to encourage a public-private involvement, taking for granted a bargaining theory aimed to manage these kinds of conflict. In this perspective, it seems that a portfolio of conservation policy options may be a powerful strategy in order to achieve the above goals, as in a portfolio there is much more scope for tradeoffs between different options.

By referring to a portfolio of conservation policy options, we mean that this approach is able to promote a sustainable conservation and urban revitalization policy within a strategic and rational framework. In such a strategy it becomes fundamental to address all values, objectives, actors involved, available means, constraints (physical, institutional and financial), and different kinds of impact, and also to define their priority and the implications of each investment in a well structured long-term perspective. In fact, most of the local problems can often only be solved by action on a wider geographical scale (Johnson, 1996).

We need, therefore, to move from a concern for individual monuments and historical centres to a cohesive focus on group values and on the wider political, socio-economic and cultural environment. In fact, if it is possible to link many urban revitalization (or CBH) initiatives, beneficial results in the whole urban context in which these assets are located may be expected.

It is plausible that good management of a portfolio of conservation policies seems to be a much more convincing mechanism in providing sufficient (and also less risky) economic returns that enables the private sector to invest in it. A portfolio of conservation policy options involves thus powerful successful factors, especially in a sustainability perspective. We may refer here to the mobilization of resources, inter-institutional coordination and cooperation, the enhancement of urban productivity, the supply of information and a more active protection of the urban environment.

The opportunity to choose the best achievable project among a range of different possibilities at a proper time and place constitutes one of the main advantages in a portfolio of conservation policy options. It concerns, in fact, a dynamic and cyclical process that allows to put off a conservation policy to a more convenient time, without dropping the chance of pursuing it in the future. Moreover, a portfolio system, characterized by a high degree of transparency and accessibility, guarantees major freedom and certainty for the private sector being involved in a conservation policy due the fact that, faced with a plurality of options, actors are able to evaluate all relevant implications (viz. costs, benefits, risks, constraints, etc) of all alternatives from their own individual interest.

We have to emphasize here the need for a strategic approach to managing the CBH and urban centres via a co-ordinated long term investment strategy by the public and the private sector. In this strategy however, it is possible to focus indifferently on single historico-cultural assets or urban centres; difference in size or value does not matter then. Then the aim may be to stimulate an incremental series of project implementations within a well structured portfolio of conservation policy options.

In fact, only by understanding and managing the indigenous problems of adopting strategic conservation policies, which have different implications and impacts on individuals and communities, will it be possible to achieve conflict reduction, or more in general, a reduction of various types of conflict: inter-actor, inter-regional, inter-temporal and intra-person (Button and Nijkamp, 1995).

Finally, we may conclude that in the perspective of a portfolio of conservation policy options, these various kinds of information transmission are present in all phases of the process. A rational and coherent information system at the local, regional and national level may be viewed as a guarantee of major transparency for the private sector in order to be involved in the portfolio implementation.

4.   A "Leverage" Perspective in the Context of a CBH Portfolio

Through a portfolio perspective it is possible to encourage the private sector propensity to be involved in integrated conservation of the urban environment in the wider context of a strategic programme, which is able to guarantee both individual and collective goals achievement. We may argue that each single policy option tends to be disjoint and likely unfeasible if it is not mapped in a broader strategic policy focused on a systemic and inter-active rationality. A network of policy options, based on complementarity and mutuality principles, has therefore to be connected with a management system, within which it is possible to select a ranking of priorities between different policy alternatives' implementation, by considering a proper time scale and common criteria. The latter issue, viz. the implementation of conservation policies, depends on both funding mechanisms and cooperative/competitive relationships between public and private sectors (Fusco Girard and Nijkamp, 1997). In effect, in order to conduct an urban revitalization policy, the portfolio must be the fruit of a comprehensive agreement between all agents involved in the implementation and management of relevant conservation policy options.

Faced with the limited availability of public funds earmarked for urban revitalization projects, we can easily conclude that the recourse to public financing represents no more than a short-term solution. This calls therefore for: (i) a more active involvement of private sectors in revitalization programmes (for more details, see Bizzarro and Nijkamp, 1996), (ii) a long-term perspectvive, and then (iii) a joint (public and private) financial body capable of supporting the implementation of the portfolio of conservation policy options. These features may be identified in a portfolio context in order to identify a win-win situation in a public-private partnership, viz. the likelihood to move from a zero-sum game to a positive sum game (Fusco Girard, 1995).

CBH conservation and urban revitalization imply diversified and often conflicting interests between public and private agents, aimed to achieve for some agents the maximization of the welfare of the community, and for others the maximization of real estate values and income opportunities. Mostly, renewal processes, within an institutional context based on private property, involve land-real estate values whose costs fall upon the community and whose benefits are appropriated by the private sector. This calls for the need both to reconsider the market role and to identify new rules aimed to alleviate the above mentioned mechanism of income privatization.

Regarding the possibility of identifying new relationships between public and private agents, it is possible to consider public-private partnerships as a crucial instrument in mobilising the financial resources and also as a more flexible way to operate in the context of both planning and implementation of the conservation process (European Commission, 1995).

In many practical cases, the partnerships that have been created in the context of urban planning were based on a synergetic relationship between public and private interests (O'Toole and Usher, 1992). Clearly, we have observed in the past years a rapid evolution in the theoretical and empirical approach to the partnership structure. Indeed, public private partnerships depend on different kinds of institutional arrangement and relationship based also on the legal, political and financial features of the external context (local, regional and national) in which they operate. Also the types of property (private, public or common) can be regarded as a determining factor for integrated conservation of the built environment.

Anyway, faced with the complexity of urban revitalization issues, it is important in a portfolio perspective to formulate a management strategy within public-private partnerships by starting to recognize proper tasks and clear responsibilities of all agents concerned.

The public sector, as promoter and actor of urban revitalization, has a specific responsibility in terms of direction, planning, control, concern for general and public interests, but also in terms of financing part of the conservation process. The main objectives of the public investments are, therefore, efficiency, equity and distribution in promoting public and social utility. With regards to financial aspects, Lichfield (1996) argued that it is necessary to resort to public resources in order to provide above all public goods and different kinds of subsidy and incentive to private agents, with the aim to render the conservation process economically less unattractive to them.

The recourse to private resources may allow the public sector to relieve its economic burdens and, at the same time, to increase financial revenues. The main objectives of the private sector, on the other hand, tend to be focused on maximization of real estate values, income privatization and promotion of higly commercial and speculative activities in architectural heritage (re)development. If we assume the above portfolio perspective as a well structured long-term profitability strategy, the private sector has to be responsible for promotion, implementation and management of the overall revitalization process, including in its objectives also a greater consideration of collective needs.

A public-private partnership has then to be based on a consensus-building of socio-economic values that are able to promote income socialization by aiming to pursue simultaneously effectiveness, efficiency and equity goals (Forte, 1995).

Besides, a public-private partnership may be widened by resorting to the involvement of other potential actors, viz. public and private institutions, non-profit conservation agencies (the third force), banks, development and construction industries, and marketing agencies. In this perspective the regulatory control policy carried out by the government may be considered as a high priority (Bizzarro and Nijkamp, 1996). This calls for the definition of new rules also in the field of cultural heritage conservation and urban revitalization, which - by overcoming the perspective of a mere real-estate value-added agency - are able to guarantee a leverage between conflicting interests and objectives.

We may conclude that, via a leverage approach, it is thus possible to guarantee the transparency of implementation and management of the portfolio of conservation policy options, and to identify mechanisms and strategies for promoting private-public partnerships by resorting to a bargaining framework, that is able to reconcile interests between all agents involved.

5.   An Integrative Bargaining Theory for Successful Conflict Management

The portfolio of policy options as the leverage perspective that have been examined in the previous sections call for a *negotiation* between public and private institutions and social and economic agents involved in the integrated conservation of the urban environment. This kind of negotiation however, has to rely on a new organization of government institutions, which implies a rational redistribution of tasks and responsibilities among governmental bodies on all levels, by identifying stable and clear rules, transparent mechanisms and support systems especially for local bodies.

One of the main targets of government institutions concerns the equilibrium in the urban market by resolving conflicting interests among all interconnected bargainers involved in the portfolio of urban revitalization policies.

From this point of view, it is essential to distinguish a set of coordination mechanisms or decision rules that are feasible, efficient and equitable for all the players involved in the cooperation process (Myerson, 1988). Indeed, notions of feasibility, efficiency and distribution play a crucial role in the perspective of urban sustainability. All these aspects, including also cultural and fairness (viz., who wins, who loses) considerations, allow to make comparisons of the strengths and weaknesses of each alternative comprised in the portfolio of conservation policies, but imply also differentiated forms of conflict. Conflicts may occur in various ways: (i) competition for scarce resources, (ii) competence on control between various management levels, and (iii) discrepancies in lateral relationships concerning work coordination and task integration (Lewicki and Spencer, 1992). In a portfolio perspective, therefore, a realistic policy to pursue a win-win situation in tackling conflict management incorporates cooperation and mutual arrangements between all actors concerned.

In the above context, it is possible to adopt three of the five conflict handling modes (competing, collaborating, compromising, avoiding and accomodating) classified by Thomas (1992): competition, compromise and collaboration. The first one, of course, represents the precondition to involve all potential actors in the urban revitalization; the compromise behavior, instead, can be viewed as an intermediate solution between a cooperative and an assertive approach, which makes due allowances for reciprocal concessions of each part; the cooperation, finally, may be able to determine the success of the portfolio implementation. Following a long-term perspective via a balanced negotiation process, it is possible to introduce new dimensions in conflict management, by trying to identify additional dimensions (i.e., qualitative and non-monetary) or alternatives that are able to overcome a rigid position for both public and private sectors.

The main purpose of conflict management lies in facilitating agreements on more favourable terms or at a higher pace. Agreement-making in negotiations reflects the need for moving from the market (real-estate income) to a negotiation form, based on clear and transparent rules, which is able to reduce real and potential interest conflicts among bargainers involved in an integrated conservation of urban heritage. Besides, a negotiation process, taken as a step in the portfolio implementation, aims to pursue not a zero-sum situation but a win-win solution.

From this point of view, we may argue that a portfolio of policy options calls for an *integrative bargaining*, or negotiation, rather than for a distributive bargaining. Whilst the latter occurs when each party tries to maximize its share of fixed-sum payoffs (Lewicki, Weiss and Lewin, 1992), determining a winner and a loser, the integrative bargaining, instead, involves a joint cooperation in order to enlarge the pie to be divided eventually by each party. All parties can gain a benefit by looking at compatibility in interests, while new possible solutions can be envisaged. This kind of bargaining can be creative and problem-solving, but implies, of course, the need for more information (see Section 3) related to a democratic and communicative participation.

6.   The Usefulness of Meta-Analysis in the Context of a CBH Portfolio

In the previous sections we have tried to investigate the nature of typical aspects and new perspectives that may arise by implementing a portfolio of conservation policy options. Our main goal is now to improve our understanding and capability in performing measures and strategies for management and conservation in urban policies in the above portfolio context. This will be done by comparing, via a meta-analytic approach, interesting European case studies in order to identify political, social, physical, cultural and economic conditions that allowed the success of a sustainable revitalization policy. These comparative studies will mainly be used as a learning model.

Meta-Analysis allows, through the application of systematic statistical methods, to pull together many empirical studies in order both to extract and organize additional information from previous results and to focus on common elements, success factors and impediments in such conservation policies. Meta-analysis will permit us to combine, compare, analyze and validate quantitative results from these studies focussing on similar phenomena under varying conditions (Nijkamp, 1996).

Notwithstanding the fact that the idea of meta-analysis has already been introduced by Glass in 1976 and has mostly been employed in social psychology and in medical and natural sciences, it has, until now, found mainly (limited) application in the context of micro-economic research (labour economics, industrial efficiency, environmental economics and transport economics). Nevertheless, we may expect that the application of meta-analysis to the sector of conservation economics will likely offer a valid contribution to guaranteeing transparency in prior CBH judgments, by either reducing the level of subjectivity in existing analyses or by providing the necessary conditions and appropriate tools to develop suitable frameworks for evaluation and assessment in the CBH field.

Rosenthal (1991) has argued that the main efforts in early meta-analytical approaches were focussed on summarising, looking at common threads and obtaining insights into aggregate links. Nowadays, several improvements can be found in this field, leading to the identification of operational possibilities offered by the application of modern meta-analysis. In our particular context, by starting from the recognition of similar case studies on urban conservation policies, we may be able:

- to compare, to evaluate, to aggregate and to rank these various studies according to specific features, criteria, indicators, relationships and practical outcomes;

- to average the prior values and to validate the reliability of the findings of the existing analyses, by identifying common elements in such studies and by linking the aggregated data of each single case to other elements of the analysis;

- to verify the validity and usefulness of specific evaluation methods applied in this particular context, allowing to promote scientific consensus on such evaluations;

- to focus on key factors in the implementation of the conservation policy (i.e, successful factors and impediments) aiming, according to the experience acquired, to propose new potential research perspectives in a sustainable CBH portfolio context.

The important issue of transversality, both horizontally and vertically, plays an important role in meta-analysis (Matarazzo and Nijkamp, 1995). We may refer here, in the first case, to the capability to identify, to select and to analyze a specific set of existing case studies by resorting to suitable meta-analytical techniques. Next, we may refer, for the vertical case, to the phase within which it is possible to move from the methodological definition of the specific issue and of the aim of the analysis carried out, to the possibility to practically apply in a policy context the findings that have been achieved.

Of course, the recourse to meta-analysis calls also for a recognition of several restraints that can seriously prejudice the outcome of an entire evaluation approach. It is, ideally, important in dealing with meta-analytic techniques to favour: (i) objectivity in collecting information in order to reduce the level of subjectivity, which characterizes the typical process of synthesizing previous studies; (ii) incorporation in the analysis of a wide spectrum of data, viz. both positive and negative results, by examining not only relevant previously published studies, but also unpublished research or technical reports undertaken in the urban conservation sector; (iii) avoidance of scientific, ethical and political bias in the available body of information; and, finally, (iv) a comparison of economic impacts when qualitative factors (e.g., environmental, ethical, historico-cultural and symbolic values) and different measurement scales play a role.

From this point of view, the choice of the operational meta-analytic approach to be adopted depends on the aim of the study; this can thus influence the outcome of the evaluation. It is possible, therefore, to identify, according to the major analytic processes of comparing and combining, different kinds of meta-analytical procedures (Van den Bergh, Button and Jongma, 1995).

7.   Rough Set Analysis

Our aim now is to explore the relevance of a rough set approach in a meta-analysis of urban revitalization studies. We will try, therefore, in this section to briefly explain the key principles of rough set theory, referring to more detailed literature on the theoretical and mathematical aspects underlying the method (see also Pawlak 1991, Slowinski 1995, Nijkamp 1996).

Rough set analysis enables us: (i) to single out the set of data (both qualitative and quantitative) and information (consistent, partial and inconsistent) available in the selected case studies, and also (ii) to find out the (ir)relevance and the potential relationships (not always observable) between these data, with the aim to create a suitable decision-making and selection framework in the context of our CBH portfolio. Rough set analysis may be viewed as a useful tool in dealing with the complexity and multidimensionality in sorting and ranking of actions or strategies coping with vagueness and uncertainty-dominated problems of choice.

The principal aim of this approach consists of both the classification of available information in classes of attributes or categories and the discovery of possible cause-effect relationships from a set of data in order to pursue a more structured and precise knowledge (Nijkamp, 1996). Rough set analysis, therefore, is able to face two particular issues of the decision-making context: (i) *explanation* (i.e., the capability to point out the critical aspects of the problem and the correlations between the data, by resorting to a set of conditional attributes and an information table) and (ii) *prescription* (i.e., the capability to evaluate the information available in a decision table, aiming to provide a comprehensive preference model as a support in the decisional and negotiation process).

The starting point of rough set analysis concerns thus the possibility to catch the differences in a set of objects U (*universe* of the discourse) and to characterize these objects by a finite set Q of *attributes* (each one is able to assume different values), in order to define distinct *equivalence classes* (or partition), within which the associated objects are *indiscernible*. Besides, the *similarity relationship*, particularly helpful in the case of quantitative attributes, allows to reduce subjective subintervals. The notions of indiscernibility and equivalence classes concur than in introducing the basic concepts of *approximation, reduction* and *dependency*.

Facing with the difficulty to represent a set X due to the available imperfect information level, it is possible to express X by resorting to other two sets, called the *lower approximation* (i.e., the maximal set of objects surely belonging to X) and the *upper approximation* (i.e., the minimal set of objects possibly belonging to X).

Therefore, a *rough set* may be defined as the family of all subsets of U, characterized by the same lower and upper approximations.

Whilst the idea of approximation allows to deal in a more accurate way with the imprecise notions of knowledge, the reduction principle concerns the possibility to eliminate all superfluous attributes in the knowledge base in order to preserve only the really significant family of elementary sets. It permits, indeed, to investigate whether all condition attributes are necessary to make decisions. This principle leads us to introduce the concepts of *reduct* (the essential part of knowledge to define all basic concepts) and *core* (the most important part of knowledge, i.e., the intersection of all reducts).

Finally, the notion of dependency, strictly related to the previous notion of reduction, refers to the relationships between basic categories in the knowledge base and implies the possibility to avails of known values of some attributes in order to express other attribute-values.

Another basic principle of rough set analysis concerns the deduction of *decision rules* (i.e., a conditional statement "if ... then") in the context of an information table, characterized by *condition* C and *decision* D *attributes*. This table may be considered as a valid support in the decision-making process, due to the fact that it outlines the kind of decisions or actions that should be undertaken when some conditions are satisfied (Pawlak, 1991). The values of condition attributes, therefore, describe the objects, while the values of decision attrbutes represent the partition of U by an expert into distinct classes, i.e. his subjective classification. The indiscernibility relations Ic and Id, generated by condition and decision attributes, involve a *decision table*, within it is possible to distinguish condition classes Xi (C-elementary sets) and decision classes Yj (D-indiscernible objects). Any row in the decision table may be interpreted as a decision rule, i.e. an implication relationship between the description of the condition class Xi and the description of the decision class Yj. However, a decision rule may be *exact* or *approximate*. In the first case it is possible to state that the decisions are uniquely determined by the conditions; an approximate decision rule, instead, is non-deterministic in individualizing the values of decision attributes which correspond to the same values of the condition attributes.

The main advantage of such rules is the validation and the better understanding of the information contained in the original decision table and the possibility to enrich this table with additional rules supplied by experts, improving the knowledge level. We may conclude, therefore, that decision rules represent the most important feature of rough set analysis. This is expecially true in a decision making process, wherein the availability of results from the analysis of known case studies may offer a valid support in tackling new problems of a choice or bargaining situation.

We will now apply this approach to a comparative analysis of 10 case studies about urban revitalization and conservation issues, aiming to identify feasible strategies which may be useful in the perspective of our CBH portfolio.

TABLE 1. Overview of selected case studies on sustainable revitalization policy

Study ID (authors, years)	Country (geographical scale)	Type of documentation	Aim of the study	Evaluation method	Interest groups	Sustainability criteria	Success factors	Policy measures	Impediments
1) Guarino 1989, F. Angeli, (Italy)	Napoli (Italy) local level	D1	A1	M1		C1 C2 C3	S1 S2 S4	P2	I2 I4
2) Ferretti 1989, F. Angeli, (Italy)	Procida (Italy) local level	D1	A2	M1		C1 C2 C3	S1 S2 S4	P1	I1 I4
3) Nijkamp, Artuso, 1995, Free Univ., Amsterdam	Bassano del Grappa (Italy) local level	D2	A1	M1	G2 G3 G4	C1 C2 C3	S1 S2	P1 P2 P3	I3 I4
4) Nijkamp, Bithas, 1995, Avebury, (U.K.)	Olympia (Greece) regional level	D1	A3	M1 M2	G4	C1 C2 C3	S1 S2 S3 S4	P1	I1 I2 I3 I4
5) Fusco Girard, Giordano, 1995, Avebury, (U.K.)	Maratea (Italy) local level	D1	A2	M1 M2 M3 M4	G1 G2 G3 G4	C1 C2 C3	S1 S2 S4	P1 P3	I1 I4
6) Coccossis, Mexa, 1995, Avebury, (U.K.)	Rhodes (Greece) local level	D1	A3	M2	G4	C1 C2 C3	S1 S2 S3	P1 P2 P3	I1 I2 I3 I4
7) Nijkamp, 1987, Free University, Amsterdam	Thessaloniki (Greece) local level	D2	A2	M1		C1 C2	S1 S3	P1	I1 I3 I4
8) Massi-mo, 1995, Avebury, (U.K.)	Lamezia Terme (Italy) regional level	D1	A2	M1	G4	C1 C2	S1 S3	P1	I1 I4
9) Ferretti, Bizzarro, 1997, F. Angeli, (Italy)	Napoli (Italy) local level	D1	A1	M1	G1 G2 G3 G4	C1 C2 C3	S1 S2	P1 P3	I2
10) Grittani, 1993, F.Angeli, (Italy)	Akamas (Cyprus) national level	D1	A3	M3	G1 G4	C1 C3	S2	P1	I2 I4

Legend.

Case Studies (titles)
1) La valutazione di progetti alternativi secondo il metodo di regime: il caso dei Quartieri Spagnoli a Napoli
2) Applicazione dell'analisi di frequenza ai piani di recupero urbano: il caso di Procida
3) Methodology and Application of Sustainable Environment Concepts for the Built Environment: a Case Study on Bassano del Grappa
4) Scenarios for Sustainable Cultural Heritage Planning: a Case Study of Olympia
5) Sustainable Conservation of the "Castello di Maratea"
6) Tourism and the Conservation of Heritage: the Medieval Town of Rhodes
7) An Application of the Regime Method to the Evaluation of Urban Monuments in Thessaloniki
8) Heritage Conservation Economics: a Case Study from Italy - Lamezia Terme
9) La scelta della destinazione d'uso per la riqualificazione di un'insula nel centro storico di Napoli
10) La penisola di Akamas

Study ID	authors, year, editor
Type of documentation	book (D1)
	research paper (D2)
Aim of the evaluation	identification of one feasible alternative (A1)
	priority ranking of alternatives (A2)
	support to the decision-making process (A3)
Evaluation methods	multicriteria methods (M1)
	impact evaluation (M2)
	direct (CVM) (M3)
	indirect (TCM, HP) (M4)
Interest groups	users (G1)
	public sector (G2)
	private sector (G3)
	government and public/private partnerships (G4)
Sustainability criteria	economic (C1)
	socio-cultural (C2)
	historico-environmental (C3)
Success factors	multidimensional approach (S1)
	conservation/socio-economic development (S2)
	long term perspective (S3)
	infrastructure policy (S4)
Policy measures	equity considerations (P1)
	management measures (P2)
	democracy and cooperation (i.e., public-private initiative) (P3)
Impediments	limited financial resources (I1)
	deficiency of local government and lack of integrated strategies (I2)
	physical-socio-economic deterioration (I3)
	externalities (I4)

8.   The application of Meta-Analysis on Sustainable Revitalization Policy

As mentioned above, following a meta-analytic approach, we will in this section try to compare, through the application of rough set method, 10 interesting European CBH policy cases in order to create a suitable decision-making framework in the context of our CBH portfolio.

In this perspective, we have firstly selected a sample of 10 homogenous case studies on sustainable revitalization policy (Table 1), focussing on a set of relevant qualitative attributes, clearly identified in the previous pages.

We have then coded these attributes into proper classes, shown in the following decision table (Table 2), with the aim to find the associated minimal decision algorithms. In this particular case we have obtained a dedcision algorithm with a consistency degree 1, but also a large number of reducts (29) due to the lack of balance between attributes and case studies.

TABLE 2 Decision table

Cases	A1	A2	A3	A4	A5	A6	A7	A8	A9	A10	d
						Attributes					
1	1	1	1	1	1	0	1	2	4	5	1
2	1	1	1	2	1	0	1	2	3	1	2
3	1	1	2	1	1	2	1	4	1	2	2
4	2	2	1	3	4	4	1	1	3	4	3
5	1	1	1	2	5	1	1	2	2	4	1
6	2	1	1	3	2	4	1	3	1	1	3
7	2	1	2	2	1	0	2	5	3	3	1
8	1	2	1	2	1	4	2	5	3	4	3
9	1	1	1	1	1	1	1	4	2	7	1
10	3	3	1	3	3	3	3	6	3	6	3

Condition attributes

A1: Country
Classes: 1   Italy
        2   Greece
        3   Cyprus
A2: Geographical scale
Classes: 1   local level
        2   regional level
        3   national level
A3: Type of documentation
Classes: 1   book (D1)
        2   research paper (D2)
A4: Aim of the study
Classes: 1   best alternative (A1)
        2   priority ranking (A2)
        3   decision-making support (A3)
A5: Evaluation method
Classes: 1   multicriteria methods (M1)
        2   impact evaluation (M2)
        3   CVM (M3)
        4   (M1) (M2)
        5   (M1) (M2) (M3) (M4)
A6: Interest groups
Classes: 0   non considered
        1   (G1) (G2) (G3) (G4)
        2   (G2) (G3) (G4)
        3   (G1) (G4)
        4   (G4)
A7: Sustainability criteria
Classes: 1   (C1) (C2) (C3)
        2   (C1) (C2)
        3   (C1) (C3)

A8: Success factors
Classes: 1   (S1) (S2) (S3) (S4)
        2   (S1) (S2) (S4)
        3   (S1) (S2) (S3)
        4   (S1) (S2)
        5   (S1) (S3)
        6   (S2)
A9: Policy measures
Classes: 1   (P1) (P2) (P3)
        2   (P1) (P3)
        3   (P1)
        4   (P2)
A10: Impediments
Classes: 1   (I1) (I2) (I3) (I4)
        2   (I1) (I2) (I4)
        3   (I1) (I3) (I4)
        4   (I1) (I4)
        5   (I3) (I4)
        6   (I2) (I4)
        7   (I2)

Decision attributes

d: **Successful strategy for a sustainable urban**
    revitalization
Classes:
1   "social complex value" maximization
2   policy for a balanced management of CBH
3   impact evaluation related to the broader
     socio-economic development

Accuracy of classification: 1.0000
Quality of classification: 1.0000

Reducts (or minimal sets) of study characteristics: 29

Frequency of attributes in reducts :
A1: Country   7 (24%)
A2: Geographical scale   8 (27%)
A3: Type of documentation   9 (31%)
A4: Aim of the study   11 (40%)

A6: Interest groups   9 (31%)
A7: Sustainability criteria   10 (34%)
A8: Success factors   7 (24%)
A9: Policy measures   15 (51%)

A5: Evaluation method	6 (20%)	A10: Impediments	8 (27%)

TABLE 3 Implementation of the analysis

Round 1	Round 2	Round 3

A2: Geographical scale
Classes: 1   local level
     2   regional/ national level
A5: Evaluation Method
Classes: 1   multicriteria methods
     2   other methods
A6: Interest Groups
Classes: 1   non considered
     2   considered
A8: Success Factors
(S1, S3 = S1; S2, S4 = S2)
Classes: 1   (S1) (S2)
     2   (S1)
     3   (S2)
A9: Policy Measures
Classes: 1   (P1) (P2) (P3)
     2   (P1) (P3)
     3   (P1)
     4   (P2)
A10: Impediments
(I3, I4 = I3)
Classes: 1   (I1) (I2) (I3)
     2   (I1) (I3)
     3   (I2) (I3)
     4   (I3) or (I2)
	A5: Evaluation Method
	Classes: 1 multicriteria methods
	     2 other methods
	A6: Interest Groups
	Classes: 1 non considered
	     2 considered
	A8: Success Factors
	(S1, S3 = S1; S2, S4 = S2)
	Classes: 1 (S1) (S2)
	     2 (S1)
	     3 (S2)
	A9: Policy Measures
	Classes: 1 (P1) (P2) (P3)
	     2 (P1) (P3)
	     3 (P1)
	     4 (P2)
	A10: Impediments
	(I3, I4 = I3)
	Classes: 1 all 3 considered
	     2 only 2 considered
	     3 only 1 considered
		A6: Interest Groups
		Classes: 1 non considered
		     2 considered
		A8: Success Factors
		(S1, S3 = S1; S2, S4 = S2)
		Classes: 1 (S1) (S2)
		     2 (S1)
		     3 (S2)
		A9: Policy Measures
		(P2, P3 = P2)
		Classes: 1 (P1) (P2)
		     2 (P1)
		     3 (P2)
		A10: Impediments
		(I3, I4 = I3)
		Classes: 1 all 3 considered
		     2 only 2 considered
		     3 only 1 considered

Reducts (or minimal sets) of study characteristics		
1. {5 6 9 10} 2. {2 5 10} 3. {5 6 8 9} 4. {2 5 8 9}	1. {5 9 10} 2. {5 8 9}	1. {6 9 10}
Core set : A5	Core set : A5, A9	Core set : A6, A9, A10
Accuracy of classification: 1.0000 Quality of classification: 1.0000	Accuracy of classification: 1.0000 Quality of classification: 1.0000	Accuracy of classification:0.6667 Quality of classification: 0.8000

We have, therefore, checked which values of condition attributes were indispensable in order to discern the values of the decision attributes, by relying both on our personal experience and by looking at the frequency of attributes in the reducts. We have decided, in effect, to focus our analysis on critical key factors of a CBH portfolio (geographical scale, valuation, interest groups, success factors, policy measures and impediments), assuming that sustainability criteria were the preconditions for the success of a sustainable revitalization policy.

Moreover, considering the large number of reducts in the first tentative analysis and the limited availability of case studies, we have introduced also some changes in the original classification.

We have therefore carried out three different runnings of the value codes in the rough set analysis (Table 3), by leaving out of consideration in the second and third round the geographical scale attribute, because we are mainly interested, in a portfolio perspective, in a broader context.

In the last round of analysis we have, instead, omitted the evaluation attribute (A5), even though it belonged to the core set of attributes in the first and second round. This particular choice was due to the fact that, since only through evaluation approaches it is possible to identify a successful strategy for a sustainable urban revitalization (i.e., "social complex value" maximization, policy for a balanced management of CBH, and impact evaluation related to the broader socio-economic development), the choise of the most appropriate evaluation methods may plausibly be considered as an intrinsic feature of the decision attribute of our analysis. In this round, moreover, the quality of core was equal to the quality of the complete set of attributes. Here, besides the policy measure attribute already identified as a core in the second round, other two condition attributes (i.e., interest groups and impediments) have been stressed. This is perfectly consistent with our previous considerations on strategic aspects of a CBH portfolio, involving also the leverage perspective and the management via the application of integrative bargaining principles.

We next passed to the analysis of the set of decision rules (Table 4), based on the core set of attributes A6, A9, A10, which can be viewed as a valid support in the portfolio implementation process. Also in this case the decision algorithm was consistent, allowing us to draw some interesting conclusions on the available information.

TABLE 4 Set of decision rules, based on the core set of attributes A6, A9, A10

N°	Rule	Class
Rule 1	A9 = 2	d = 1
Rule 2	A9 = 3 , A10 = 3	d = 1
Rule 3	A6 = 1, A9 = 3, A10 = 2	d = 1
Rule 4	A9 = 3, A10 = 1	d = 2
Rule 5	A9 = 1, A10 = 2	d = 2
Rule 6	A6 = 2, A9 = 3	d = 3
Rule 7	A6 = 2, A9 = 1, A10 = 1	d = 3

In particular, we have focused our attention on rule 7, which involves the overall aspects concerned with a portfolio perspective (i.e., interest groups, equity considerations, management measures, democracy and cooperation, and, finally, various kinds of impediments) in order to pursue an integrated sustainable conservation by looking, therefore, at the impacts related to the broader socio-economic development. It is easily seen however, that the policy measures attribute showed up in each rule, implying that it in any case represents the precondition for a successful strategy for a sustainable urban revitalization.

9.   Conclusions

In this paper, by presupposing that urban CBH revitalization is able to provide a significant contribution to socio-economic development, we have stated the necessity to envisage this process in a wider policy context. We have tried, indeed, to single out the strategic benefits of a sustainable urban revitalization, pursued through a systematic, multi-layer and sequential process of conservation policies. A portfolio of policy options for integrated conservation of cultural built heritage has then been identified in order to create the conditions for a synergy of economic, social and political interactions.

Moreover, by introducing the notion of a leverage approach, we have pointed out a potential alternative to guarantee the transparency of implementation and management of such a portfolio and, at the same time, to identify mechanisms and strategies for promoting private-public partnerships in the context of an integrative bargaining framework.

Our main interest has been in discovering the necessary conditions which have to be fulfilled by a given portfolio in order to be effectively and successfully implemented in a policy context. In this perspective, the recourse to a meta-analytic approach, through the application of rough set analysis to a sample of 10 interesting European CBH policy cases, has offered us the opportunity to evaluate the information available in the selected case studies, aiming to provide a suitable framework as a support in the decisional and negotiation process within our CBH portfolio analysis.

References

Bergh, van den J.C.J.M., Button, K.J. and Jongma, S.M. (1995) A Meta-Analytical Framework for Environmental Economics, *TRACE Discussion Paper*, TI **82**,Tinbergen Institute, Amsterdam.

Bizzarro, F. and Nijkamp, P. (1996) Integrated Conservation of Cultural Built Heritage, in *Serie Research Memoranda*, 1996-**12**, Vrije Universiteit, Amsterdam..

Button, K. and Nijkamp, P. (1995) A Typology of Environmental Assessment Issues and the Usufulness of Meta-analysis, *Tinbergen Institute Research Paper*, Amsterdam.

Coccossis, H. and Nijkamp, P. (eds) (1995) *Planning for our cultural heritage*, Avebury, Aldershot

Dal Piaz, A. and Forte, F. (1995) *Interessi Fondiari, Regole Perequative*, Clean, Napoli.

Forte, F. (1995) Una Nuova Intelligenza nella Progettazione del Piano Regolatore Comunale attraverso la Perequazione, *Urbanistica Informazioni* **140**, pp. 10-12.

Fusco Girard, L. (1995) L'utilita` dei Beni Culturali nella Citta` Moderna, *Restauro*, **131**.

Fusco Girard, L. (ed.) (1993) *Estimo ed Economia Ambientale: le Nuove Frontiere nel Campo della Valutazione*, F. Angeli, Milano.

Fusco Girard, L. and Nijkamp P. (1997) *Le Valutazioni per lo Sviluppo Sostenibile della Citta' e del Territorio*, Franco Angeli, Milano.

Hubbard, P. (1993) The Value of Conservation, *Town Planning Review*, **64**(4), pp. 359-373.

Johnson, J. (1996) Sustainability in Scottish Cities, in Jenks M., Burton E., Williams K. (eds.), *The Compact City: A Sustainable Urban Form?*, E&FN SPON, London.

Lewicki, R.J. and Spencer, G. (1992), Introduction and Overview, *Journal of Organizational Behavior*, **13**(3), pp. 205-207.

Lewicki, R.J., Weiss, S.E. and Lewin, D. (1992), Models of Conflict, Negotiation and Third Party Intervention: A Review and Synhesis, *Journal of Organizational Behavior*, May 1992, **13**(3), pp. 209-254.

Lichfield, N. (1996) *Towards Successful Strategies for the Economic Revitalisation of the Cultural Built Heritage in Europe*, Draft Report R. 6 (Paris 12/3/96), Council of Europe, Strasbourg.

Matarazzo, B. and Nijkamp, P. (1995) Methodological Complexity in the Use of Meta-Analysis for Empirical Environmental Case Studies, *TRACE Discussion Paper*, TI **141**,Tinbergen Istitute, Amsterdam.

Myerson, R.B. (1985) Analysis of Two Bargaining Problems with Incomplete Information, in Roth A.E. (ed), *Game-theoretic Models of Bargaining*, Cambridge University Press, Cambridge.

Nijkamp P. and Artuso, L. (1995) Methodology and Application of Sustainable environment Concepts for the Built Environment, *Serie Research Memoranda* **48**, Free University, Amsterdam.

Nijkamp, P. (1987) Culture and Region: A multidimensional Evaluation of Monuments, *Serie Research Memoranda* **71**, Free University, Amsterdam.

Nijkamp, P. (ed) (1996) *META-Analysis of Environmental Strategies and Policies at a MESO level*, FINAL REPORT (May 1996), Commission of the European Communities, Environment Programme, DG XII (Science, Research and Development), Bruxelles.

Nijkamp, P. and Vleugel, J. (1994) *Missing Transport Networks in Europe*, Avebury, Aldershot.

Nijkamp, P. and Voogd, H. (eds) Fusco Girard L. (1989) *Conservazione e Sviluppo: La Valutazione nella Pianificazione Fisica*, F. Angeli, Milano.

O'Toole M. and Usher, D. (1992) Editorial, in Healey, P., Davoudi, S., O'Toole, M., Tausanoglu, S., Usher, D. (eds.), *Rebuilding the City*, E&FN SPON, London.

Pawlak, Z. (1991) *Rough Sets. Theoretical Aspects of Reasoning about Data*, Kluwer Academic Publishers, Dordrecht, The Netherlands.

Pearce, G. (1993) Conservation as a Component of Urban Regeneration, *Regional Studies*, vol. **28** (1), pp. 88-93.

Rosenthal, R. (1991) *Meta-Analytic Procedures for Social Research*, SAGE publications, Beverly Hills.

Slowinski, R. (1995) Rough Set to Decision Analysis, *AI Expert* **3**.

Thomas, K.W. (1992) Conflict and Conflict Management: Reflections and Update, *Journal of Organizational Behavior*, May, **13**(3), pp. 265-274.

Tisdell, C.A.(1991) *Economics of Environmental Conservation*, Elsevier, Amsterdam, The Netherlands.

Woltjer, J. (1995) Consensus Building in Infrastructure Planning, *9th AESOP Congress*, Glasgow.

EVALUATION AND EQUITY IN ECONOMIC POLICIES FOR ENVIRONMENTAL PLANNING

M. L. CLEMENTE, G. MACIOCCO, G. MARCHI,
F. PACE, F. SELICATO AND C. TORRE*

1. Introduction

Since the extension of concepts like carrying capacity, non-market resources, sustainability has taken a wide place in the scientific debate among scholars of planning, of estimating and of environmental economics, the practices on equity planning and evaluation have assumed a new importance; at the same time, the interest in their operational issues is increasing.

The consciousness of the need to consider natural resources as a "value of the community" leads to sharing the advantages and the disadvantages which derive from the use of those resources and, consequently, leads to assuming new mechanisms and methods for assessing their value and for identifying the correct thresholds to their *consumption*.

The importance of the environmental needs has given a new dimension to the concept of equity planning. It is clear that equity planning can not realise in practice the principle of *giving to each individual or each community their own equal part*.

In a modern vision equity planning can be considered the planning practice which derives from the consciousness that (i) the existence value of environmental goods belongs to the whole community, not only to individuals, and (ii) the existence value is related to the ethical dimension of planning (MacLaren, 1996).

What is explored now is the possibility of creating a mechanism of balance by the use of planning/economic instruments on a wide spatial scale, in order to make the need for profit (of individuals and the community which use the land) compatible with the preservation of nature: when defining this balance, the assessment of those environmental resources and those natural amenities which exist significantly in a regional context, is an important factor of effectiveness (Blomquist and Whitehead, 1995).

---

* Paragraph n. 4 is due to G. Maciocco and M.L. Clemente ; paragraph n. 5 is due to G. Marchi.; n. 2 is due to F. Pace; paragraph n. 3 is due to F. Selicato.; Paragraphs n. 1 and n. 6 are due to C. Torre

N. Lichfield et al. (eds.), Evaluation in Planning, 213–227.
© 1998 *Kluwer Academic Publishers. Printed in the Netherlands.*

A first approach to the question comes from urban economics. The monetary evaluation of environmental amenities has been a main subject in the literature of economics since the late '70s, when Rosen (1977) developed and applied a methodology of assessment based on the contingent valuation method. In this context, a direction for innovation could be the updating of the methods by the construction of models which can take into account more aspects of the value of socio-environmental resources than in the past.

A second approach is related to those economic instruments which are implemented in planning policies. The regulations based on the taxation of the use of natural resources, or on the influence of planning policies on the market have also found a place in the scientific debate of environmental economics (Tietenberg, 1992; Hanley, Shogren and White, 1996).

Even though the idea that it is hard to obtain effectiveness and equity at the same time in planning policies is quite frequent among researchers, the demand for renewal of planning practice imposes investigation of the methodological and the operational issues of the evaluation of natural resources related to equity planning in a regional context. It is possible to read the need for this renewal in some events :

i) the need for a new planning system which takes into account a concept of equity related to the sustainability of the development;

ii) the difficulty in implementing the planning instruments in variously characterised wide spatial contexts [1];

iii) the necessity to adopt new methodological criteria to assess natural resources and natural risks, in order to make the application of the environmental planning instruments (provided by earlier legislation) more effective.

In the Acsp-Aesop Conference in Toronto, for instance, it was underlined in several papers [2] that the need for environmental protection deals with the question of equity, especially when, in regional planning practices, attention is placed on those typologies of land use (tourism industry, recreational uses of landscape) which may have a heavy impact on the "natural capital". Another example is the attention placed in the INU (Italian Town Planning Institute) Conference of 1995 on the need to introduce some mechanisms of equity in the new planning system (Dal Piaz and Forte, 1995).

Nevertheless studies of the last decade, in the field of estimating and environmental economics, increase the awareness about non-market resources evaluation (Fusco Girard, 1993); the researchers involved in studies on future applications of operational planning have emphasised the relation between environmental impact and regional planning. According to the economic disciplines, instruments would be market regulation and taxation of land-uses; according to planning disciplines, instruments are regulation and the limitation of land-uses. The question of the effectiveness of environmental policies passes through the interaction of these two approaches.

2.   The Use of Resources in Environmental Policies

2.1.   SOME QUESTIONS

Although studies on the concept of sustainability have reached a sufficient level of diffusion and of acceptability in the scientific world, they seem to face a difficulty in being accepted as a real instrument of investigation by the public planning agencies or in being used for the orientation of governmental directives.

Some developed countries quite recently have begun to show a real interest in environmental problems: in most cases environmental protection is the scope of some studies commissioned by governmental agencies, which are not immediately available for the aims of political economy or for legislative action, due to their partial character tending to verify the presence of impacts, or to evaluate the effect of a project or/of a policy from a defined point of view (for instance atmospheric pollution, air quality and water quality, natural environment etc.).

In those studies the question of how to share advantages and disadvantages related to the use of resources appears not to be discussed enough.

For a long time the scholars of regional planning disciplines (and/or economics, politics, finance etc.) have dealt with the question of distributive justice and the localisation of resources, when they are involved in investigations on decision-making procedures in policy more than in the market.

In the ambit of different theories, usefulness, contracts, rights and dialogue are used as concepts, taking into account that policies have to be evaluated in terms of effectiveness and not only in terms of efficiency, and that the same policies have to be accepted by relative social components.

The relationship between planning and economic-.evaluative disciplines appears quite complex (economic disciplines already exist as a support in planning in terms of financial plans, feasibility studies, assessment of expropriation value, etc.).

In the 80's, economics and estimating were interested by the debate about the principle of rationality, which is a base for both disciplines. The new awareness of the (natural and built) environment and of social groups poses the question of environmental evaluations; at the same time the ethical question of inter-generational relationships assumes new relevance.

The consideration of market and ownership as the only factors regulating the optimum level of externality (Coase, 1960) gives rise to a common criticism on the missing correspondence between theoretical models and reality.

This criticism can be summarised as follows:

i)   the real absence of an ideal market and of an ideal competition;
ii)   the difficulty in negotiating, due to the difficult identification of actors and costs;
iii) the difficulty in negotiating when the polluter and/or the polluted refer to resources owned by private or by public bodies.

The existence of pollution, considered by the environmental economists as an example of market failure, leads to the creation of modalities of intervention aimed at the control of the pollution itself, starting from various regulations to the reduction of emissions in different industries, to taxes on polluting waste, to incentives and fiscal facilities and to influence on the market.

Starting from the 80's, the real tendency has been based on the search for "satisfactory" solutions to be obtained by negotiation and/or public agreements, more than the optimal solutions of marginalist theory (according to this theory an "optimum" exists as a result of a negotiation between polluter and polluted). There is, moreover, a relevant shift of interventions in the production process, in terms of prevention, and integration by instruments which are able to regulate the choices in economic policy.

These choices deal with wider tasks and scopes, because of the necessity to respond to criteria of coherence between aims and methods, of effectiveness in sharing advantages and disadvantages and of efficiency in economic and environmental policies.

Especially when different communities (and their different administrative bodies) are interested in the management of resources, negotiations and pacts are necessary. In this case, for instance, the pollution produced in a territory moves to other territories. An analysis of costs that are necessary for pollution control (and therefore of costs of benefits coming from that control) can be inadequate to establish thresholds and standards to be respected, if there is not a common definition of quality.

In the same way it is possible to establish some standards regarding the use of some defined areas (like a conservation area or a heritage area) or to establish common requirements for sustainability. Therefore, due to the necessity of maintaining the amount of "natural capital" constant, the total loss of resources should be balanced by an increase of natural value.

There is in any case a problem in evaluation (in terms of attribution of value) which inevitably crosses the expression of judgements and preferences, by individuals and by communities, and which deals with the need to maintain the ecosystem. Therefore an ethical problem remains.

2.2.   SOME EXAMPLES

In some cases, even though the importance of concepts of scarcity of resources and of sustainable development have been accepted, the determination of costs of environmental policies still remains a question for the public administrations. At the beginning of the Eighties the Environmental Agency of the Federal Republic of Germany (Schultz and Schultz, 1991) promoted ten pilot studies for the evaluation of the benefits deriving from the measures of environmental improvement adopted in special areas, in order to demonstrate how the economic evaluation of environmental impact was really possible. Before that, the Ministry of the Environment had provided a research programme looking at the monetary costs of environmental damage, organised in modules, each one according to a different perspective; the special character of these modules consists in considering the non-material impact, such as the psychological costs of pollution, or the costs connected to loss of non-use value like the option value or the existence value, (which are typical of environmental goods); the programme, still in progress, studies the environmental damage from different starting points: the specific environmental component (air, water, noise), the social groups involved (customers, institutions, business men), the production sector (agriculture, industry, tourism). One of these studies has the purpose of evaluating in economic terms the "conservation of nature"; the approach is the prevention of damage, the costs of which is put before the judgement of the community by the use of interviews. These interviews describe different scenarios, on the basis of different expected degrees of conservation (and the different determination of the willingness to pay, in connection with which different motivations are also investigated).

The willingness to pay approach is used also for an investigation into the demand for environmental quality, which has the purpose of finding a relationship between welfare and improvement of environmental quality, and the value which citizens would pay to obtain those effects. Research has been carried out with the use of 3000 interviews.

The principle of the willingness to pay, despite its many limitations of chance, moves from an acceptable need, that is to give an indication of the "global" type of the degree of welfare or, at least, of the value given to such a concept, or to goods considered in connection with this concept. A problem seems in fact to be how to include in a comprehensive model the possible inter-relationship existing among all the effects of a defined policy, including the distributive character of the same policy.

Regarding this, the introduction in the USA of the Analysis of Impact Regulations (AIR) has required (with the Executive Order 12291) since 1981 for all the main norms issued in the federal states, seems to be of some interest; this kind of impact has to be estimated by using the maximisation of the net benefit for society. This provision follows the well-known and more partial NEPA (National Environment Policy Act) of 1969, and its rules, that requires the setting-up of Environmental Impact Assessment for specific programs and projects (Froehlich, Hufford and Hammett, 1991).

The utilisation of estimates of environmental benefits also within studies and provision for an economic policy not only regards the AIR, but also the Environmental Impact Assessment applied to projects in studies by the U.S. Environmental Protection Agency (1987) which carries out research with an environmental character which provides useful indications of economic policy (for instance a classification of different sources of environmental contamination, depending on the different types of effect caused to health and/or the ecological system) , besides giving criteria and directives for the application of the EIA at the federal level and for the explanation of laws issued by Congress.

3.   Economic Instruments in Environmental Policies

The evaluation/distribution instruments in environmental policies, still now directed especially at the control of pollution factors, can be classified in three typologies (Allègre, 1990).

The first typology (which is more frequently used in real praxis) is direct regulation. It means the use of a set of limitations and prohibitions by governments, which appear as "haesogenous compulsory action". In this first category all those regulations and norms defining the behaviour which can be adopted or avoided are included.

Behaviour regarding the maximum level of polluting substances and the ban on using phosphates in agriculture, for instance, are included.

In this approach the most important limitation (not easily acceptable) is the compulsory character of actions on individuals. If the real tendency is addressed - as it appears - to a vision of environmental protection as a cause of limitation and prohibition, the defence of the environment will probably not create a necessary public consensus.

The second typology of instrument (often complementary to the first) is derived from an economic instrument, and is based on the imposition of new taxes, and above all, on the destination of corresponding economic resources for expenses related to environmental protection. In this second case the instrument is powerful as an indirect regulation. In a more ambitious perspective the fiscal instrument, when considered as a way of implementing environmental policies, should internalize not only the social cost of environmental changes, but also some externalities which are usually forgotten in planning. Those externalities can be classified in two groups: negative externalities due to the action of public operators, and positive externalities which produce favourable effects to the community.

The environmental damage caused by public goods like power stations, plant for waste treatment and highways belong to the first category of externalities : they can be evaluated in episodic environmental impact assessments, that is to say not in normal regional planning procedure.

Social services provided to the community by historical cities, rural areas, parks, belong to the second category of externalities : their character as public goods is recognisable by the imposition of limitation and prohibition of building in those areas, without any   compensation for private owners. In a wide spatial context those positive externalities constitute a system, a network of function and inter-relations. The characterisation as a system makes the determination of existence value and of the use value of those social services in the balance of regional economics unavoidable. Planning policy is to give a response on the correct use of territorial sources, and contemporary economic disciplines produce models of evaluation of this use, in order to find a way for defining balance, as will be shown below.

Also in this case it is necessary to underline that the imposition of new taxes is a difficult and a unpopular policy: when the taxation is not directed only at those actors that find a source of profit in those positive externalities, all the community could be negatively involved without distinction. In fact, a necessary condition to apply a fiscal instrument which can be useful to promote incentives for environmental protection is the possibility of "identifying, calculating and sharing social costs" (Danielis and Giacone, 1991) : those costs should regard - in particular - promoters of territorial changes and - in general - those people which receive a benefit from the services offered.

The third instrument (that is an indirect instrument) relates the environmental question to economic policy, by considering the costs of nature protection as costs belonging to the "expenses list" of society. Surely this solution is the most difficult (nobody owns the air or the water). This instrument can be considered as a sort of "payment for the use of the environment". "Each oil-tanker that crosses the ocean should pay, for instance, a price proportional to the dimension of the oil loads and to the duration of use of the sea ; the money will be used to adopt solutions to the problem of the pollution of the Sea.(...) Similarly, each fishing enterprise should pay a rent for the Ocean, on the basis of the species of fish to be protected".(...) "It would be possible, moreover, to establish a price for pollution and, with the money obtained, to adopt a solution to eliminate the pollution itself."

Therefore, the possibility - necessity of choice between direct norms and indirect economic instruments exists. The choice is a constant dilemma in the policies of various sectors (Weitzman, 1974). In environmental policies, in spite of the economic theory which suggests the use of indirect instruments, it happens quite often that only the direct norms have been applied (OECD, 1990). The use of heavy direct norms has moreover shown its weakness. In many countries the severity of the regulations has been proportional to the loss in control and the loss in effectiveness of the sanctions (Dente, 1984).The advantage of using indirect economic instruments instead of direct norms is not, however, absolute. Direct regulation still plays a fundamental role in those situations in which the damage depends on the character of the polluting substance, on the sensitivity of economic operators, on the environmental character of the area. In conclusion, experience leads us not to completely substitute the economic instruments with direct regulations, to promote the experimentation of economic instruments wherever useful.

An interesting instrument created by economists (Crocker, 1996 ; Dales, 1968) and introduced in American legislation in 1977, is the permission of exchangeable emission (PEE). The permission is a certificate that allows the emission of polluting substances in a well-identified space and a well-defined period. The use of PEE is articulated in two phases : in a first moment, the public body determines the "level of environmental quality" to be maintained in a defined area and, consequently, the level of emission to be allowed and the number of permission to be given. In a second moment, the companies are allowed to utilise the PEE on the basis of the defined amount of emission. In this way the PEE allows private entities to use in well defined terms the public good represented by the capability of absorption by the environment, and to create a "market" in which this use has an "equal price". Altough PEEs are indirect regulations which use the market processes, they require the intervention of public bodies which are able to control the environmental impact of economic activity (Cellerino, 1985).

Effectiveness, incentives and flexibility are peculiarities required for economic instruments in environmental policies.

Effectiveness in sharing the costs of reduction of pollution, to obtain a defined level of environmental quality, is a more recognised advantage of PEE, than of direct regulations (Montgomery, 1972; Baumol and Oats, 1975). Each company when costs are higher (lower) than the costs of PEEs, will decide to require (offer) more PEEs, and, consequently, to cause an increase/decrease of price of PEEs. The result is a balance in which the cost of the reduction of pollution is shared among the companies (Danielis and Giacone, 1991).

The effect of incentives is due to the fact that PEEs internalise the damage. It is an interest of private operators to find better technologies for reducing the amount of pollution. It is believed that this mechanism may encourage the improvement of technology, with economic benefits for companies and environmental benefits for society.

An important feature of PEEs is also flexibility. Flexibility means easiness in maintaining environmental quality standard, also when changes occur. In theory PEEs give advantages in terms of flexibility, in comparison to direct regulations and taxes, because the very market of PEEs creates arrangements to maintain the "Standard objective" (Danielis and Giacone, 1991).

It is, however, necessary to underline that the possibility of outcome deriving from the use of a fiscal instrument depends on the institutional context in which the same instrument is utilised. In USA, in detail, PEEs are in context characterised by the existence of a technical-administrative body which, for a long time, had experimented with the control of the level of quality by the previous legislation. The use of PEEs in Italy has to deal with the ineffectiveness due to the burocracy that characterises real environmental policies.

The main causes of this ineffectiveness can be resumed as follows:

i)   confusion and conflicts as regards the definition of administrative competencies;

ii) lack of cooperation between different public bodies;

iii) low level of skill as regards clerks and technicians in technical public offices;

iv) impossibility to monitor and create an environmental control (Bresso, 1989);

v)   scarce willingness to address the efforts towards improvement, more than the maintenance of environmental quality.

It is clear that it is necessary to improve administrative efficiency and the ability to govern through a continous process of experimentation and test of the actions implemented.

4.   Equity in Environmental Planning.

Studies and research works referring to case studies of city and town environments only recently have dealt with the wider field of land rent equity of environmental planning. In this field, the question of environmental protection, safeguard and transformation of *environmental public* [3] *goods* plays a strategic role in defining and controlling conditions for *sustainable development.*

The historical-morphological analysis of territory and of the landscape structure; the landscape scenarios at various enjoyment scales; the definition of the spatial compendia for integral conservation, transformation, environmental recovery and repair, must absolutely constitute the fundamental reference elements for the planner's activity, through an interdisciplinary approach.

We assume the activity we are dealing with is aimed at building a landscape plan.

Regional landscape planning recognises the "landscape-environment" as "immaterial public goods [4]".

On the other hand, these "immaterial goods" can take an economic value only if a "strong public agreement" were reached between the planning actors[5] on its medium- and long-run strong options [6] on territorial invariants.

Such invariants identify the *integral-conservation areas* in their strongest binding degree.

In Italian legislation, *environmental bonds* on private ownership do not require compensation [7] since they have a *merely declarative* [8] character, though.

In this framework, it is extremely difficult to reach a strong public agreement among planning actors to jointly recognize *environmental values.*

Certainly this does not come from the awareness of these values- therefore from their consistent recognition from the scientific point of view; instead, such a situation is generated by the assessment of the economic consequences of their identification, which implies different - and extremely diversified - operational possibilities for the various economic actors involved in the plan's implementation.

For example, in the case of "integral-conservation areas" the environmental bond determines a strong limitation of their theoretical operational capacity [9] for their owners.

On the other hand, the same environmental bond constitutes an economic advantage for those subjects who are willing and able to exploit its positive externality at least in terms of *immaterial value.*

In the Italian normative framework, the approval of a landscape plan - e.g.: a plan of a regional park - can generate a true shock on the land use market, and the system of those area which are directly usable for tourist settlements, when exploiting the immaterial value generated by conservation areas, becomes able to express a higher land value.

5. A model of Fiscal Imposition in Planning Instruments for Environmental Protection

An example of how the regional economy may be re-balanced by fiscal policies which are related to wide area planning is now illustrated.

A perspicuous case may be represented by the environmental planning activity of a regional public administration [10].

Generally and at the regional scale, two systems can be identified:

- the first, Y, is a public system which from the landscape-environmental plans' approval and the consequent definition and protection of natural resources - e.g.: the regional parks - produces *environmental public goods* [11];

- the second, X^I, is a private system which consists of construction industry entrepreneurs who operate in the coastal tourist zones.

Let us assume land as the only factor of production.

In the Y system the *integral-conservation areas*:

- can be assumed to be an indirect production factor which consists of a positive externality which is exploited by the system X^I [12], (but also a contemporary negative externality against private citizens who are land-owners since this bond prevents them from implementing certain uses of their land);

- express an absolute maximum ($=V_{iUPA}$) for the environmental value function;

- express an absolute minimum ($= V_{ma}$) for the land value function [13].

In the X^I system the *areas directly usable for tourist settlements*:

- can be defined as a direct production factor;

- express a relevant intrinsic land value with regard to their building potential ($=V_e$);

- take advantage of the positive externality ($=V_{iUPA}$) produced by the system Y.

As previously affirmed, the process of approval of a regional park plan generates a shock in the land market in economic terms, since in the areas subject to the bond [14] of *integral conservation* the already small *land economic value* ($= V_{ma}$) tends towards zero since no real market for these areas can exist.

On the other hand, system X^I - the *areas directly usable for touristic settlements*- exploits the immaterial value generated by Y and becomes able to express a total land value, $V_{TOT(X^I)}$, which can be expressed as follows:

$$V_{TOT(X^I)} = V_e + V_{iUPA} \tag{1}$$

To solve the puzzle and find a solution which re states market conditions in terms of equity, we propose to make reference to the Pigouvian [15] principle opportunely qualified [16]. Our solution is shown in Figure 1.

Let us suppose the shock occurs as a consequence of the approval of a plan for a regional park (in the current regional legislative framework): production system X^I - the *areas directly usable for tourist settlements*- once it owns the immaterial value generated by Y, can be represented:

- through the demand curve D - D;
- through the marginal cost curve MCo [17].

The MCs curve represents the social marginal cost generated by system Y, which is negative for owners of the areas where bond Y is established, but certainly positive for

system X^I that takes advantage from it. This advantage - an increasing advantage in fact - reaches its optimal value at point b, where the productive system X^I tends naturally to identify its output X^I_1, since this very point b - where the demand curve (D - D) and the ordinary marginal cost curve (MCo) of firm X^I intersect - represents the maximum benefit of X^I.

The MCs curve is the marginal social cost curve of productive system Y; a portion of this cost - that is V_{iUPA} - represents a benefit for system X^I.

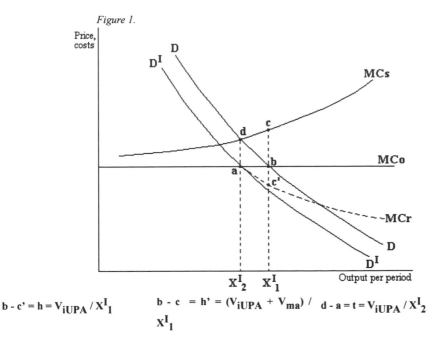

Figure 1.

$$b - c' = h = V_{iUPA} / X^I_1$$

$$b - c = h' = (V_{iUPA} + V_{ma}) / X^I_1$$

$$d - a = t = V_{iUPA} / X^I_2$$

From the public interest point of view, the optimal output of productive system X^I is equal to X^I_2, since here the marginal benefit from output of system X^I [18] coincides with the marginal social cost (this is point d in Figure 3). The productive system X^I - without any public intervention - tends towards overproduction with respect to the optimal quantity represented by the public interest, at point X^I_1, however.

A suitable per-unit indirect[19] taxation t would push the demand curve downward up to the new position D^I - D^I.

As a result of this taxation - this is the segment d-a in Figure 3 - equal to the following:

$$t = V_{iUPA} / X^I_2 \qquad\qquad (2)$$

the new demand curve D^I - D^I. determines the following facts:

- firm X^I is pushed to produce X^I_2, since at point a - as a result of taxation - it cannot enjoy the surplus value generated by activity Y [20] any more;

- quantity X^I_2 is the social optimal level of output, since at point d the marginal social cost curve (MCs) and the demand curve (D - D) intersect.

Basically, output X^I_2, induced through taxation t allows the maximisation of profit of private entrepreneurship X^I at normal market conditions and the optimisation of output from the social point of view. Of course, consumers would pay not only the direct cost from MCo but also the indirect cost of taxation t [21].

Of course, what would be earned from taxation should be utilized to compensate the owners of the land with a particular environmental importance (park) which are subject to the environmental bonds (the *integral-conservation areas*); these areas are necessary for the operational activity of Y[22].

This could be achieved through the association of the environmental bond- which cannot receive compensation- with a legal expropriation procedure to compensate the original owners of these areas at least for the agrarian value of their properties V_{ma}, which from nominal becomes real again.

The above model can be considered a case of trade-off between the fiscal imposition and the use of land for the tourism industry, but a more relevant trade-off could be considered between the consumption of coastal resources and the conservation and the valorisation of the natural resources inside the protected areas. The second trade-off could represent a typical case of application of the principle of *weak sustainability,* according to which the loss of an amount of a specific natural resource can be balanced by the increase in the general natural stock value.

The possibility to find a balance in the second trade off could be considered a way of measuring the equity of the implemented planning policy. The more the natural stock value can be increased, the more the social costs deriving from the coastal consumption can be balanced.

6.   Concluding Remarks.

Each example shown in the various paragraphs of this paper has shown how the questions of equity and environmental effectiveness in planning are referring to trade-off models, which should take into account the balance between individual profits and social costs, as a result of fiscal intervention, and the balance between the use of natural resources and their consumption. According to some scholars of green economics (Pearce, Markandia and Barbier, 1989), this solution can be an acceptable result.

The VIA methods adopted in Germany, the tradeable pollution permission represented by PEEs and the fiscal taxation of the tourist use of coastal areas represent an application of   the   principle   of   weak   sustainability.   The   possibility   that environmental planning can use an effective determination of value of natural resources still remains a limited option. Even though it is always possible to assess the impact of a regional policy on the environment in non-monetary terms (by the use of appropriate methods), it is still difficult to evaluate the monetary cost of the conservation of nature. The application of monetary evaluation methods, although they are imperfect, is a useful support for decision in a process like policy making, which is also imperfect (Blomquist and Whitehead, 1995).

Maybe the increase in participation and public representation in planning policies can be a way to shift the question (and the solution) of equity to an upper level of discussion.

---

[1] For Instance two different situations such as the British and the Italian ones. Cliff Hague, during a seminar in Italy in February 1997, noticed that the structure planning practice in England has been difficult in recent times; at the same time, regional plans are still missing in many Italian regions.

[2] For instance, Campbell, 1996 ; Glickman, 1996 ; Kuiper, 1996 ; Ismail, 1996 ; MacLaren and Benson, 1996.

[3] Goods is defined as "pure public" if it has the non-excludability and non-rivalry characteristics. Furthermore, non-pure public goods, such as environmental public goods, could lose, at least to some extent, their non-rivalry characteristic. "Environmental public goods" such as a "charming landscape" are intrisically indivisible and non-rival since consumption of a single consumer does not imply a decrease in the availability of the goods for other consumers.

[4] The landscape environmental units (UPA) of a landscape territorial plan include the agrarian value of their land (V_{ma}, which constitutes a material value) and an immaterial value (V_{iUPA}) expressed through their environmental general qualitative characteristics. These qualitative characteristics can be, as an example, considered as the *trade mark immaterial value* of a firm that produces quality goods; this trade mark characterizes and differentiates this firm from others and gives it an incremental value with respect to its production's intrinsic value.

[5] Actors whom initially we considered as divided into the following two categories:
-    public actors: Y;
-    private actors: X.

[6] G. Maciocco, 1991.

[7] Italian Constitutional Court, Sentence n. 56/68.

[8] The environmental bond is not a further burden for the normal utilization of these goods; it simply recognizes as characteristics of the goods some qualities which are instrinsically embedded in them.

[9] The limitation of the subject's theoretical operational capacity comes from the absence of any surplus value of the land, since it will not be ever used for anything else but its agricultural potential and furthermore under the established environmental bonds.

[10] This case is even more significant if we consider an island as a closed system.

[11] Parks are the most important environmental resource of Sardinian internal zones which are characterized by a strong economic downturn with respect to coastal zones.

[12] This advantage is due to the *immaterial value* (V_{iUPA}) which the park- as goods- transfers to the coastal-tourist entrepreneurship.

[13] The intrinsic economic value of land is equal to its agrarian value (V_{ma}, which constitutes a material value). Moreover, the very agrarian value of land constitutes a merely nominal value since the bond of integral protection- which is not to be compensated - makes it really impossible to sell this bounded land.

[14] This bond- which does not imply an expropriation procedure- is not to be compensated (see note 24); therefore the original owners keep the property right.

[15] Pigou, 1946.

[16] Nicholson, 1972, p. 808.
"The government could impose a suitable excise tax on the firm generating the external diseconomy. Presumably, this tax would cause the output of Y to be cut back and would cause labor to be shifted out of the production of Y. This classic remedy to the externality problem was first put forward lucidly in the 1920s by A.C. Pigou; although it has been somewhat modified it remains one of the 'standard' answers to the externality problem given by economists".

[17] The MCo curve represents the marginal costs system X^1 would incur in normal conditions. The real curve of the marginal costs is represented by the dotted line MCr. In fact, it can sustained that the maximum benefit of system X^1, equal to V_{iUPA}, can be considered as a proportional reduction of marginal costs following the regime operational activity of system Y.

[18] The consumer's willingness to pay.

[19] That is a taxation charged directly on the final consumer.

[20] It has shown that at point a the real marginal cost curve (MCr) coincides with the normal marginal cost curve (MCo) of firm X^1; this firm behaves here as if it were in the normal competitive market without receiving a direct benefit from Y's activity.

[21] Really, taxation t- a true indirect cost for the consumer- will become an added direct cost transferred from entrepreneurship X^1. In fact, the current price system of the tourist market supply is rapidly evolving from a "cost-plus" to a "price-minus" condition due to international competition. this indicates that market prices-which up to the recent past were based on the production costs and on profit margins of the producer- will be increasingly determined through the consumer's willingness to pay for these particular goods; profit margins and production costs follow this first moment. This will seemingly imply internalisation on behalf of the producer of the costs which should indirectly be charged to the consumer as a consequence of taxation t.

[22] The following example shows how to determine the size of the tax (t) to acquire through expropriation the *integral-conservation areas* of the Park of Gennargentu.
The hypothesis of the park's organisation studied by the Province of Nuoro takes a comprehensive area of 61,826 ha, 10% of which- this is about 6,182 ha- are destined to be an integral reserve because of their very high naturalistic value; 5% of the total area- this is about 3,091 ha- is destined to services and infrastructure.

References

Allègre, C. (1990) *Economiser la Planète*, Libraire Arthème Fayard, Paris

Baumol, W. J. And Oates, E.(1988) *The Theory of Environmental Policy*, University Press, Cambridge.

Blomquist, G.C. and Whitehead, J.C. (1995), Existence value, contingent valuation, and natural resources damages assessment, *Growth and Change*, **26**, fall, pp.573-589.

Bresso, M. (1989) Environmental policy in relation to territorial distribution and productive activities, in Archibugi, F., Nijkamp P., (eds) *Economy and Ecology : Towards sustainable Development.* Kluwer Academics, Dordrecht.

Campbell, H. and Marshall, R. (1996) Ethical issues and Planning Practices : has planning changed its spots ? Paper presented in the Aesop-Acsp Conference, Toronto.

Cellerino, R. (1985) Diritti di inquinamento: lo strumento e le sue applicazioni. *Rivista di Diritto Finanziario e Scienza delle Finanze* 1, pp.352-384.

Coase, R.H. (1960) The problem of the Social Cost, *Journal of Law and Economics*, n. 3 October 1960, pp. 1-44.

Crocker, T.D. (1966) The structuring of atmospheric pollution control systems, in Wolozin, H. (ed.) *The Economics of Air Pollution*, W.W. Norton, New York.

Dales, J. (1968) *Pollution, Property and Price*, Toronto U.P. Press, Toronto.

Danielis, R. and Giacone, P. (1991) Politiche ambientali e politiche urbanistiche: nuovi strumenti a confronto, in Bonacci, F. and Gorla, G. (eds.) *Economie Locali in ambiente Competitivo*, Angeli, Milano, pp. 191-217.

Dente, B. et al. (1984) *Il Controllo dell'Inquinamento Atmosferico in Italia: Analisi di una Politica Regolativa*. Officina, Roma.

Dal Piaz and F. Forte, (1995) *Piano urbanistico: interessi fondiari e regole perequative*, CLEAN, Napoli.

Froehlich, M., Hufford, D.J.C. and Hammett, N. (1991) The United States: legislative and executive requirements for Benefit Analysis, in Barde, J. P. and Pearce, D.W. (eds.) *Valuing the Environmental*, Earthscan, London.

Fusco Girard, L. (1993) ed., *Estimo ed Economia Ambientale: le nuove frontiere nel campo della valutazione*, Angeli, Milano.

Glickman, N. (1996) Does economic development 'cause' inequality ? Aesop-Acsp Conference, Toronto.

INU, Istituto Nazionale di Urbanistica (1995), Regime degli Immobili e fiscalità per la nuova legge urbanistica, INU Conference, sept. 1995, Turin.

Ismail, A. (1996) The environmental consequences of societies in transition. Analysis of the paradigms, economic systems, and istitutional linkages. Paper presented in the Aesop-Acsp Conference, Toronto

Kuiper, G. (1996) Compensation for environmental degradation. The case of the Highway. Paper presented in the Eindhoven-Oss, Aesop-Acsp Conference, Toronto.

Hanley N., J. F. Shogren and B. White (1996) *Environmental Economics in Theory and Practice*. MacMillan, London

Maciocco, G. (1991) Introduzione, in G. Maciocco (ed.) *La pianificazione ambientale del paesaggio*. Angeli, Milano.

MacLaren, V. (1996) Voluntary agreements: an effective alternative to environmental regulation. Paper presented in the Aesop-Acsp Conference, Toronto.

Montgomery, D. (1972) Markets in licenses and efficient pollution control programs, *Journal of Economic Theory* 5, pp. 395-418.

Nicholson, W. (1972), *Microeconomic Theory Basic Principles and Extension*, sixst edition, The Dryden Press, Harcourt Brace College Publishers, Fort Worth.

Pearce, D.W., Markandia, A. and Barber, E. (1989) *Blueprint for a Green Economy*. Earthscan, London

Pigou, A.C. (1946) *The Economics of Welfare*, 4th. ed. Macmillan & Co. London.

Schultz, W. and Schultz, E. (1991) Germany concrete areas of application with environmental policy relevance, in Barde, J. P. and Pearce, D.W. (eds.) *Valuing the Environmental*, Earthscan, London.

Tietenberg, T. (1992), Environmental and Natural Resource Economics, third ed., Harper Collins, New York.

U.S. Environmental Protection Agency. (1987) EPA Use of Benefit-Cost Analysis: 1981-1987. Report of the Office of Policy Planning and Evaluation.

Weitzman M.L. (1974) Prices vs. Quantities, *Review of Economic Studies*, (XLI) 128, pp. 477-491

ON THE EVALUATION OF "WICKED PROBLEMS"

Guidelines for Integrating Qualitative and Quantitative Factors in Environmental Policy Analysis[1]

H. GLASSER

1. Introduction

> There are moments of history when we simply must act, fully knowing our ignorance of possible consequences, but to retain our full rationality we must sustain the burden of action without certitude, and we must always keep open the possibility of recognizing past errors and changing course. (Arrow, 1974: 32)

> The contemporary challenge is to recognize that policy is a learning process and to build on the knowledge revealed by past error. Politics and policy do change. One of the lessons of the past half-century is that intelligent and accurate criticism will make a difference. (Ingram, 1988: 51)

Individuals and society face a litany of decisions involving competing interests and conflicting ends. In surprisingly many cases these decisions entail substantial environmental trade-offs and carry with them significant environmental consequences. In this paper I focus on one very narrow element of the overall planning process: the generic decision problem of evaluating, *ex ante*, alternative policy or project proposals. While I do not address the design phase of planning, in which the alternatives themselves are conceived and developed, I take creating a framework that is capable of evaluating a broad array of qualitatively distinct alternatives to be critical to the overall planning process[2]. My emphasis is on creating a conceptual framework for evaluating diverse alternatives with significant environmental and societal implications. The overarching goal is to explore the potential of forestalling environmental problems through anticipatory, adaptive planning.

The interrelated, coevolving effects of poor environmental planning sometimes announce themselves dramatically, but often they appear only after a significant separation between cause and effect. Consider for instance the effects of deforestation and poor agricultural practices on ancient civilizations (Marsh, 1865; Glacken, 1967;

N. Lichfield et al. (eds.), Evaluation in Planning, 229–249.
© 1998 *Kluwer Academic Publishers. Printed in the Netherlands.*

Ponting, 1991) and the effects that may result from contemporary anthropogenic greenhouse gas emissions, synthetic hormonal mimics, or loss of critical habitat for endangered species. While the secondary effects of projects and policies have often been poorly understood or plainly misunderstood[3] and what understanding there is has rarely been systematically analyzed or widely communicated, the environmental consequences of these projects and policies have seldom been totally unforeseen or unforseeable.

The pattern of overexploitation and exiguous conservation practices that has insidiously degraded the environment, extinguished species, and toppled civilizations appears to be, *inter alia*, a result of the unexamined implications of human actions - not an arrant ignorance of these implications (Glasser, 1995). The importance of an anticipatory approach is underscored by the high social and environmental costs of mistakes - mistakes that an exploration of historical environmental crises suggests might be avoidable (Glasser, 1995). As is well documented, however, it is particularly onerous to evaluate decision problems that are plagued by incommensurable criteria and multiple, often conflicting, goals. Because such complex, dilemma-laden social choice problems can never be fully characterized, and because resolving them requires subjective judgments, they have been termed "wicked" by Rittel and Webber (1973).

2.   "Wicked Problems" and Comprehensive Rationality

The comprehensive rational model of choice presumes that efficacious *means* can be found to satisfy previously defined *ends*. In its archetypal, idealized form, this model of rationality rests on four key assumptions[4]. First, it assumes that ends and means are separable. Second, it assumes that decision-makers can, and do, identify all of the relevant alternatives. Third, it assumes that decision-makers can, and do, survey these alternatives with respect to all of their possible consequences. Finally, it assumes that decision-makers, fully cognisant of their objectives and with a completely defined preference function, which is capable of aggregating these objectives, mechanically choose the alternative or sets of alternatives that maximise their overall utility.

To consider *all* of the facets of a planning situation, however, is clearly impossible. Because of its untenable assumptions, comprehensive rationality has received devastating criticism (Rittel and Webber, 1973; Lindblom, 1959, 1965; Self, 1974; Webber, 1983; Simon, 1972). Despite these critiques and despite visionary efforts by Lichfield (1968, 1996) and Hill (1968) to develop and promote descriptive overview methods for *ex ante* evaluation, the influence of comprehensive rationality is still widely felt. This is particularly so in the United States. Comprehensive rationality is the model of rationality that underpins neoclassical economics (Samuelson and Nordhaus, 1992), benefit-cost analysis (Mishan, 1988; Prest and Turvey, 1965), and environmental economics (Baumol and Oates, 1988; Pearce and Turner, 1990). It also underpins the orthodox, substantively rational approaches to project evaluation (Dasgupta et al., 1972) and policy analysis (Stokey and Zeckhauser, 1978). It even underpins some of the most farsighted approaches for evaluating the sustainable management of resources (Costanza et al., 1997). Because the presence of comprehensive rationality is widely felt in the tools for environmental policy analysis, I will survey eight of its major "flaws"[5]

The first flaw with comprehensive rationality is that complex policy problems are multicriteria problems. Policy analysis usually requires satisfying an array of competing and conflicting objectives that are assessed with criteria that are often incomparable (Rittel and Webber, 1973). These problems are mathematically ill-defined - global maximization is usually not feasible and utility, as a highly aggregated "super"-criterion is unobservable and immeasurable - without gross oversimplification, multicriteria problems are simply not amenable to substantively rational solution strategies, even assuming the existence of perfect information.

Second, complex policy problems do not satisfy the condition of perfect information. They involve inherently incomplete information, mixed data, immeasurable factors (e.g., intuition and experience), uncertainty, and ambiguity (Lindblom, 1965; Dakin, 1963; Zadeh, 1973).

Third, information is not costless, so there will always be a tradeoff between action and reflection (Churchman, 1968). Even if it were feasible, such an approach would represent an inefficient use of time and thus constitute poor planning because it neglects balancing the need to reflect with the need to act.

Fourth, individuals adjust their aspirations to their perception of what possibilities are available and to their evolving understanding of the consequences associated with these alternatives (Elster, 1982; Nelson and Winter, 1982; Lindblom, 1965). Preference functions are not complete and latent, but fragmentary and revealed dynamically through adaptive learning (March, 1978; Gregory, 1993).

Fifth, individuals often evaluate their preferences for different objectives with incomparable, unaggregatable criteria and plural rationalities. Moral, economic, political, social, legal, scientific, and ecological considerations all enter into the evaluation process and each encompasses a unique set of standards for making judgments (Diesing, 1962; Dryzek, 1987; Sagoff, 1988). Even if we could guarantee comprehensive rationality we would not be able to guarantee ecological sustainability, because it requires a particular value-inspired preference ordering that places ecological rationality above other conflicting objectives (Glasser, 1998 forthcoming).

Sixth, policy evaluation is a normative pursuit where ends and means are intertwined. The idea of generating objective "best" alternatives by applying logical reasoning to valid premises and clear, quantifiable objectives is a fallacy. The best we can hope for is to search for "preferred-compromise" alternatives by making informed, well-considered judgments. Values, principles, and ethical commitments all play a crucial role in decision-making processes (Anderson, 1979; Sagoff, 1988). Just as no policy or project can be value-free (Gilroy and Wade, 1992), no methodology for choosing between alternative policies or projects can be entirely value-free.

Seven, the selection of alternative policies and projects is a social process. Kuhn (1962, 1977) argued that there is no single criterion sufficient for justifying the selection of alternative scientific theories. Lack of consensus on appropriate criteria, their measurement, and their value orderings is sure to generate conflicts that cannot be resolved by appeals to "objective" science. In a similar vein, Norgaard (1992) argues that environmental science - comprising an even greater variety of perspectives and individual disciplines including climatology, hydrology, geology, toxicology, environmental engineering, conservation biology, and environmental economics, to name only a few - cannot be combined into a coherent unity. There is no "single logical paradigm, set of assumptions, and data set from which conclusions can be deduced" (p. 96). There is no unified interpretation of science from which to generate policy consensus on scientific grounds (Norgaard, 1992; Ludwig et al., 1993). Furthermore, additional information cannot be presumed to resolve conflicts; it might even sharpen conflict (Caldwell, 1977). The notion that informed argumentation and open debate are indispensable for resolving complex policy issues within democracies has been emphasized by many other authors (Fischer, 1990; Sagoff, 1986; Mumford, 1964; Churchman, 1968; Ellul, 1992; Majone, 1989).

Eight, rationality is bounded: there are limits to human information processing capabilities. Psychological experiments demonstrate that individuals and organizations do not, in practice, employ anything close to comprehensive rationality (Miller, 1956; DeSoto, 1961; Shepard et al., 1961; Parat, 1969). Rather, they seem to be guided by a host of incomplete and fragmentary multicriteria impressions, which they overwhelmingly attempt to collapse into a single "good versus bad" dimension (Shepard, 1964). The consequence of this simplification is a substantial loss of information. If we were not limited by our ability to acquire information or the quality of our information, comprehensive rationality would be impossible simply because of what Simon (1972) refers to as the "bounds" of rationality, our limited ability to *store* and *process* the information available to us.

Using evaluation frameworks that draw on idealized comprehensiveness to generate planning decisions is both ill-advised and dangerous; it proffers a false sense of objectivity and it misrepresents the capabilities of substantively rational tools. Unfortunately, practitioners and theoreticians alike in economics, policy analysis, and project evaluation reflect too infrequently on the dangers of employing decision models that rely on comprehensive rationality (Breheny and Hooper, 1985)[6].

3.   Critical Rationality: A Middle Ground Between Comprehensive Rationality and "Muddling Through"

While most orthodox approaches to policy analysis seem to naïvely presuppose comprehensive rationality, we are not forced to view comprehensive rationality as a realizable possibility. Neither are we forced to adopt Rittel and Weber's pessimism regarding our rational problem solving capabilities or Lindblom's (1965) disjointed-incrementalism. We need not, and should not, be too quick to dismiss some of the benefits of comprehensive rationality as an *ideal*. We ought not abandon rationality as a critical process for considering the future consequences of possible actions. As Popper advised (Popper, 1966, Vol. II: 232):

> [A]rguments [alone] cannot *determine*... fundamental moral decision[s]. But this does not imply that our choice[s] cannot be *helped* by any kind of argument whatever. On the contrary, whenever we are faced with a moral decision of a more abstract kind, it is most helpful to analyze carefully the consequences which are likely to result from the alternatives between which we have to choose. For only if we can visualize these consequences in a concrete and practical way, do we really know what our decision is about; otherwise we decide blindly.

We are inescapably bounded by our prior experiences, our creativity, and our limited information acquisition and processing capabilities, but by focusing on the *procedural* or process oriented aspects of rationality we can respond with foresight by demanding increased accountability, considering a broader range of perspectives, expanding the range of alternatives to be explored, and subjecting these alternatives to greater deliberation and debate. Gupta (1981) emphasized that the social and environmental impacts of development projects are often not revealed in the simple accounts of disbursed funds and transferred technologies. Comprehensive review of policy proposals, often inspired by public criticism of initial pro-development proposals, has often uncovered many previously unaddressed issues and generated improved policies. A careful review of the environmental considerations figured prominently in the pro-development policy reversals on the supersonic transport (Primack and von Hippel, 1974), the Tock's Island Dam (Feiveson et al., 1976; Tribe et al., 1976), the Naramada Dam (Morse, 1992), the Storm King Mountain pumped storage hydroelectric project (Carroll, 1973), the development of Point Reyes Seashore (Eaton, 1995), and the clearcutting of Clayoquot Sound (Berman et al, 1994; Matas, 1995). A model for early intervention that shows great promise can be gleaned from the report of an independent review panel, which evaluated the potential environmental impacts of the proposed 3,400 kilometer Paraguay-Paraná Navigation Project (Moore and Galinken, 1997).

A recent cognitive anthropological study on the environmental values of Americans, by Kempton, Boster, and Hartley (1995), reveals that while the public has strong environmental sentiments, its understanding of policy issues is often riddled with misunderstandings, which tend to skew its support for particular policies. Stuart Hill (1992), in an insightful study of citizen attitudes on the Diablo Canyon reactor issue, confirms that the public can form coherent, reasoned positions on complex issues, given access to information, sufficient guidance, and a reasonable amount of time to weigh the alternatives and evaluate the various "benefits" and "costs" of the alternatives. Other recent developments indicate not only that the public can form well-considered policy stances, but it can also be the driving force behind innovative policies (Yuba Watershed Institute et al., 1994; Hochschild, 1995; Kemmis, 1990; Berkes, 1989). The primary challenge of the contemporary era is to develop proactive strategies for considering the environmental implications of technological choices that can facilitate our potential for making informed, well-considered decisions, which reflect a respect for environmental constraints.

I contend that appropriate reasoning processes for decisions involving environmental tradeoffs must faithfully reflect the complexity of the problems while recognizing that such decision-making is part of a social process, in which plural rationalities dominate. In the face of limited and imperfect information, uncertainty, lack of scientific consensus, alternative value perspectives, multiple objectives, and the theoretical impossibility of comprehensiveness, a commitment to rationality must become a commitment to work toward creating informed, publicly deliberated, and accountable decisions. Substantively rational techniques can be employed with great benefit as long as they are a subset of a wider approach that recognizes their limitations. Such an interpretation of rationality seems more consistent with everyday experience in which goals are not etched in stone but act more as tentative guidelines to be revised as information is revealed and predicted consequences are deliberated. This more realistic notion of rationality has been at the heart of Lichfield's Community Impact Evaluation Approach (1996). The alternative to employing some form of procedural rationality is, at best, to make decisions myopically and, at worst, as Popper admonishes, to make them blindly.

Adequate consideration of complex policy analysis problems requires an adaptive, deliberative process that recognizes ethical commitments and community interests along with scientific data and intuition. Procedural rationality leads to descriptive (as opposed to prescriptive) decision aids, which, because of the limited relevance of any single approach, demand pluralism in methodology (Norgaard, 1989). In public policy the aim is not simply to make new prescriptive laws and rules, but to make ones that can be coherently justified so as to incline individuals to carry them out, politicians to follow through on their legislative responsibilities, and legal authorities to enforce them.

When alternative selection problems are straightforward, when they are dominated by a single criterion or a single "best" alternative is obvious, or when the consequences of the individual and collective decisions are insignificant, there may be no need to develop elaborate and systematic decision frameworks. In such cases an incremental or "muddling through" strategy is entirely appropriate. When, however, problems are complex and conflict ridden, this strategy is no longer appropriate. In such situations it may be desirable to search out the unforeseen, but not unforeseeable, impacts and attempt to systematically explore these along with the more readily foreseen impacts. The level of formalism and comprehensiveness used to approach an alternative selection problem should be dictated by the problem at hand, not simply the tools in our hands. Good judgment is not a substitute for inadequate methodology and good methodology is not a replacement for exemplary judgment.

I have been arguing that environmental problems are not merely cybernetic failures of the policy process. While it is certainly necessary for information flow and processing to be improved, it does not follow that a comprehensively rational strategy, with its tendency toward hierarchical organization, highly abstract analysis, quantified, fungible parameters, and mathematically "optimized" solutions, is the appropriate framework for approaching wicked problems.

What wicked problems demand is a new decision framework that expands both our empirical and our epistemological systems boundaries. Such a framework must connect science to values and ends to means. It must allow us to pay more attention to the non-quantifiable, non-separable, and non-decomposable properties in policy analyses. And it must not homogenize conflict by focusing on a single objective.

In contrast to pure optimizing approaches to policy analysis, which are teleological (exclusively focused on narrow ends), I argue for a deontological, multicriteria policy theory, which emphasizes wider consequences by considering the interrelatedness of means and ends. As Lindblom contends, we may be limited to muddling, but I argue that we can make this muddling much more deliberate and systematic. Next, I outline a series of ten postulates that specify general considerations for developing a systematic, critically rational framework for policy analysis.

4.   "Ten Tenets" for "Wicked Problem" Resolution

The following tenets are descriptive and normative postulates directed at making planning decisions more consistent with treasured social values, ecological exigencies, analytic constraints, and pragmatic political realities. They are intended to exploit the human potential for exploring the implications of our actions, in advance of acting. They address the normative meta-issue of what we *should* consider when deciding how to decide. One example is that while searching for a single "best" alternative we should also allow for a collection of "acceptable" or "preferred" alternatives and the possibility of "no acceptable" alternatives. The essential premise of this strategy is that informed choice is superior to uninformed choice; clarifying our assumptions and drawing out and exposing the consequences associated with different alternative courses of action may help us make both better individual and better collective choices. The tenets are intended to serve as both guidelines for assessing frameworks for responding to complex, conflict-ridden policy analysis decision problems and as propositions for guiding the development of procedural reforms to better respond to such decision problems. As such, they represent a tentative outline of minimal considerations that viable approaches for making such decisions should address.

1. *Individuals and society both have multiple objectives. These objectives, which, among other things, reflect ethical, social, aesthetic, environmental, scientific, economic, spiritual, and political considerations, are usually competing and are frequently conflicting.*

Reflective decision-making rests on taking great pains to tentatively outline a representative set of objectives. These objectives must faithfully mirror the complexity of the problem at hand, but they need not and cannot be exhaustive. When choosing between complex multicriteria alternatives the existence of competing and conflicting objectives creates dissonance both within individuals and amongst individuals inescapable. Realistic and defensible decision frameworks must be capable of incorporating this multidimensional conflict from the outset.

2. *Humans perceive and experience a wide range of values. To reflect these multiple senses of value and plural rationalities, individuals and societies operationalize their multiple objectives and gauge their degree of satisfaction by employing a broad spectrum of incommensurable criteria.*

A circumspect representation of the foreseeable consequences resulting from a collection of alternatives will often necessitate a host of qualitative and quantitative criteria. These criteria might reflect instrumental, non-instrumental, and even intrinsic values. The set of criteria need not, and from the stand-point of manageability should not, be exhaustive, but it must reflect the set of objectives in a manner considered to be sound by all of the relevant actors. The choice of criteria should be orchestrated so as to facilitate the exploitation of accessible and information-laden indicators. Some of the criteria will be expressible in a definite and precise manner and some of them may only be expressible in a fuzzy or imprecise manner. This, at least, ostensible lack of fungibility makes for significant difficulties when attempting to compare alternatives. These procedural difficulties, however, do not justify attempts to artificially restrict individuals and societies from incorporating their full range of value considerations into the evaluation process. Reflective decision-making rests on both outlining a set of criteria to faithfully represent the set of tentative objectives and working to exploit the information potential of this set of disaggregated, qualitative and quantitative criteria.

3. *In complex alternative selection problems, in which multiple objectives and multiple criteria predominate, the notion of an "objective," scientifically optimized, "best" alternative is a fallacy.*

Transforming a multicriteria problem into a single objective function problem will give rise to a global optimum, but it usually results in excessive oversimplifications and loss of problem complexity. Such an approach is methodologically inappropriate and inadequate to the task at hand. However, even if we were to assume that our complete spectrum of values could be represented quantitatively by a true, multiobjective, multicriteria vector optimization problem, such formulations almost never yield unambiguous global optima. This result is not a function of our limited analytical capabilities, it is an ineradicable characteristic of the multiobjective, multicriteria vector optimization problem. The sea of relative extrema cannot be culled without providing additional normative criteria to weight our relative preferences for the various objectives and criteria. These fundamental difficulties are only exacerbated if one addresses the collective choice problem by attempting to aggregate individual preferences to seek out even more elusive multicriteria potential pareto-optimal alternatives.

4. *The relative importance of different objectives and criteria cannot be fully characterized in advance of systematically investigating how the various criteria, which characterize the different alternatives, tradeoff.*

Learning can and does occur during the process of policy evaluation. Preferences are not fixed and underlying, as is often presumed, but revealed dynamically, through adaptive learning as new information is made available. Policy evaluation frameworks can foster this adaptive learning process and provide feedback for recalibrating our analyses. While an initial exploration of objectives is a prerequisite to evaluation, these objectives, and our expectations, are often tentative, unstable, ill-defined, and conflicting. This is an important reason why prescriptive approaches, which result in unjustifiable selections of "best" alternatives, are often poorly received and why a descriptive approach is more well-suited to improving decision outcomes. Pursuing this strategy requires investigating and describing the foreseeable consequences of the alternatives so that conflicts and tradeoffs between the alternatives can be systematically and interactively assessed to understand critical, "emergent" properties.

5. *The impossibility of reflecting non-fungible social objectives with "value free" mathematical optimization techniques clarifies why the process of choosing multicriteria policies and projects amounts to a normative exercise of judging most acceptable or best compromise alternatives.*

Substantively rational decision approaches, by themselves, are inadequate for the task at hand. We cannot generate justifiable "best compromise" alternatives only by applying logical reasoning to valid premises and clear, quantifiable objectives. An adaptive, deliberative process that recognizes ethical commitments and community interests along with scientific data and intuition is needed. Procedural rationality leads to descriptive (as opposed to prescriptive) decision aids and pluralism in methodology. Procedural rationality recognizes that the process of choosing alternative policies is a social process that should not be artificially separated into technical and political components. Plural rationalities must be combined into one integrated process whose primary objective is to carefully consider, within the constraints of reason, prudence, and feasibility, the broadest range of perspectives.

6. *So as to not a priori bias the decision-making process toward a narrow subset of alternatives before the decision problem has been adequately explored, the widest possible range of alternatives should be sought out and considered.*

This proviso is both to ensure that we do not foreclose our options and to ensure that we allow for surprises. Such a strategy is particularly important when enduring impacts are possible, there is potential for irreversibilities, and the range of impacts are qualitatively diverse. It suggests, for instance, that when considering alternatives that will increase supply we should also consider those that might reduce demand. When considering alternatives we must attempt to suspend our implicit biases. This does not suggest an illusory quest for traditional objectivity, but rather a conscious awareness of its practical impossibility. Alternatives must be pruned, but they should be eliminated only when careful analysis demonstrates that they are clearly dominated by other alternatives, not simply because of preconceived notions or prejudices.

7. *Striving for idealized comprehensiveness in planning, however, is both unwise and dangerous.*

There are two important ways in which comprehensiveness is limited. First, there are limits to human information processing capabilities. All relevant alternatives, objectives, and criteria cannot be considered, no less considered simultaneously. Second, even abbreviated comprehensiveness comes at the cost of significant time and financial investment. Sound planning requires finding an appropriate balance between conflicting objectives such as action *and* reflection and specificity *and* generality. We must act creatively and cautiously to limit problem complexity while acting to enhance our ability to comprehend complex problems.

8. *Ultimately, decision criteria must be aggregated, at least in some sense, to allow for inter-alternative comparisons. The methods and dynamics of aggregating procedures will affect the decision outcome.*

There is an important process distinction between assuming the viability of fungible criteria and single objective functions and recognizing the specious nature of such assumptions, while still conceding the pragmatic importance of making inter-alternative comparisons. A more defensible strategy, that incorporates this second perspective, calls for thoroughly exploring the consequences of alternatives, in their representative criteria, before making an attempt to weight and aggregate the criteria. Such a strategy is more consistent with an iterative, interactive, and dialectical approach for weighting criteria than a prescriptive or optimizing approach. When comparing alternatives we must consider the potential for zero-sum results, in which every gain is balanced by an equivalent loss and the possibility for non zero-sum results, in which there can be both winners without losers and losers without winners. In the case of non zero-sum results, no assertions about the balancing of wins and losses is offered. Flexible, staged analyses that iteratively hone in on the most troublesome issues, without expending undue effort on less consequential issues, facilitate such a search.

9. Because the selection of multicriteria policies and projects is a social process it should encourage public participation and support equal access for all stakeholders.

Social values can only be communicated by citizens, acting as both experts in their field and as active members of society. The public's interest cannot be adequately served by a small group of experts acting as surrogate representatives. Open democratic decision processes help to: combat against technocratic decision-making, prevent oversight, guard against decision-making that caters to narrow subgroups, and garner broad support for the final outcome. From the earliest stages of analysis, proactive as opposed to reactive citizen involvement must be encouraged. Citizens must be able to actively participate in the policy formulation process, not just react to decisions that have already been formulated by "experts." Limiting citizen participation to commenting on proposed policies or speaking at public hearings after the decision process is far along does not adequately address this concern.

10. Because highly technical analyses are not easily digested and because they often send obfuscating, unintended signals, a viable decision framework must convey its results in a manner that is transparent to experts and citizens alike.

So as to facilitate open and free debate, the models and assumptions that undergird our analyses must be intelligible and accessible. The tradeoffs between the alternatives must be explicit and clearly delineated, rather than hidden. The sensitivity of the analysis to important assumptions and uncertainty must be considered. And the conclusions must be justified with evidence and argument, keeping specialized language to an absolute minimum. Even if citizens do not wish to actively participate in the evaluation or selection process they have a right to have access to the decision-makers' assumptions, analyses, and justifications for pursuing a particular policy or project.

5.   Outline of a Descriptive, Participatory Approach to Environmental Policy Analysis

The notion of "best" or optimal policies needs considerable explication. The idea of an objective "best" alternative, agreed on unanimously, is usually illusory. The notion of a socially determined "best compromise" or optimal policy, however, as in a wise choice made through careful deliberation of the options and weighing of the tradeoffs, in an open process that cultivates public participation, is both imaginable and feasible.

As Frank Fischer emphasizes efforts to improve policy by promoting collaborative learning and discourse face two primary epistemological questions (1990: 375). First, there is the difficulty of integrating qualitative and quantitative knowledge. Second, there is the problem of integrating the expert's scientific-technical knowledge with the citizen/client's "ordinary" knowledge.

In response to the first question, I call for a multistage, multicriteria decision framework that incorporates mixed data, compensatory and noncompensatory criteria, and plural rationalities. The general concept is to view the process of policy analysis as a learning exercise - as an experiment based on an emerging set of "what-if" questions and criteria weighting scenarios[7]. To use anticipatory analysis to avoid problems, we must constantly test and revise our assumptions, recognizing that our preferences emerge as we learn more about the various consequences of alternatives.

In response to the second question, I argue for an open policy process that solicits public participation from the earliest stages of analysis. A public deliberation process based on open debate can facilitate the explicit discussion of values and underlying premises. By doing so it helps to reveal and test hidden assumptions as well as avoid misunderstandings and clarify areas of more serious conflict. Furthermore, the only feasible way to incorporate intangibles is through discussion and debate. In this task "experts" and "citizens" must be seen as collaborators and equal partners.

12 STEPS OF A DESCRIPTIVE, PARTICIPATORY APPROACH TO POLICY ANALYSIS

1. Develop a tentative outline of the assessment problem

The first step involves clarifying the boundaries of the assessment problem. Broad objectives for assessing the project or policy are defined in terms of eight general criteria categories that reflect the range of plural rationalities[8]. These categories include: ecological considerations, spiritual/aesthetic considerations, ethical considerations, social considerations, legal considerations, political considerations, scientific/technical considerations, and economic considerations. In many cases these criteria will be able to be further decomposed and represented qualitatively or quantified. In other cases nominal expressions for each alternative, relative to each criteria, will need to be defined.

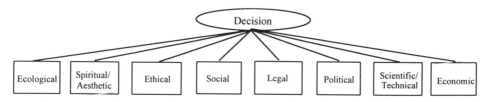

Figure 1. Eight Primary Criteria Categories

2. *Identify a set of feasible alternatives*

The primary objective here is to identify the widest possible array of alternatives for future consideration. The purpose of this step is not to eliminate alternatives, but rather to seek out a broad array of structurally unique and contrasting alternatives. No endogenous limitations are placed on the number of alternatives that may be considered, although the cost of data gathering and analysis must certainly be taken into account. The approach outlined is particularly well-suited for structuring alternative selection problems when there are 10 alternatives to consider.

The best way to create the possibility of a sustainable future is to sample the available possibilities in an organized fashion. We should not exclude any plausible alternatives simply because, upon first blush, they appear inconsistent with our worldview or, worse yet, because they do not fit into our framework for analyzing alternatives.

3. *Identify a tentative set of criteria*

The objective of step three is to exhaustively (within reason) identify a representative, disaggregated criteria set for the alternatives in question (see Figure 2, following page). The conjunctive constraints or noncompensatory (nontradeoffable) criteria should be identified. When appropriate, an effort should be made to exploit hierarchical properties of the systems in question to explore if representative high-level surrogates exist (cf. x_1 and x_4). In the case of economics, one single, high-level surrogate exists in the form of a benefit-cost ratio. In the case of the law, a selection of legal requirements may need to be satisfied. Similarly, in the case of science, a collection of climatological, air quality, mass balance, and energy balance criteria might be relevant.

4. *Prune criteria set*

The purpose of this step is to characterise a minimal criteria set that while certainly not exhaustive or complete, adequately represents the complexity of the assessment problem. The idea here is that there may be surrogate indicators that reflect a host of disaggregated sub-considerations (see Figure 2). Because conjunctive constraints are noncompensatory, great care must be taken when searching for high-level surrogates. Ensuring that the final attribute set is "minimal" involves checking to see that redundant objectives have been culled and that easy to evaluate surrogate attributes, which may represent a collection of other attributes, replace the disaggregated, lower-level attributes. This element of the fourth step is critical for reducing the level of analytical computation that will be necessary in the evaluation stages.

Note, however, that x_1 and x_4 are not entirely separable because they share x_{1131}, x_{1132}, and x_{1133}. As an example of a hierarchical surrogate consider fossil fuel consumption, which is a good surrogate for CO_2 emissions, particulate emissions, and reactive organic gas emissions from automobiles. In other cases, particularly where nominal criteria are involved - like Boulding's honor, justice, pride, and vice[9] or beauty, resilience, stability, love, and intelligence - further separability and decomposability are unlikely to be feasible. Following Miller's (1956) experimental results on human information processing limitations, and for reasons of accessibility and transparency, I recommend limiting the number of compensatory criteria to 10. When criteria are conjunctive constraints, like inelastic legal requirements or fixed pollution emissions standards, there are no restrictions on the number of these criteria that may be considered.

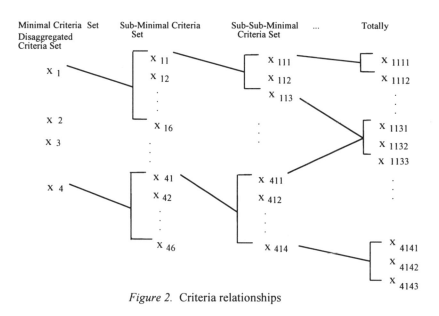

Figure 2. Criteria relationships

5. *Form effects matrix*

After the minimal criteria set has been chosen the task of data gathering, modeling, ethnobotanical surveys, social impact analysis, economic analysis, risk analysis, environmental impact analysis, etc., for building the effects matrix begins. Some very complex and detailed technical analysis may be necessary to acquire the effects matrix elements.

6. *Evaluate noncompensatory criteria*

Many objectives (criteria) are noncompensatory. There are two main types of noncompensatory criteria, deontological and satisficing. Deontological criteria are nonnegotiable and nontradeoffable. Endangered species cannot (at lonat not legally in the United States) justifiably be traded off against development, similarly minimal water quality standards cannot be traded off against economic efficiency. Satisficing criteria must fulfill minimum standards or not exceed maximum standards. A firm evaluating new pollution control technologies may be indifferent to further pollutant reductions as long as fixed criteria pollutant levels are not exceeded. Satisficing criteria must always be examined before applying methods for evaluating compensatory criteria because decision-makers may be indifferent to marginal increases (or decreases) in the criteria values once the initial constraints have been satisfied. In this case it would be inappropriate to further prune alternatives based on their being dominated with respect to these criteria. The conjunctive constraints approach can be used at this stage as a screen to evaluate noncompensatory criteria without requiring any preference information to be stipulated. Conjunctive constraints can also be established for deontological criteria by treating them as binary criteria (go/no go).

It may also be feasible to perform additional screening at this stage by iteratively and interactively making the conjunctive constraints successively more stringent. All noncompensatory, nominal criteria must be evaluated before performing additional pruning, lest feasible (globally nondominated) alternatives might be inadvertently pruned. The conjunctive constraints method might also be applied to compensatory criteria too. In this case, real-time graphics showing three-dimensional projections of the alternatives in normalized criteria space might be particularly helpful for illuminating appropriate cutoff levels.

After conjunctive constraints methods have been applied, it may also be appropriate to apply dominance relations at this point, although they frequently will not reduce the nondominated set substantially.

7. *Compare the remaining alternatives interactively under various criteria weighting scenarios*

The remaining criteria that must be considered are now all compensatory. A variety of methods for evaluating compensatory multiple criteria have been developed and reviewed (Lichfield, Kettle, and Whitbread, 1975; Goicoechea, Hansen, and Duckstein, 1982; Voogd, 1983; Nijkamp, Rietveld, and Voogd, 1990). To increase opportunities for learning, so as to facilitate adaptive preference formation, it is particularly appropriate to apply ideal point methods at this stage (Steur, 1986; Jannssen, 1991). These techniques are suitable for real-time analysis of the alternatives under different criteria weighting scenarios. They are also particularly suitable because their algorithm is quite transparent and thereby supports mathematically unsophisticated decision-makers in performing parametric scenario games.

As another alternative, ELECTRE methods (Benayoun et al., 1966; Hwang and Yoon, 1981; Roy, 1968, 1978; Roy and Bertier 1973; Roy and Bonyssou, 1986) are also particularly well-suited for being applied at this stage because, by relaxing the dominance criterion with concordance and discordance relations, they provide a selection of non-dominated alternatives that can be subjected to more detailed examination and public discussion and debate.

8. *Revise criteria, constraints, or alternatives and iterate, if necessary*

At this stage, depending on the size of the set of remaining non-dominated alternatives and the desired decision outcomes (single alternative, small set of non-dominated alternatives, etc.) any one or a combination of the above mentioned steps might be repeated.

9. *Perform sensitivity analysis and assess tradeoff relations*

This step may not be necessary if sensitivity analysis has been built into step seven. If interactive parametric studies under different criteria-weighting scenarios were not performed earlier, than they must be performed at this point.

10. *Choose one or more "preferred" options from the remaining non-dominated alternatives*

11. *Subject the set of "preferred" options to discussion and debate*

At this stage the transparency of the decision method plays a crucial role. In order for the remaining alternatives to be considered and critiqued the weighting assumptions and implicit assumptions built into the assessment algorithm must be readily discernible to decision-makers, multicriteria analysis experts, and citizens alike.

12. *Choose a single alternative, a collection of acceptable alternatives, no alternative, or iterate again*

6. Conclusion

An integrated policy framework, incorporating all varieties of mixed data, with compensatory and noncompensatory decision methods is necessary so that in our analytical search for "best" solutions we do not inadvertently eliminate second-best solutions in a technical screening. The aim is not to prescribe particular "best" decisions, but to offer an alternative, descriptive paradigm for *guiding* the comparative evaluation and public assessment of alternative projects and policies based on accessible, well-defined, and highly certain effects data. In particular, I focus on explicitly incorporating into the evaluation and assessment process detailed information about the interconnected, coevolving causes of environmental problems. The motivating premise has been that, from the standpoint of hindsight, many historical decisions that led to environmental crises often appear to be avoidable. The linchpins, however, seem to be in the process of exploring tradeoffs, making them understandable, and maybe most importantly in how we address the inevitability of conflicts in the process of social decision-making.

The focus of my descriptive, participatory framework is clearly not *a priori* prescriptive, to find a single "best" policy. I have argued that such a goal is at best illusory, and at worst disingenuous. My argument, rather, is that for complex policies and projects a circumspect and lucid outlining of the foreseeable consequences of a broad range of feasible alternatives is a necessary prerequisite for both informed discussion and debate and for effective decision-making. The framework has four key elements that distinguish it from orthodox approaches to policy analysis, which are grounded in comprehensive rationality. First, it incorporates multiple objectives and all forms of mixed, qualitative and quantitative criteria. Second, it incorporates noncompensatory and compensatory methods in successive stages. Third, it is based on the principle that social systems are "inquiring," and that this dynamic, adaptive learning component can be tapped to generate improved policy. Finally, it is compatible with several different methods for weighting and aggregating multiple criteria. While critics might argue that they could identify the same preferred policy with a single objective function optimizing approach, assuming they had all the necessary information, this is only partly true. They could most certainly arrive at the same preferred policy by chance, but to argue that they could identify it by analytic reasoning, taking the *same* considerations into account, is to invoke *petitio principii*. These processes and their assumptions preclude the accessing of information gleaned through the dynamic, adaptive learning process, that is critical to make these decisions. The novelty of this framework is that it integrates all of this information from the outset, While also incorporating other factors, particularly qualitative information, that may not be accessible to other more restrictive frameworks.

References

Anderson, C.W. (1979) The Place of Principles in Policy Analysis, *The American Political Science Review* **73**, pp. 711-723.

Arrow, K.J. (1974) *The Limits of Organization*, Norton, New York.

Banfield, E.C. (1994) Ends and Means in Planning, in Faludi, A. (ed.) *A Reader in Planning Theory*, reprint, Pergamon Press, Oxford, pp. 139-149.

Baumol, W. and Oates W. (1988) *Theory of Environmental Policy*, Cambridge University Press, Cambridge.

Benayoun, R., Roy, B. and Sussman, N. (1966) *Manual de Reference du Programme Electre, Note de Synthese et Formation*. Direction Scientifique SEMA, Paris.

Berkes, F. (ed.) (1989) *Common Property Resources: Ecology and Community-Based Sustainable Development*. Bellhaven Press, London.

Berman, T., Ingram, G.B., Gibbons, M., Hatch, R.B., Maingon, L. and Hatch, C. (1994) *Clayoquot and Dissent.* Ronsdale, Vancouver, B.C., Canada.

Breheny, M. and Hooper, A.J. (eds.) (1985) *Rationality in Planning: Critical Essays on the Role of Rationality in Urban and Regional Planning*, Pion Limited, London.

Caldwell, L.K. (1977) *Quantifying Values: Four Intractable Problems.* A discussion paper, prepared for the Workshop on Environmental Values Research, held at the Institute on Man and Science, Rensselaerville, NY.

Carroll, J.D. (1973) Participatory Technology, in ed. Barbour, I.G. (ed.) *Western Man and Environmental Ethics: Attitudes Towards Nature and Technology*, Addison-Wesley, Reading, Mass., pp. 204-224.

Charnes, W. and Cooper, W.W. (1964) Constrained Extremization Models and Their Use in Developing Systems Measures, in Mesarovi´c, M.D. (ed) *Views on General Systems Theory: Proceedings of the Second Symposium at Case Institute of Technology*, John Wiley, New York, pp. 61-88.

Churchman, C.W. (1968) *The Case Against Planning: The Beloved Community.* Space Sciences Laboratory, Social Science Project, Internal Working Paper 80, University of California, Berkeley.

Costanza, R., d'Arge, R., de Groot, R. Farber, S., Grasso, M., Hannon, B., Limburg, K., Naeem, S., O'Neil, R.V., Paruelo, J., Raskin, R.J., Sutton, P. and van den Belt, M. (1997) "The Value of the World's Ecosystem Services and Natural Capital." *Nature* **387**, pp. 253-260.

Dakin, J. (1963) An Evaluation of the 'Choice' Theory of Planning., *Journal of the American Institute of Planners* **29**, pp. 19-27.

Dasgupta, P., Sen, A. and Marglin, S.(1972) *Guidelines for Project Evaluation*, United Nations, Vienna.

DeSoto, D.B. (1961) The Predilection for Single Orderings, *Journal of Abnormal Social Psychology* **62**, pp. 16-23.

Diesing, P. (1962) *Reason in Society: Five Types of Decisions and Their Social Conditions*, University of Illinois Press, Urbana.

Dryzek, J.S. (1987) *Rational Ecology: Environment and Political Economy*, Basil Blackwell, Oxford and New York.

Eaton, J. (1995) From the Archives: A Master Plan to Develop Point Reyes, *The Wilderness Record* 20 January, pp. 3-7.

Ellul, J. (1992) Technology and Democracy, in Winner L. (ed.) *Democracy in a Technological Society*,. Kluwer Academic Publisher, Dordrecht and Boston, pp. 35-50.

Elster, J. Sour Grapes - Utilitarianism and the Genesis of Wants, in Sen, A. and Williams, B. (ed.) *Utilitarianism and Beyond*, Cambridge University Press, Cambridge, pp. 219-238.

Feiveson, H.A., Sinden, F.W. and Socolow, R.H. (eds.) (1976) *Boundaries of Analysis: An Inquiry into the Tocks Island Dam Controversy*, Ballinger, Cambridge, MA.

Fischer, F. (1990) *Technocracy and the Politics of Expertise*, Sage, Newbury Park, CA.

Gilroy, J.M. and Wade, M. (eds.) (1992) *The Moral Dimensions of Public Policy Choice: Beyond the Market Paradigm*, University of Pittsburgh, Pittsburgh.

Glacken, C. (1967) *Traces on the Rhodian Shore: Nature and Culture in Western Thought from Ancient Times to the End of the Eighteenth Century*, University of California Press, Berkeley.

Glasser, H, Craig, P. and Kempton, W. (1994) Ethics and Values in Environmental Policy: The Said and the UNCED, in van der Straaten, J. and van den Bergh, J. (eds.) *Concepts, Methods, and Policy for Sustainable Development*, Island Press, Washington, D.C., pp. 80-103.

Glasser, H. (1998) Ethical Perspectives and Environmental Policy Analysis, in van den Bergh, J.(ed.) *Handbook of Environmental and Resource Economics*, Edward Elgar, Cheltenham, U.K., forthcoming.

Glasser, H. (1995) *Towards a Descriptive, Participatory Theory of Environmental Policy Analysis and Project Evaluation*, Ph.D., Department of Civil and Environmental Engineering, University of California, Davis.

Goicoechea, A., Hansen, D.R. and Duckstein, L (1982) *Multiobjective Decision Analysis with Engineering and Business Applications*, John Wiley, New York.

Gregory, R. (1993) Valuing Environmental Resources: A Constructive Approach, *Journal of Risk and Uncertainty* 7, pp. 177-197.

Gupta, A. (1981) Monitoring of Rural Projects Through People's Participation, *Ekistics* **291**, pp. 434-442.

Hall, P. (1992) *Urban and Regional Planning.* 3rd ed., Routledge, London.

Hill, M. (1968) A Goals-Achievement Matrix for Evaluating Alternative Plans, *Journal of the American Institute of Planners* **34**, pp. 19-29.

Hill, S. (1992) *Democratic Values and Social Choice*, Stanford University Press, Stanford, CA.

Hochschild, A. (1995) Amazon Nation: In the Heart of the Colombian Rain Forest, Indians are Returning to Their Traditional Way of Life, *San Francisco Magazine*, July 23, pp. 14-17, 26-32.

Holling, C.S. (1978) *Adaptive Environmental Assessment und Management*, John Wiley & Sons, . New York.

Hwang, C.L. and Yoon, K. (1981) *Multiple Attribute Decision Making*, Springer-Verlag, . New York.

Ingram, H. (1988) Comments, in Rosen, H. and Reuss, M.(eds.) *The Flood Control Challenge: Past, Present, and Future (Proceeding of a National Symposium, New Orleans, Louisiana, September 26, 1986)* Chicago: Public Works Historical Society, pp. 49-51.

Janssen, R. (1991) *Multiobjective Decision Support for Environmental Problems*, Ph.D., Vrije Universiteit Te Amsterdam, Amsterdam.

Kemmis, D. (1990) *Community and the Politics of Place*, University of Oklahoma Press, Norman, OK.

Kempton, W., Boster, J.S. and Hartley, J.A. (1995) *Environmental Values in American Culture*, MIT Press, Cambridge, Mass.

Kuhn, T.S. (1977) Objectivity, Value-Judgment, and Theory Choice, in Kuhn, T.S. (ed.) *The Essential Tension: Selected Studies in Scientific Tradition and Change*, University of Chicago Press, Chicago pp. 320-339.

Kuhn, T.S. (1962) *The Structure of Scientific Revolutions*, University of Chicago Press, Chicago.

Lee, K.N. (1993) *Compass and Gyroscope: Integrating Science and Politics for the Environment*, Island Press, . Washington, D.C.

Lichfield, N., Kettle, P. and Whitbread, M. (1975) *Evaluation in the Planning Process*, Pergamon Press, Oxford.

Lichfield, N. (1968) Economics in Town Planning, *Town Planning Review* **38**, pp. 5-20.

Lichfield, N. (1996) *Community Impact Evaluation*, University College of London Press, London.

Lindblom, C. (1959) The Science of Muddling Through, *Public Administration Review* **19**, pp. 78-88.

Lindblom, C. (1965) *The Intelligence of Democracy*, Free Press, New York.

Ludwig, D., Hilborn, R. and Walters, C. (1993) Uncertainty, Resource Exploitation and Conservation: Lessons from History, *Science* **260**, 2 April, pp. 17-36.

Majone, G. (1989) *Evidence, Argument, and Persuasion in the Policy Process*, Yale University Press, New Haven.

March, J. (1978) Bounded Rationality, Ambiguity, and the Engineering of Choice, *Bell Journal of Economics* **9**, pp. 587-608.

Marsh, G.P. (1965) *Man and Nature: Or Physical Geography as Modified by Human Action* (reprint of 1865 ed.), Harvard University Press, Cambridge, MA.

Matas, R. (1995) B.C.'s Logging Decision Celebrated, *The Globe and Mail*, July 17.

Merchant, C. (1980) *The Death of Nature: Women, Ecology, and the Scientific Revolution*, Harper and Row, New York.

Miller, G.A. (1956) The Magical Number Seven, Plus or Minus Two: Some Limits on Our Capacity for Processing Information, *Psychological Review* **63**, pp. 81-97.

Mishan, E.J. (1988) *Cost-Benefit Analysis*, 4th ed., Unwin Hyman Ltd., London.

Moore, D. and Galinkin, M. (eds.) (1997) *The Hidrovia Paraguay-Paraná Navigation Project: Report of an Independent Review*, Environmental Defense Fund and Fundação Centro Brasileiro de Referância e Apoio Cultural (CEBRAC), Washington, D.C.

Morse, B. and Berger, T. (1992) *Sardar Sarovar: Report of the Independent Review*, Canada, Resources Futures International, Ottawa.

Mumford, L. (1964) Authoritarian and Democratic Technics, *Technology and Culture* **5**, pp. 1-8.

Nelson, R.R. and Winter., S.G. (1982) *An Evolutionary Theory of Economic Change*, Belknap Press of Harvard University Press, Cambridge, MA.

Nijkamp, P., Rietveld, P. and Voogd, H. (1990) *Multicriteria Evaluation in Physical Planning*, North-Holland, New York.

Norgaard, R.B. (1992) Environmental Science as a Social Process, *Environmental Monitoring and Assessment* **20**, pp. 95-110.

Norgaard, R.B. (1989) The Case for Methodological Pluralism, *Ecological Economics* **1**, pp. 37-58.

Parat, A.M. and Haas, J.A. (1969) Information Effects on Decision-Making, *Behavioral Science* **14**, pp. 98-104.

Pearce, D.W. and Turner, R.K. (1990) *Economics of Natural Resources and the Environment*, Johns Hopkins University Press, Baltimore.

Ponting, C. (1993) *A Green History of the World: The Environment and the Collapse of Great Civilizations*, Penguin Books, London.

Popper, K.R. (1966) *The Open Society and Its Enemies* (two volumes), Princeton University Press, Princeton.

Prest, A.R. and Turvey, R. (1965) Cost-Benefit Analysis: A Survey, *The Economic Journal* December, pp. 683-735.

Primack, J. and von Hippel, F. (1974) *Advice and Dissent: Scientists in the Political Arena,* Basic Books, New York.

Rittel, H.W. and. Webber, M.M. (1973) Dilemmas in a General Theory of Planning, *Policy Sciences* **4**, pp. 155-169.

Roy, B. and Bonyssou, D. (1986) Comparison of Two Decision-Aid Models Applied to A Nuclear Power Plant Siting Example, *European Journal of Operations Research* **25**, pp. 200-215.

Roy, B. and Bertier, P. (1973) La Methode Electre II: Une Application au Media-Planning, in Ross, M. (ed.) *Dublin VIIeme Conference Internationale de Recherche Operationelle,* North-Holland, Amsterdam, pp. 291-302.

Roy, B. (1968) Classement et Choix en Presecé de Points de Vue Multiples (La Méthode ELECTRE), *R.I.R.O.* **2**, pp. 57-75.

Roy, B. (1978) ELECTRE III: Un Algorithme de Classement Fondé sur une Représentation Floue des Préferences en Présence de Criteres Multiples, *Cahiers du Centre d'Etude de Recherche Opérationnelle* **20**, pp. 32-43.

Sagoff, M. (1986) Values and Preferences, *Ethics* **96**, pp. 301-316.

Sagoff, M. (1988) *The Economy of the Earth,* Cambridge University Press, . Cambridge and New York.

Samuelson, P.A. and Nordhaus, W.D. (1992) *Economics.* 14th ed., McGraw-Hill, , New York.

Self, P. (1974) Is Comprehensive Planning Possible and Rational?, *Policy and Politics* **2**, pp. 193-203.

Shepard, R.N., Hovland, C.I. and Jenkins, H.M. (1961) *Learning and Memorization of Classifications.* Psychological Monographs: General and Applied vol. **75**(13), American Psychological Association, Washington, D.C.

Shepard, R.N. (1964) On Subjectively Optimum Selections Among Multiattribute Alternatives, in Shelly II., M.W. and Bryan, G.L. (eds.) *Human Judgments and Optimality,* John Wiley, New York, pp. 257-281.

Simon, H.A. (1972) Theories of Bounded Rationality, in McGuire, C.B. and Radmer, R. (eds.) *Decision and Organization,* North-Holland, , Amsterdam, pp. 162-176.

Steuer, R.E. (1986) *Multiple Criteria Optimization: Theory, Computation and Application,* John Wiley and Sons, New York.

Stokey, E. and Zechauser, R. (1978) *A Primer for Policy Analysis,* W.W. Norton, New York.

Tribe, L.H., Schelling, C.S. and Voss, J. (eds) (1976) *When Values Conflict: Essays on Environmental Analysis, Discourse, and Decision,* Ballinger, Cambridge, MA.

Voogd, H. (1983) *Multicriteria Evaluation for Urban and Regional Planning,* Pion Limited, London.

Walters, C. (1986) *Adaptive Management of Renewable Resources,* Macmillan, New York.

Webber, M.M. (1983) The Myth of Rationality: Development Planning Reconsidered, *Environment and Planning B: Planning and Design* **10**, pp. 89-99.

Yuba Watershed Institute (1994) Timber Framer's Guild of North America, Bureau of Land Management. *The 'Inimim Forest Draft Management Plan.*

Zadeh, L.A. (1973) Outline of a New Approach to the Analysis of Complex Systems and Decision Processes, *IEEE Transactions on Systems, Man, and Cybernetics* **3**, pp. 28-44.

---

1. The present paper draws liberally from Chapters 2 and 5 of my Ph.D. dissertation, Towards a Descriptive, Participatory Theory of Environmental Policy Analysis and Project Evaluation (Glasser, 1995).

2. The evaluation phase is weakly related to the design phase that produces the alternatives. I assume that before reaching the evaluation phase the different alternatives go through a systematic critical design analysis. Unlike evaluating discrete alternatives, this critical design analysis is a continuous problem because the decision parameters can be adjusted freely. An example would be to determine the appropriate amount of electricity to supply and the best mix for the year 2000 under competing demand, environmental, population, and cost constraints. This problem is one of continuous evaluation because the particular supply mix and quantities of supply from each form in the mix have not been pre-specified. If all of the alternatives are rejected in the evaluation phase, the insights from this evaluation can be fed back to improve the re-design process. For an overview of urban and regional planning and the planning process itself, see Peter Hall (1992).

3. Examples include: the perception, in Roman times, that reduced agricultural productivity was due to senescence (Glacken, 1967: 134, 136); the belief, in ancient Greece, that metals reproduced themselves "through small metallic seeds" (Merchant, 1980: 29); and the more recent tacit assumption that the environment is a limitless sink for residuals.

4. A variety of interpretations of this model have been offered (Lindblom, 1965: 137-138; Banfield, 1994: 140; Rittel and Webber, 1973). None of these authors, however, addresses my first consideration. Their interpretations are all relatively consistent with assumptions two, three, and four; they differ only in form and precision rather than substance.

5. A more thorough discussion of these flaws of the comprehensive rational model is developed in Glasser (1995), Chapter 2.

6. For a positive example of such reflection by high-level policy analysis practioners see Glasser, Craig, and Kempton (1994).

7. This idea of policy analysis as an experiment and learning exercise is drawn from Holling (1978) and Walters (1986) work on adaptive environmental management. See also Kai Lee (1993) for a more recent extension of the principles of adaptive management to sustainable development.

8. The first category, ecological considerations, is drawn from Dryzek's (1987) concept of ecological rationality. The last five categories are drawn from Diesing's (1962) concepts of plural rationalities. I have included two additional categories, ethical considerations and spiritual/aesthetic considerations.

9. See Boulding's poem on the limitations of using single-objective function optimizing models for complex public policy problems (Charnes and Cooper, 1964: 61).

EVALUATION IN THE DIGITAL AGE

M. BATTY

1.   Introduction

As the convergence of computers and communications continues, every aspect of the city is becoming computable. We face the prospect of a world in which the science that we use to analyze and plan is simultaneously changing that very world. In an era of such immediacy, the role of science and of planning begins to change for we can contemplate involving many interests hitherto excluded, previously peripheral to the design of the future. Digital communications, the net and virtual reality systems where the emphasis in on linking all to all through the visual media herald a multitude of opportunities for the many to interact and decide where few have done so in the past. We chart this emerging world where seemingly unlike phenomena and interests can be juxtaposed. To give some sense of where we might be heading, we demonstrate three possible pointers: first how the science of cities built around mathematical models developed by the very few might be made accessible to the many through VR systems; second how we can develop network-based systems where the many can interact remotely and discuss planning and design online within a visual context; and third, how the many can develop their own analyses though simple software which enables them remotely to explore and manipulate data through their own GIS.

In all these possibilities for applying computers to planning, evaluation is central. Evaluation is the process of assessing the relevance of plans but it is also central to the way diverse interests who are not usually involved in the scientific design of planning proposals bring their interest to the process. In this, the dramatic developments in ways of communicating using new digital media are central and although computers can be used at every stage of the planning process, it is in evaluation that the greatest opportunities for using the new media lie.

As we approach the millennium, it is not computers or communications systems per se that will dominate the way we will understand, plan and build our cities of the 21st century but the very activity of digital computation. It is digital computation that represents the singly most important transition from the industrial to the postindustrial world. Computation does not simply mark the divide between mechanics and

N. Lichfield et al. (eds.), Evaluation in Planning, 251–272.

electronics but between the world of the tangible, hard, material, and the emerging world of the intangible, soft, ethereal. Over half a century ago, the great pioneers of computing, Turing and von Neumann, and before them, Godel, Boole, and back to Leibniz, realized if phenomena could be represented using binary distinctions, then there was the prospect of building a single machine which might manipulate phenomena of very different kinds. These pioneers invented the basic logics which showed that digital computation was possible but they did not invent the machine - the computer. These came from a very different source, and during the last 50 years, society has been dominated by the development of ever more powerful computers. In this headlong rush, the notion that everything might ultimately become computable using these same machines has, at best, remained implicit. Apart from some prescient speculation such as the insight of Vannevar Bush (1945), only now is there a dawning realization that the age of the universal machine is almost upon us.

Computation is about manipulating phenomena which can be represented digitally. Until quite recently, most computation was concerned with manipulating digital artifacts on single computers where the purpose has been to produce some result or product, which is then input to some other set of activities, often nondigital. But in the last decade and especially since the rise of the internet as a global phenomena, the emphasis in computation has changed from manipulation to communication of data and ideas. The quest is increasingly communication and it is the process rather than the product which is all important. This has, of course, become possible from a convergence of computers with communications systems whereby digital information can be transmitted at high speeds across networks, thereby enabling computation to take place 'at-a-distance'. Computation is becoming a strange mix of manipulating data and ideas but communicating the same for diverse purposes. And some products particularly those that relate to abstract concepts such as scientific ideas, financial transactions, and various forms of entertainment might never leave the digital media, existing entirely in digital space, in cyberspace as it has come to be called (Whittle, 1997).

It is the notion of 'communication' that raises the prospect that cities might be computable. Cities exist to facilitate communication, first and foremost as market places where trade is initiated and goods are exchanged. These markets have hitherto always been places where some physical transaction has taken place even though the media might have been symbolic, through money, for example. But the prospect is before us that these markets might not longer reside in the material realm, that there now exist goods and services which will remain entirely in the ethereal realm and whose manipulation will inevitably be mediated through computation. This notion was raised nearly 40 years ago by Richard Meier (1962) in an early prelude to the computable city, and it was foreseen in Mel Webber's (1964) nonplace urban realms, and in the concept of the transactional city (Gottman, 1983). Of course, the idea that cities might be entirely computable is far fetched and it is difficult to foresee a time when cities will not be places where the material and the ethereal will interact in subtle and confusing ways.

But the ethereal realm is fast expanding, and within a generation, it will be significant in terms of the physical infrastructure of the city. In a lifetime, it will be commonplace.

The phenomenon of the computable city can be summed up in the following way. With the convergence of computers and communications and the generalization of computation to the manipulation of digital activity 'at-a-distance' so-to-speak, computers are becoming part of the very fabric which we are seeking to understand using those same computers. We are beginning to use software on computers to understand the use of that same software on those same computers for quite different and often contrary purposes. In one sense, planners have always been part of the system they have sought to plan and some have identified planners and plans as being part of the problem rather than its solution. Nevertheless, we have never faced a time when the very methods which we seek to enhance our understanding are the same as those used to confound the very phenomena we seek to unravel. This poses a conundrum. Combined with a multitude of subtle interactions between the material and the ethereal world, these new ways of understanding are paralleled by new levels of complexity in the city which these new ways of understanding are themselves generating. This is the phenomena of the computable city (Batty, 1997a).

Almost 20 years ago, James Martin (1978, 1981) coined the term 'Wired Society' in a book of the same name. In the intervening period, computer and communications systems have converged to provide a platform for the realization of this image. For a long time, computers simply stood alone. Users were connected to them remotely as early as the late 1960s but neither users nor computers interacted with one another. It was not until the late 1980s that computation in the traditional sense of the word began to drift onto networks and away from single computers. At first, most interaction between machines was in terms of data and information, much of it of immediate import such as file transfer and email. But increasingly computation itself has begun to take place over the net and information far from being passively communicated is being changed continually as it is communicated over and over again to differ-ent users in diverse contexts. The real challenge in fact is to understand how traditional forms of computation are being merged with new forms of computation through communication, and it is here that the biggest impacts on urban planning and decision making will take place. In this essay, we will address this question of communication as computation and relate it to various aspects of the planning activity, showing how communication at every level is providing new and innovative ways of addressing urban problems and their solution through plans.

First we will discuss new forms of computation as they are being manifested in software across nets and within computers. These will provide radically different kinds of opportunity for planning in the early 21st century (Kelly, 1994). We will examine the role of communication in planning, indicating how different processes of communication are likely to be affected by the transition to a digital world. We will then develop three different scenarios which combine participation and communication in different ways within the planning process. We will outline how traditional planning science using computers can be aided by virtual reality systems, how virtual worlds might be created within which users can interact with each other for a variety of tasks from professional to political, from passive to active, and how network systems can be used to provide data and analysis which a wide variety of users can manipulate. The examples that we will show involve the use of virtual reality software over network systems into which geographic information, traditional planning models, and digital varieties of user/public participation are embedded (Batty, 1997b). At present digital planning is characterized by a confusion of media and models: every kind of data and science as well as fact and value are capable of being juxtaposed within the digital world. We will allude to this confusion. In conclusion, we speculate on how digital planning will evolve in the short and medium term futures.

2.   Accessible Computation: VR and the Net

The history of 20th century computing is one which reflects the evolution of its elements - hardware, software, and users - coming closer together at every stage. In one sense, this is a process of integration but in another, it is one where every more users from an increasingly wide array of lifestyles and roles are finding uses for computers and computation in more and more diverse ways. This process has taken place in waves as computers themselves have swept into the infrastructure. Its magnitude is hard to measure but at present the accessibility which the population in general have to computers is probably proportional to the actual number of personal computers. In specialist groups, the accessibility to machines is much higher, for example in small research groups, the number of machines is likely to be two or three times the number of users but once we scale to the organization, then the number of PCs is as good a measures as any. This is particularly so in that virtually all network links are made by connecting such standalone machines.

Computers stood in splendid isolation from one another until quite recently. Because computers are universal machines, they have always attracted more than one user or at least have had the potential to execute more than one task, and from the beginning, a user community existed around single installations. But it was not until the mid-1970s, that users became physically connected to their computers from remote sites across dedicated network lines but using dumb terminals to communicate. These star networks enabled users to interact with machines for the first time, *en masse* so-to-speak, and gradually the idea of the file server existing at the central hub of the network caught on. The PC however blew this notion apart. The 1980s saw the massive growth of standalone machines which were by and large assumed to be "personal" and thus not "connectable". However as soon as PCs were introduced, they were connected to main frame and minicomputers through modem technology and although for most of the 1980s PCs were configured to act either as dumb terminals to remote hosts or as standalone machines, the notion of the client and its server was borne. In fact, there is still very little interactivity between machines which are connected together. The norm is that software and data are downloaded from servers to their clients while most processing takes place locally. In the workstation environment, more ingenious swapping is engendered but we are only just beginning to enter an era where computation is across many linked machines, on the net. This is a prospect for the early 21st century.

Software too shows increased accessibility to users. The whole notion of user-friendliness is built around the idea that visual and textual interfaces need to be constructed to make software more accessible, hence more usable. It was only after the advent of the PC, however, that the idea of software itself really took off. Before then, computation was the act of running a computer program but once PCs emerged it became immediately clear that the concept of a program which had always remained in the image of a calculation, was too restricted a notion. Inexorable miniaturization led to ever more memory begin devoted to graphics and eventually windows interfaces emerged in which different software running on the same machine became visually apparent. But there is still very little integration between software running on the same computer. On PCs, it is not quite possible to run more than one process simultaneously apart from various low-level operating system tasks and as yet, there is little communication between different software packages. However within the next decade, communication between software running on the same machine, on the same desktop, will be possible although this may related more the developments in hardware architectures than the needs for integrating different software from a user perspective.

Users themselves have begun to connect up across the internet. The sudden rise of the net is due to the convergence of modest computers with modest networks, both used for different purposes in the first instance - for standalone computing such as word-processing and for communication using telephones, for example. For a decade, professional users have shared files and mail across the internet, and a variety of other communications based on emergent cyber communities such as bulletin boards, internet gaming and so on have developed (Rheingold, 1993). Since the early 1980s, it has been possible to communicate directly with users logged onto the same machine but the concept of full interactivity is still on the drawing board. What the internet and the layer on top of this which is most used to communicate and view information - the world wide web (WWW) - has done is to broaden routine communication and to make remote access to other processors possible while extending the concept of client and server over much wider distances. It has not yet generated full interactivity for this depends upon users dedicated to the same tasks working on projects across the net and developing systems which enable group and team problemsolving. All this, of course, is under rapid development. Although various demonstrations of online working are available through the idea of cyberlabs and internet games, most communication at present is non interactive and involves single users communicating remotely in a passive way.

Developments in graphics and network software are likely to make true interactivity using computers possible, through new concepts in virtual reality (VR) systems in network and other interactive contexts. For the first time ever, the net provides *'complete connectivity'*. In principle and in practice if no constraints are put on communication, everyone can interact with everyone else. There has much been written on the notion that telecommunications and networking computing is annihilating distance but the real issue is connectivity. Already with the growth of the web, available information is being massively multiplied. Many strange worlds are already colliding in cyberspace and images of anarchy and chaos which pervade descriptions of the net are not far wrong at this early stage. Complete connectivity of course will never be the reality. Cyberspace is being tamed, and the net is being structured in various ways which reflect the power and influence of groups in society which have an influence over its form. But for the first time, there is a medium through which it is possible to reach very large numbers of people, to disseminate what has remained rather specific and somewhat esoteric, to communicate new ideas, and of course to educate. The power of the net in communicating science, for example, to many different kinds of audiences, cannot be underestimated.

VR represents the contemporary development of visual software, of techniques for visualization and of representation using computer graphics. The graphics interface has dominated the way we interact with computers for over a decade but as machines have got faster, bigger in memory, and smaller in size at exponential rates of growth, then the ability to process graphics fast enough to provide real time motion has engendered the development of graphics environments which are referred to as 'virtual'. VR began in a rather narrow context with single users immersing themselves within a visual environment and interacting with software which enabled them to move graphical objects in real time with various devices, often coordinated with their own body movement (Kawalsky, 1993). This is the world of the headset where users can see themselves and their own motion within the graphics world that they create. However, such experiments are highly restrictive in terms of interaction with other users. Theaters are now being built in which many participants can interact in conventional as well as virtual terms. For example, using the same software technologies, users can casually converse as they themselves interact with the virtual world, manipulating the world directly and observing changes in the world in complete coordination with their actions. This is not immersive VR but a half-way house involving a mix of real and virtual communication.

Two related developments must be noted. First, rather special kinds of theaters called CAVES are being built which represent immersive environments in which several users can be present, see each other in real terms but be surrounded by a complete screen environment - on all sides - where the projection of the graphics is onto all the surfaces of the room. This makes possible very different views of the phenomena which might involve both real and abstracted windows on the world. Different areas of the CAVE can be given over to these windows showing the way data, models, predictions might be communicated to all these users, and of course changed in real time. Second, the more pervasive development is the emergence of VR on the net for most of what stands alone on computers in terms of computer aided design and animation has now moved to the net. *VRML* (Virtual reality modeling language(s)) and related visualization software makes possible very fast real time motion across the net. Users can log on remotely from anywhere in the world and with appropriate hardware and software can run VR-like animations which others have prepared. With the interactivity that will break in the early 21st century, users will be able to change these models as they run. The latest developments bring VR on the net into the CAVE, the theater and the headset. The possibilities for different types of communication are endless but to make sense of what is possible, there must be some sense of what is required.

3.   Communicable Planning: Participation, the Net and VR Come Together

When planning was first institutionalized as part of modern western democracies, the process was entirely technical, operated by architects supported through philanthropists and reformers but implemented as top-down *fait accompli*. This model was only rarely modified for it was simply assumed that insofar as the public had any relevance to the process of preparing plans, this was purely to inform and educate, rather than involve in any direct way. Patrick Geddes' (1915, 1949) civic exhibitions were the nearest that the process ever came to participation and in the early years of this century when planning first became a function of the state, this kind of communication was regarded as radical. For the next 60 years, the process of involving groups other than the experts who prepared the plan slowly widened. By the time planning was being operated comprehensively in Britain at least, formal procedures for involving the public-at-large or at least those who were directly affected by any planning proposal were in place. This process had come to be known as public participation by the 1960s but it was conceived as something which was tacked onto the technical process of plan preparation rather than as an activity which suggested involvement of the wider public in the actual process of plan making.

The model that had emerged by the 1960s was one in which problems were defined and analyzed largely by planners, data was collected and analyzed and plans designed again by planners. Then came the process of evaluation which at first was technical, then narrowly political through elected representation, and finally open to public scrutiny from whence came objections and a limited form of involvement. The process was one-shot, seen as starting with scientific analysis of the problem, leading to design of solutions and then providing opportunities for participation. Schematically it iswhere **Pr/Da** provides the driving force from problems and data, **S** the science that interprets and suggests the meaning and impact of the problems on the city, **De** the process of designing plans, and **PP**, the process of involving wider political interests and the public-at-large. Some limited cycling of the process might occur in practice but the entire sequence was often only accomplished once. The process returned of course on a

entire sequence was often only accomplished once. The process returned of course on a slow cycle which was mandated by legislation but a return to intermediate technical stages which participation should require was rarely achieved.

This top-down process was attacked for its limited role in communicating plans during the widespread questioning of institutionalized planning which began in the 1960s (Davidoff 1965). Various alternative models were suggested based on bottom-up schemes in which various interests drove the planning process with technical help and advice but with the decision making firmly in the hands of a wider public. This advocacy planning as it was called lay at the basis of community action and community architecture. No formal models were suggested but idealizations of the process built around actual conflicts which usually pitted labor against capital, authority against the individual and so on, were developed in which openness through communication at all stages of the process was advocated. In one sense, the top-down rational decision model in which communication was an after the fact activity and the bottom-up community model in which communication dominated all stages of the process are ideal types. The actuality of decision making in planning and how different groups and constituencies are involved is somewhat different (Forester, 1989). The process is manifestly less technical autocratic and there are many more opportunities for the interaction of diverse groups (Healey, 1996). Classification of these groups is difficult but to give some sense of what is possible a crude typology will be suggested. The key stages in the process which require communication are identified above as those involving problem definition and data, scientific analysis, design and the evaluation of the plan itself. Distinct groups of interest involve scientific experts, planning professionals, the immediate political representatives in the process, and the wider public. Arraying these stages against these groups gives an idea of where communication might be engendered within the process: in fact, it is everywhere as we will suggest.

There is no clear view as yet as to how the new media might be embodied in the planning process. In one sense, it exists on all levels which communicate technical and highly professional concerns through to more prosaic and superficial issues, no less worthy. Enabling scientific insights to be communicated to non-experts is a clear possibility while deriving new insights for experts themselves is another. For every possible activity, there is communication across the spectrum from informed to less informed although it is unlikely that the new media will be homogeneously distributed throughout the entire process. Some structuring will occur, particularly where these possibilities are grafted onto institutional structures. Moreover different types of communication to different groups will exist in different contexts for much of what is now possible will be deployed differentially and not in the coordinated fashion which the rational planning process might imply. In short, participation and the communication with planners, the planning process and plans *per se* suggests that the one-shot model in which such participation occurs at the end of the cycle, be replaced with a model in which participation is central. This can be portrayed as where several different kinds of participation can be developed at the various stages of the process.

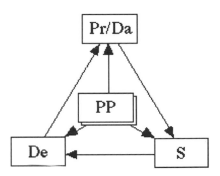

All of the technologies we are talking about are visual or rather enter the digital world through the display of phenomena in visual terms. This in itself is forcing us to juxtapose unlike data and ideas alongside, with abstractions of their systems and problems being directly comparable with realistic images. As we have implied, we have little idea as to the predominant applications of these new communications media which will define digital planning in the early 21st century but there are pointers to what is likely. Rather than attempt a comprehensive review, we will choose three which we suggest will develop rather quickly in the medium term and become commonplace in the fully wired world. First we will demonstrate how we can make the science of cities in terms of mathematical models traditionally used for analysis and forecasting intelligible as devices for portraying alternative plans. Then we will show how public participation might be engendered over the net with interests - individual or collective - coming together remotely in digital scenes from which meaningful discussion of alternative designs might take place. And third, we will show how data and its analysis can be effected at a simple but professional level by software which is available over the net through 'GIS for everyone'. These are but pointers in a wind of change which will define planning in the emergent digital world.

4. VIRTUAL URBAN MODELLING:THE OLD SCIENCE OF CITIES COMES ALIVE

Simulation models to predict spatial economic activity such as population, employment, retail turnover at shopping centers and such like were developed extensively in the 1960s when planning problems were conceived as problems of optimizing the efficiency of cities through manipulating spatial locations. These models were developed in the tradition of spatial economics and transportation behavior. They were computable using computers of those times and they provided forecasts of the impact of changing land use and transportation infrastructure on the distribution of demographic and other activities. Their development marked a watershed in thinking about a science of cities but the experience was plagued with problems. Computers were too small, data requirements too large and the policy focus too diffuse. A reaction set in and the world turned but before it did the seeds of their future development were planted (Lee, 1973; Harris, 1994).

Typical amongst these kinds of models were those which were orientated around the concept of spatial interaction or movement. Most of these were transportation models whose sole purpose was to predict the origins and destinations of trips and assign the resulting traffic volumes to the network so that demands for new infrastructure could be assessed. But a special class of such models were concerned not only with predicting interaction but also with summing such interaction at origins or destinations to provide measures of locational activity (Wilson, 1974). The most significant of these types of models simulated retail location as a direct function of the size of consumer expenditure at the origin of the trip and size of shopping center at the destination with distance or travel time acting as an inverse function. These shopping models were quite widely applied to predict the impact of carving up existing shopping hinterlands by locating out of town centers which were a comparatively new phenomena both in the US and western Europe in the 1950s and 1960s (Lakshmanan and Hansen, 1965).

Although the science of these models made slow progress as their initial problems were understood and resolved, they did in fact creep out into practice. Development companies and large retailing chains began to use them in helping to locate new stores, while planning agencies began to use them to marshall the case against the decentralization of shopping from town centers (Penny and Broom, 1988). In essence, these models were used in the crudest of ways. Usually sites for centers were identified using criteria other than that pertaining to the hinterland catchment population and its potential expenditure and then shopping models were used to compute the potential turnover of these locations. Very little "what if" types of analysis were accomplished for the models were dinosaurs in terms of their data and computing requirements. Slowly but surely as computers and data improved, their running became easier and in the last decade, they have become an important part of the emerging area of geo-demographics (Longley and Clarke, 1995). They also play an important part in the wider science of

GIS although much of this focus is data rather than model driven with exceptions (Birkin, Clarke, Clarke, and Wilson, 1996).

In parallel to these developments, progress with the theoretical structure of these models was made. From early work on their calibration to real situations (Batty, 1976), a great deal of attention was given to their structure. Spatial interaction models had been formulated as optimization problems as far back as the mid 1960s where the problem was one of maximizing some function such as accessibility or spatial utility subject to constraints on what was allowable and not allowable in terms of interaction. In the 1960s, Alan Wilson reformulated such models in an entropy-maximizing framework and the die was cast to examine these models as formal optimization problems. Such problems are nonlinear and thus not easy to solve although common versions of them have convex objective functions, hence unique solutions (Wilson, Coelho, Macgill, and Williams, 1981). But when cast into a semi-dynamic context where size of the center depends on turnover and vice versa in the shopping model context, for example, such problems become insoluble using standard algorithms. For example, if the problem is cast as one where the model is one which predicts turnover as a function of shopping center floorspace and distance between origins and destinations and if floorspace is also a function of turnover, then the equilibrium of such models is problematic. Moreover if this is now cast into an optimization framework in which it is required to optimize some function such as turnover or accessibility subject to appropriate constraints which embody the model, such problems can only be solved heuristically.

Consider formulating this problem visually. We have a map of where shopping centers exist with a fixed size of floorspace, and we have a retail turnover surface draped over the map, where each point on the turnover surface corresponds to the sales generated at the point on the map below. The surface is of course an interpolation from discrete points where turnover is generated in discrete centers. Distance is embodied in the map itself so all the key variables of the model are part of the visualization. As we change the independent variables of size and distance (location), then the turnover will change. In the past what we could do was to pick off sample points on the map surface and run the model for specific shopping center changes in size or location or both and examine the turnover generated. This is still the method used by the geo-demographics constituency. As we have implied, the problem is hard to solve in optimization terms so it is not possible to identify the best location for a given size and location of center. We can see this in visual terms. Imagine that we are trying to locate a new center where it would maximize turnover. If we place the center at the point where turnover seems least, then what happens is that the system redistributes its sales in such a way that the center which is placed at the best point is no longer optimal. In short, as we drag the shopping center across the map, the best position changes which is reflected in a forever changing retail turnover surface. This image of the problem can be given a much more definite form using virtual reality techniques.

What we are able to do is to set up the optimization visually so that the user of the system is literally able to drag the shopping center across the map and see the retail turnover surface changing. As the center is dragged towards the 'best position', that position is no longer 'best', but the user is able to 'feel' a way to the best position by continual experimentation of placing the center at many positions on the map. This entire approach depends upon having a computer and visualization system which is fast enough to recompute the model at motion picture rates, and to redisplay the map and surface at these same rates. As the shopping center is picked up by the mouse say, and dragged, the model is recomputed for every point where the center is dragged, the turnover surface interpolated and the screen refreshed at a rate around 30 frames per second. In short the shopping center and its location are the independent variables, the dependent variable is turnover, and the model connects them both. The optimization process is the visual act of dragging the independent variables into different positions. The process stops when the user has explored the retail surface and its possible best points or when a 'best' point has been arrived at. In more structured form this process can be represented as

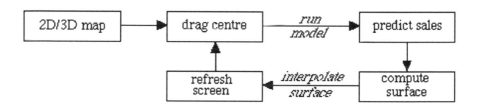

To do this in real time for a non-trivial problem where one might have hundreds of origins and destinations requires a very different computing environment from those on which such models have been run. We are fortunate at CASA to have a virtual reality system based on a SG Onyx Infinite Reality Engine 2 which renders 10 million polygons a second. Thus the visual refresh problem is immediately soluble. The modeling problem is severe but we are at work on this; gravitational models of the kind developed in this field are separable in some measure and small change can be approximated without recourse to massively parallel processing. We believe we will be able to implement such a system quite effectively using the Onyx.

However the real issue is who uses the visual model and in what context. The VR theater which is able to project these images contains large screen projection around which decision- makers and scientists of many kinds can gather and discuss the problem. We can project the problem into this theater and various users can pick up any shopping center in the scene and move it, exploring the surface and finding the "best" location. This can be done for every center in the problem and of course for every point of consumer expenditure and for every distance measure as it is reflected in the highway system and all clusters of these variables thereof. In short any independent variable or group of such variables can be altered on-the-fly and the problem visualized in real time. Of course, there are many aspects of the problem which are more abstract than this visualization and we can easily add these to the interface. So while the centerpiece might be the virtual reality of the real map, we would have various more abstract visualizations of the problem alongside to aid in exploration and decision-making. Extending this kind of modeling into the CAVE environment provides an obvious next step.

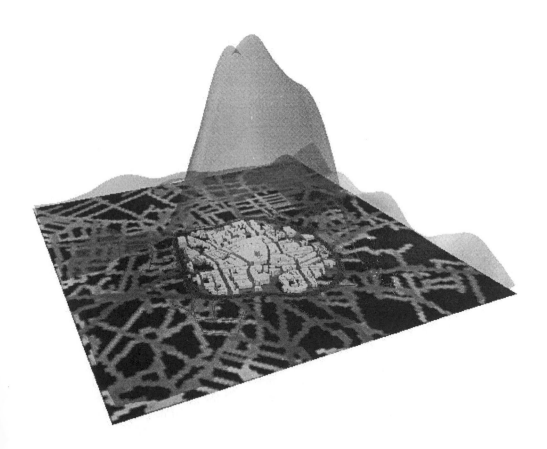

Figure 1: Visual Solutions to Complex Location Problems: Retail Turnover Associated

A picture of what is possible is illustrated in Figure 1 where we show the retail surface of Wolverhampton draped over a 2D map onto which a block model of the town center has been extruded from the basic 2D map geometry. We can fly around this surface at any speed we like using the Onyx, that is the speeds we can fly at are faster than the speeds we can push the mouse, assuming we are the only users of the system and that the link is directly into the VR Theater using a *VRML* representation within the *Cosmo Player (A Netscape* Plugin). We consider that once we move to dedicated VR software, then we will have no difficulty developing the system that we have speculated upon here. A generation or more ago in a prescient and somewhat premature project in Allende's Chile, Stafford Beer (1971) attempted the same kind of visualization for the macro economy. The 'operations room' that he established was meant to be the intelligence hub of the economy with the Map and Block Model of Wolverhampton's Town Centre just as our use of the VR theater in which to display and manipulate this retailing problem provides an environment to enable intelligent locational decision-making. 30 years on it will all be commonplace.

5.   Virtual Urban Design: Avatars in Whitehall

Participation in planning in the institutional context of the public domain is largely passive with an emphasis on providing relevant interest groups with information and interpretations of that information. There is minimal opportunity for interaction in the process apart from the statutory procedures of representation and appeal. In contrast, advocacy planning is quite the opposite in that plans and designs emerge from the grassroots but these processes are usually dominated by interests which are highly structured and do not allow much *ad hoc* participation. In both contexts and the continuum between, there is considerable effort involved in organizing the participative process which mitigates against the involvement of the public in general. Because the media involve face to face interaction, this precludes *ad hoc* response and often leads to the charge that such participation is elitist and narrow, from whatever perspective.

Communication using digital media opens up a world of potential participation which is unprecedented. Since the beginning of the electronic age, there has been wild speculation about push-button voting on issues through media such as simple interactive television. But the wider context of information and information systems for all has only become a possibility since the rise of the net and the convergence of computing with low cost, extensive connectivity. In the first instance, the net provides a major source of low-cost upto-date information which can be targeted at many audiences with diverse interests in the planning process. In developing the idea of the computable city, using the net for retrieving possible information is the first stage of communicable computation. Examples range from scientific involvement in retrieving data from remote hosts to learning about the problem through various analytical systems. In terms of involving a wider interested public, many sites specializing in disseminating planning information organized by public agencies at all levels have appeared during the last two years. Electronic journals in planning, bulletin boards, sites listing where development and planning permits have been sought and issued are fast becoming part of the infrastructure of the digital age. But all these with the possible exception of those used for downloading data are passive. Of course, information can be passed interactively through various kinds of forms or email but fully interactive exchange between remote users is in its infancy.

To demonstrate what is now possible and what will be commonplace, we suspect, in a fully wired world, we will illustrate the kind of virtual urban information which makes possible fully interactive participation through dialogue and design. Urban problems are usually represented visually through 2D and 3D maps and models but pictures, video clips and other visual media are important to communication and response. In CASA, an early project involved building a visual information system about the buildings and rooms in the college which was interfaced through a digital map and video interface. The UCL Navigable Movie as it was called was essentially a map from which pointers to videoed scenes viewed in *Quick Time* were hotlinked through a *Hypercard* stack. The power of hotlinking is of course at the basis of the world wide web and the system was easily convertible to a web browser. Bits and pieces can still be viewed at http://www.ge.ucl.ac.uk/ The idea of hotlinking 2D to 3D scenes and vice versa but in the wider context of linking to other sites remotely on the web is a significant notion which can be widely exploited in the construction of virtual urban information systems.

Andy Smith has taken this idea into a modest but potentially powerful product which has many applications in a participative context (Smith, 1997). He has built such an interface to information about places of interest in central London, in the government and entertainment heart of the capital. His "Wired Whitehall" provides a 2D map interface to several 3D videoed scenes in Trafalgar Square, Horseguards Parade, Parliament Square, Whitehall itself, Covent Garden and related areas. If you connect to http://www.casa.ucl.ac.uk/vuis/ you can experience this information system which provides a "gateway" to massive amounts of information about the Whitehall area. Scenes load as 360 degree panoramas from which buildings are identified as you point but more important are the web sites that are accessible from these pointers. Point and click on places like the National Gallery and you will visit their homepages, from which you can web crawl your way to the ends of the earth. Wired Whitehall is surrounded by information about where to stay and how to travel to and within central London and as such it is part of the emerging area of virtual tourism.

But Wired Whitehall was not primarily built as a passive information system through which you could journey as virtual tourist. It was built and is being used as a platform for online participation in the planning process. There are two very significant developments of the system which turn it into a powerful means for connecting diverse interests interactively. First and perhaps most important, because there may be many participants remotely connected to the system simultaneously, it is a straightforward matter to log this and to identify exactly where they are within the information system. If they are viewing a scene in 3D for example, then they will be moving around the scene, zooming and panning and there will be some sense of their virtual location. This virtual location can be translated into a pictorial representation within the scene - as an avatar usually. If there is more than one person within a scene at any one time, then they will appear together and it is easy to initiate dialogue. A dialogue box will open and the participants can begin to converse across the net, about any matter they wish (Dodge, Smith and Doyle, 1997). It is very easy to see how this can be extended to verbal conversation if the appropriate sound technology is in place and how the digital video might be replaced by real time video from web cameras on site. So everything is in place for anyone, anywhere to log onto any webbed scene and talk to each other through these virtual locations. The prospects for meaningful participation concerning planning issues which are primarily visual are remarkable. Of course any dialogue might be supplemented by many other representations of a more abstract nature such as those we alluded to above. Figure 2 illustrates what is possible.

The second feature of this interface to Wired Whitehall is the ability to redesign what you see remotely and with others. Simple objects can be introduced into the scene and then manipulated by the avatars. For example, we have introduced objects of street furniture such as telephone boxes into the scene. The user can manipulate these using cursor keys or mouse buttons to position the box. The basic notions can be represented as

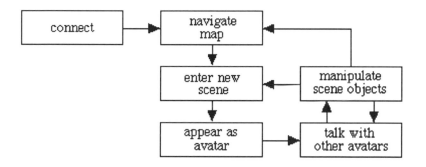

Two technologies - the video scene displayed in *QTVR* and the box in *VRML*, say, are being merged in such a context. If there are many avatars in the scene all trying to position their own virtual telephone boxes, there will be chaos. But the point is made. Proper CAD models on the web supplemented with real time scenery and digital video can ensure that basic notions of collision control and appropriate protocols for the behavior of the avatars can be invoked. What cannot be ensured are the protocols for effective participation which opens another new world of questions involving how to take part in meaningful dialogue. A million other issues are raised by such possibilities with many ways of interaction using many different types of visual representation. But the fact that we are now able to ensure presence in any scene or set of information viewed visually is a basic prerequisite for effective dialogue, and of course for better problem-solving and design.

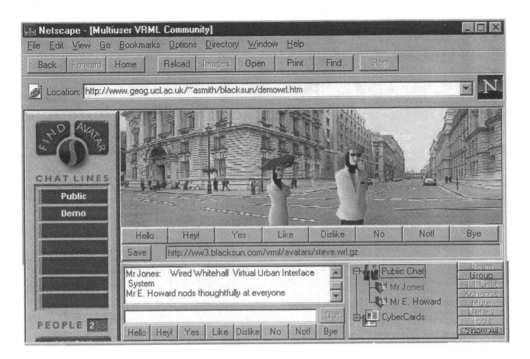

Figure 2: Avatars Converse in Wired Whitehall

6.   Data Across the Net: GIS for Everyone

Part of the inexorable move to a fully wired world is the drift of software out of standalone or even networked computers themselves onto the very network which links them: "The computer is the network, the network is the computer" which has been Sun Microsystems slogan for a decade points to the vanguard of change where computers are so widely distributed to be virtual networks in their own right. Software itself is increasingly likely to be formed across the net and to operate across the net. Yet this move to the net is still surprising: it is of course the most effective of delivery systems - what proportion of users load *Netscape* browsers from media other than the net ? - but it is probably demand economics that is the most persuasive issue in the massive growth of net-based software. The costs are much lower to users but the dramatic increases in the number of users raises profitability by an order of magnitude. Dissemination is important in all this but the valueadded to software comes from many potential users being developers. There are so many more people exposed to the use and development of software through the net that its use and growth is irresistible. Products diversify in novel and surprising ways through such use.

The development of geographic information systems has followed the same path as the development of computers and computation: first mainframes and minis, then workstations, with the present focus on the desktop, and the emergent focus on the network. In the faster more sophisticated environment of the workstation, the emphasis in GIS is upon elaborate toolkits for developing geographic models and analyses. On the desktop, the emphasis is more modest, being based on higher level functionality of a more generic, integrated kind which many different types of users may adapt. In the emergent network environment such diversification is going in two directions. Sophisticated toolkitlike map servers are being developed for large scale applications such as delivering extensive digital data sets, for example Yellow Pages, while the most popular use is developing very simple network GIS but more elaborate than the passive non interactive server applications which simply deliver canned products such as maps and their various renditions.

Most proprietary GIS software is following this route. For example, ESRI are continuing to develop workstation *ARC/INFO* as an ever more elaborate toolkit environment while *ArcView* is fast becoming a universal desktop platform for a variety of extensions which are tailored to particular applications. On the net, *Internet Map Objects* is forming the basis a professional toolkit while the slimmest and most popular GIS is *MapCafe* which is rather like a five-button *ArcView* for net users. From this, one can develop only minimal functionality but it can be developed over the net. Users can zoom and pan but also set up queries and in time are likely to be able to engage in transparent but nevertheless relatively sophisticated applications. In a sense, this is educating a very wide set of potential users in the value of geographic and spatial thinking as well as in products which enable such thinking to be applied.

Just as all the software we have referred to in this paper can be linked to multimedia, it is eminently feasible to use net-based GIS as the basis for the 2D component of the virtual environments that we introduced in the previous section. We have already used ESRI's *Internet Map Server* to construct a prototype demonstration for central London which uses scenes from Wired Whitehall. An illustration is given in Figure 3 where the user can develop basic functionality in the map but at the same time hotlink to 3D scenes which at present are recorded in digital video but might be delivered in *VRML* from the GIS itself. Such links are being pursued in another project in CASA where Martin Dodge is developing simple *VRML* scripts within *ArcView* so that urban designers can visualize the edits they make to the plan forms which are embedded within the GIS. A more elaborate form of linking has been achieved by

Simon Doyle who is using *MapCafe* to deliver panoramic scenes of UCL to the desktop within the browser in the manner of Wired Whitehall. A project for developing a Virtual Prospectus for UCL is using a large data base of digital photos and video and *MapCafe* contains hotlinks to this which in turn of course are hotlinked to many other scenes and webpages. An illustration of this application is presented in Figure 4.

Figure 3: An Internet GIS for Wired Whitehall

We noted the importance of developing protocols and structured procedures for participation over the net when we dealt with the design of virtual environments in the previous section and many of the same issues pertain here. But of somewhat different import is the issue of how naive GIS users might use simple GIS to serve their needs. Panning and zooming are straightforward but structured queries are harder to design for naive users. Other forms of functionality that we might decide are useful to general users, also require much thought in terms of interface design. We are about to begin a project in CASA which is to develop a web-based GIS for defining town centres boundaries for many town centres in the UK and these will inevitably require local input of data, albeit these definitions being largely based on national data sets. In this, the way in which users upload data and download results is all important and the way in which users might then interrogate the system (the GIS) and develop their own applications needs considerable thought. It is in these areas that basic research is required and experimental systems need to be developed and tested. The technology is less a problem that the way it is adapted and developed for specific applications, and this will be the key challenge in developing digital planning which ensures effective participation and communication.

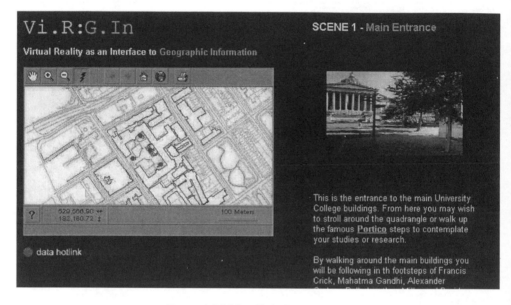

Figure 4: Multimedia in Internet GIS

7.   The Digital Century

Paul Kennedy (1993) in his book *Preparing for the Twenty-First Century* points the conventional wisdom that if the 19th century belonged to British, the 20th belongs to the Americans while the 21st will belong to the Japanese. But in the 21st century, this is likely to be turned on its head in the realignment of power which the world will face as nations breakup and merge into radically different alliances. Globalization changes everything. The kind of fully wired world in which the applications we have described here will flourish at the modest local levels will belong to everyone and no one. The key point is that total connectivity provides opportunities which will change the world in ways that we simply cannot imagine.

It is hard to classify the array of applications which will define digital planning, but it is likely that the key activities of problem-definition and data analysis which focus largely on communicating information, on modeling and forecasting, and then on design will form the basic set of possibilities. However these are generic activities which are being applied largely to the material world of the city although there exists the parallel world of cyberspace to which they might also be applied. For example, it is entirely possible to build models and make forecasts of how individuals are manipulating and building cyberspace. Take the online worlds which are emerging where participants log on, communicate, colonize their own space and build their own worlds, in a way which is not dissimilar to the way Wired Whitehall is being developed. We can build models of such processes where the data of how such worlds are built is entirely recorded in the digital media. In the same way, the process of design within such worlds can be studied and simulated. Cyberspace exists alongside real space and everything that we have been doing in real space is mirrored in cyberspace. These virtual realities can be studied through virtual reality and so on into infinite regress.

Yet the most important problem facing digital planning is providing protocols as to how one might best behave using the new media. Goals and objectives, ends and means must be closely scrutinized and the ergonomics and psychologies of how users might interact with information and with each other in the digital media must be well prescribed. As we stand at the beginning of this new world, it is not yet possible to state in any categorical way how the media will be used at various stages of the planning process which is inevitably set against a background of continual public interaction with planning problems in the most general sense. Nevertheless we have pointed to the most obvious applications, ones which are at present under development. Three years ago, these example were barely formed, and doubtless by the beginning of the 21st century there will be many new realizations within the broad array of possibilities in digital planning. The transition to a fully wired world proceeds apace and urban planning itself might be very different within a generation. Whatever happens, this will be intrinsically bound up with the emergence of the digital world.

8.   Acknowledgements

Thanks to Martin Dodge, Simon Doyle, Chiron Mottram, and Andy Smith for developing some of the applications described and for help on the graphics. The author can be contacted at m.batty@ucl.ac.uk and at www.casa.ucl.ac.uk.

9.   References

Batty, M. (1976) *Urban Modelling: Algorithms, Calibrations, Predictions*, Cambridge University Press, Cambridge, UK.

Batty, M. (1997a) The Computable City, *International Planning Studies*, 2, 155-173.

Batty, M. (1997b) Virtual Geography, *Futures*, 29, 337-352.

Beer, S. (1971) *Designing Freedom*, John Wiley and Sons, Chichester, UK.

Birkin, M., Clarke, G., Clarke, M., and Wilson, A. (1996) Intelligent GIS: Location, Decision and Strategic Planning, *GeoInformation International*, Cambridge, UK.

Bush, V. (1945) As We May Think, *Atlantic Monthly*, 176, 101-108.

Davidoff, P. (1965) Advocacy and Pluralism in Planning, *Journal of the American Institute of Planners* 31, 331-338.

Dodge, M., Smith, A., and Doyle, S. (1997) Urban Science, *GIS Europe* 6 (10), 26-29.

Forester, J. (1989) *Planning in the Face of Power*, University of California Press, Berkeley, CA.

Geddes, P. (1915, 1949) *Cities in Evolution*, Williams and Norgate, London.

Gottman, J. (1983) *The Coming of the Transactional City*, Institute of Urban Studies, University of Maryland, College Park, MD.

Harris, B. (1994) The Real Issues Concerning Lee's 'Requiem', *Journal of the American Planning Association* **60**, 31-34.

Healey, P. (1996) The Communicative Work of Development Plans, in S. J. Mandelbaum, L. Mazza, and R. W. Burchell (Editors) *Explorations in Planning Theory*, Center for Urban Policy Press, Rutgers University, New Brunswick, NJ, pp. 263-288.

Kawalsky, R. S. (1993) *The Science of Virtual Reality and Virtual Environments*, Addison-Wesley, Wokingham, UK.

Kelly, K. (1994) *Out of Control: The New Biology of Machines*, Fourth Estate, London.

Kennedy, P. (1993) *Preparing for the Twenty-First Century*, Vintage Books, Random House, New York.

Lakshmanan, T. R., and Hansen, W. G. (1965) A Retail Market Potential Model, *Journal of the American Institute of Planners* **31**, 134-143.

Lee, D. B. (1973) Requiem for Large Scale Models, *Journal of the American Institute of Planners* **39**, 163-178.

Longley, P. and Clarke, G. (Editors) (1995) *GIS for Business and Service Planning*, GeoInformation International, Cambridge, UK.

Martin, J. (1978, 1981) *The Wired Society* (republished as *Telematic Society*), Prentice-Hall, Englewood-Cliffs, NJ.

Meier, R. L. (1962) *A Communications Theory of Urban Growth*, MIT Press, Cambridge, MA.

Penny, N. J., and Broom, D. (1988) The Tesco Approach to Store Location, in N. Wrigley (Editor) *Store Location, Store Choice, and Market Analysis*, Routledge, London.

Rheingold, H. (1993) *The Virtual Communities: Homesteading on the Electronic Frontier*, Addison-Wesley, Reading, MA.

Smith, A. (1997) Realism in Modelling the Built Environment using the World Wide Web, *Habitat* **4**, 17-18.

Webber, M. M. (1964) The Urban Place and NonPlace Urban Realm, in M. M. Webber *et al.* (eds) *Explorations into Urban Structure*, University of Pennsylvania Press, Philadelphia, PA, pp. 79-153.

Whittle, D. B. (1997) *Cyberspace: The Human Dimension*, W. H. Freeman, New York.

Wilson, A. G. (1974) *Urban and Regional Models in Geography and Planning*, John Wiley and Sons, Chichester, UK.

Wilson, A. G., Coelho, J. D., Macgill, S. M., and Williams, H. C. W. L. (1981) *Optimization in Locational and Transport Analysis*, John Wiley, Chichester, UK.

PART III

**Linking Practice
To Theory**

LINKING PRACTICE TO THEORY

Introduction

D. BORRI

Evaluation in Rational vs. Communicative Planning Models

It is well known how the key role of evaluation in planning can be traced back to the development of rational planning methods, with their claim for the implementation of planned actions to be based on the systematic evaluation of alternative courses of action and the choice of the best among them (Faludi, 1973).

It is also well known how this rational style has become more a form of rhetoric to get social consensus and valorisation for planning experts than a real model for plan-making; if the above is particularly true for routine plans, literature also shows plenty of criticisms for the ways in which evaluation has been misused in strategic plans and projects.

If rational theory emphasises the analyse-evaluate-choose imperative in the planning procedure, very little is known about aspects and outcomes of the rigorous adoption of this method in practice.

The turn, in recent years, to a communicative style of planning (Khakee, 1996) has revealed the bias of expert planning rationality -both instrumental and substantive- in the light of the perspectives offered by a more promising social learning based on collective knowledge and in particular on the virtuous interaction of expertise and common-sense: in this view; planners perform the subtle roles of community listeners and interpreters, at the same time bringing to the game their ability to suggest proper courses of action and to draw on social learning creativity.

Environmental concern, too, has led to the introduction of a more sophisticated approach to planning, warning against deterministic and simple assumptions of planning rationale and asking for sound consideration of the eco-systemic aspects of reality (Soderbaum, 1976).

All these events and processes have meant that evaluation -far from loosing its flavour as a potential means to improving the quality of public and private policies- has been changed into an integrated and continuous mechanism of theory-in-action able to take into account a complex set of qualitative and quantitative variables and expert and non-expert knowledge.

N. Lichfield et al. (eds.), Evaluation in Planning, 275–281.
© 1998 *Kluwer Academic Publishers. Printed in the Netherlands.*

The evolution of decision theory and operations research -with the increasing importance in it of cognitive aspects and fuzzy reasoning- parallels the new things which are happening in the evaluation field (Munda, 1995); the attention to qualitative and linguistic variables leads to the adoption of scenario-like approaches and makes mainstream planning evaluation always more sensitive to uncertainty and complexity issues.

The characteristic interactionist feature of communicative planning calls for continuous and flexible, multi-objective and multi-criterion evaluation approaches, in the context of plural reasoning and non-monotonic logic (Borri, 1997).

However, as public policies become more and more market-oriented and economic privatisation of interventions makes progress, a strong resistance is raised to the introduction of the new integrated and pluralistic evaluations; in the meantime cost-benefit like approaches continue to be supported especially for their use in strategic planning and projects. Even if merely economic approaches to the evaluation of impacts are distrusted by sophisticated experts and communities, the apparently clear message of budgetary appraisals remains unequalled from the normative point of view and everywhere old-fashioned administrative contexts demand rough and speedy estimates of policy performance.

Moreover, it is well known how rational planning did fail in developing evaluation methods of its own emerging from its typical also moral background -perhaps with the only exception of the Hill's Goal Achievement Matrix method and later of some original MCA approaches- with the consequent importation from other fields of methods mostly elaborated to serve economic accounts and profit calculations.

No wonder, then, if many people in communicative planning circles and grassroots organisations dislike formal evaluation exercises, rejecting their normative assumptions and claiming for a new, more democratic and sensitive style of decision-making, in which the positions of the less politically represented -both humans and non-humans-can be fully acknowledged and respected; there is also a growing idea, on this side, of how any monologic statement turns to be cognitively vane and ineffective in practice -at least in the long run- and how interactional or even conflict-maintaining hybrid (expert-commonsense) approaches are more creative and resilient for the sake of society and the environment (Barbanente et al., 1996).

The theory-practice gap for evaluation exercises does not differ from what has been largely recognised as one of the major troubles of planning: in the planning domain, the need of sound theory -particularly for decision analysis- crashes into the limitations of knowledge and forecasting in the face of huge systemic and probabilistic problem domains; more adaptive, evolutionary, and interactive models of knowledge and action seem today to be good challengers of these difficulties, in particular when a model of social learning -with its typical positive feature of thinkers-actors open to self-reflection and continuous restructuring of their own previous positions- is adopted.

Back to strategic vision: integrating local and global decision making

To what extent can the following contributions cope with the primary problems and dilemmas of contemporary evaluation and more in general of contemporary planning: limits of classical rational models of evaluation and planning in coping with the envisaged new categories of phenomena at hand; inadequacy of recent communicative and social learning approaches in dealing with the institutional, often aggregate, demands which continue to be raised against formal evaluation and planning systems, now coexisting with the micro-analyses, micro-policies, and subjectivities determining those approaches; apparent impossibility to reduce environmental and social phenomena and their ecosystemic complexity and uncertainty to the trivial schemes and anticipations of rational models; difficult coexistence of economy rights and community rights in a complex ethical perspective which includes moral argumentation.

Lichfield and Prat going through the development of the British planning system indicate its basic stability around a core of local plans led and controlled by central government; they also underline the persistent rational problem solving inspiration of that system. However, in the face of this apparent rationality, there would be a reality of practice much less linear, because of uncertainties about facts, conflicts about values, localisms, organisational constraints, differences in policy and professional behaviour.

L&P in principle accept the traditional dichotomy between ex ante evaluation (dealing with the two major decisions making problems of uncertainty in respect of the future and efficiency, effectiveness and equity of the decision) and ex post evaluation (concerning understanding the -anticipated and unanticipated- changes produced by the policy and the process that has induced them, and judging the degree of success), as a direct consequence of the start-and-end structural features of the action; but they call for adding a third, intermediate, stage of evaluation (in itinere, monitoring policies and socio-economic changes and evaluating ongoing policies), to grant continuity to the whole process.

L&P note the increasing tendency, in UK, to evaluate implementation and success of plans, which obviously means to emphasise monitoring and ex post approaches and to look for plans more updated, responsive, and reactive to external changes; they make a point, however, on the good theoretical and institutional frame of ex ante evaluation, currently offering a lot of well established methods, raising a strong criticism about the ways in which this new season of in itinere evaluation has been started in UK: here monitoring, far from being a continuous activity, with a robust technical and financial organisation, is developing in a context of improvisation, with a patchwork of methods inconsistently used by public and private bodies at different levels, often diverging from ends; the same holds, according to L&P, for methods of plan review.

To improve this situation, L&P proposes the introduction of a PMLS (Planning and Management Learning System), as a continuous evaluation environment for planning, well grounded on integration of methods and resource organisation; they see this system as also useful for more plural rationality and enhanced participation in plan making.; the aim of PMLS is to go beyond the articulation -typical of any positivistic social research- in descriptive (knowledge oriented) and prescriptive (value oriented) components, and to integrate facts and values in order to make policy more informed and reflective.

Alexander looks at the current highly rational and normative planning model in Israel, with its typical statutory and hierarchical features; in this model, the evaluation of alternatives is largely institutionalised, both for serving the evaluation as a basis for judicial review according to the administrative law and for the effects of the introduction of EIS in 1992. The still prevailing instrumental rationality of these evaluations is now obviously under criticism while some form of multi-objective or multi-criterion decision analysis is appearing at the local or project scale.

Alexander says that most of these evaluations, however, are poor and arbitrary, hiding predetermined choices: a sensitivity analysis applied by him to the case of Trans-Israel Highway no. 6 - where the three alternatives of government-, authority and franchise-based action were evaluated through a MCA - shows macroscopic methodological errors and systematic bias in favour of privatization. Alexander strongly criticises the abandonment of substantive rationality in this exercise and the undue use in it of rhetorical argumentation and communicative style where in reality it is strategic planning and decision making which is involved; if MCA has undoubted merits, for it makes explicit evaluation assumptions and integrates empirical and experiential knowledge about the alternatives with a systematic account for goals and priorities, it is also true that often MCA is misused, as the cited Israeli case shows: here communicative and interactive rationality, putting aside classical rationality, has been perverted and changed into a vane rhetorical expedient.

The point here is not to fight against communicative exercise and style of planning but to make clear the risk of biased uses of MCA: otherwise MCA, far from virtuously facilitating political discourse and consensus in always more complex environments - y informing all those involved about the alternatives and the criteria for choosing the best or more satisfactory among them - becomes a subtle instrument for empowering professionals and bureaucrats serving public and private vested interests.

Nash, in his brief history of the pioneering application of CBA in transportation studies in Britain points out the active role of central government in this development and the tendency, instead, of local authorities to use more speedy methods when not asking for some financial help from the state; still later, the spread of privatisation in the British transport system made CBA less relevant because of the increasing informality and diffusion among many actors of evaluations However, it seems that a more consistent approach is growing out of these developments: a more strategic and multi-objective vision, less concerned with mere economic efficiency, in which local and state positions can finally converge.

To be sure, the British Department of Transportation still makes large use of monetary evaluations in the CBA tradition (see its computerized program COBA), distrusting any innovative method for considering environmental and distributive aspects. It is well known that multi-criterion evaluations satisfactorily deal with these be conceived -as Nash suggests- as an integration to CBA in the last stages of the decision process, because of their offer of a well disaggregated and weighted analysis; this integration of methods is particularly consistent, then, when all the objectives-criteria relate to factors affecting the welfare of the population concerned and achievement and weighting measures are based on the preferences of those people and hopefully subjected to equity considerations.

Even if the outcomes of the Leitch Committee gave rise, in the late 1970s, to a new evaluation approach for transportation issues, still enforced with some refinements by the DoT, based on monetary and non-monetary, quantitative and qualitative assessment, Nash points out the limitations of this approach, in particular regarding the distribution of incidence groups and effects and for the absence of strategic understanding - including scale problems- of phenomena at hand: the recent decision of the central government to bring trunk road evaluation into the framework of regional and structural planning could be seen as a positive novelty to meet this need for integrated and strategic evaluation. However, increasing privatisation of transport, in Britain, does not contribute to integration of methods and visions, for it determines a strong tendency to resort to financial evaluation, underscoring social considerations.

Millichap in his paper says that ex ante evaluation methods generally fail in the anticipation of the negative impact of development projects, with effects of damaged environment and deligitimization of planning; moreover, resorting to ex post remedies is also problematic in UK, as the "discontinuance" powers of local planning authorities in the case of damages emerging from the implementation of a project are limited by the requested economic compensation.

In principle, according to Millichap, improved forecasting could be given by insuring its social components against the risk of failures following wrong decisions: this mechanism, in fact, would probably lead to more cautious forecasts. Millichap invokes, also, a number of other innovations for evaluation in a condition of structural uncertainty: clear argumentation and respect of "community rights" by decision makers, monitoring (comparing reality with forecasts), introduction of a temporal dimension in planning, more detailed analysis; compensation procedures similar to those used in EIA.

The market, in substance, becomes a good partner in this monitor-and-mitigate strategy to cope with uncertainty: reimbursements for failure in forecasts will lead to more virtuous behavior in uncertain decision making.

Corkindale starts from the basic mechanism of regulation of public and private interest on land and its transformation performed by the British planning system to observe the absence in that system of any provision for measuring the impact of planning permission; Corkindale, in particular, stresses the reluctance of planners to adopt economic concepts when evaluating their own work: the high cost of operation of the British planning system and the often negative impact it has on public and private economies are also underscored.

Planners, however, combat similar criticism with the argument that a planning system would be highly beneficial for the society which uses it; what, then, about these benefits? Corkindale says that answering this question would mean comparing objectives of plans and real outcomes from them, but that in general the absence of a clearly stated objective in the plans makes it impossible to follow this way; applications of CBA to planning problems with their typical content of public policies show all the range of these difficulties, with the absence of good and accessible data and poor orientation towards distributional issues.

Corkindale sees internalisation of technological externalities by planning (accounting for the environmental impact) as crucial, for it balances competing public and private interests at a level adequate to the current policy challenge; to get this internalisation, relevant actors have to be involved in negotiation and monitoring; a normative statutory frame, alternating liability rules and property rules, would give a good chance of success on this terrain. An extension of liability (dispositive) rules could be particularly beneficial in the case of the presence of high amounts of potentially conflicting subjects, as recent legal trends on privatisation and deregulation in planning show; a wise mix of granted development rights for investors and landowners in areas where environmental conservation is not required and less market-oriented policies in sensitive areas could be the best strategy to adopt, as the successes of Tradeable Development Rights in Usa and elsewhere indicate.

Evaluating non market values: creativity and self-reliance in planning from below

Problems raised by the contributions collected in this section deal, on the one hand, with the planning and evaluating system's general ability to answer demands for more effective selection of environmentally and socially sustainable actions and, on the other hand, with the specific ability of that system to improve and integrate the rationale of the operational methods -in Britain as yet largely based on the CBA tradition- in the light of the increasing complexity and uncertainty of involved phenomena.

In the first case, all the authors underline the importance of argumentable plan choices, in a perspective spanning from the routine need for policy transparency and justification (Alexander, Corkindale) to communicative and community approaches (Lichfield, Millichap) or to the potential creativity of radically transformative solutions (Hall); the integration of different evaluation methods -in particular multi-objective and multi-criterion evaluations to traditional cost/benefit, always highly appreciated in institutional environments- and the enhancement of open monitoring and control are presented as possible solutions to this prime need. Again in this case, which claims for argumentability and robustness of plans and their choices -in the light also of the complex scale and theory/practice problems posed by emerging environmental and distributive affairs- we have to locate the following contributions -notwithstanding their explicit reference to the UK case- when they ask for strategic recomposition (in a wider spatio-temporal and policy perspective) of key actors behaviour and interests at different levels, from local planning authorities more concerned with implementation and consensus problems to central government authorities more concerned with aggregate accounts of their own economic, social, and environmental policies.

The need to bring back selected parts of policies to the market, in a more articulated and cautionary framework if compared with the general deregulation of the 1980s, also seems to be a leitmotiv in the positions of these authors, in their search for those practicalities and automatic compensations and mitigations which can be provided - within certain limits- by the invisible hand of the market.

In the second case, that is the adoption of more evolved forms of rationality, the following contributions do not abandon, in general, the sequential procedures of optimal or satisfying problem solving typical of rational planning and decision making models, searching for the integration and consolidation of those different stages: here again the emphasis on objectives and criteria to be defined in the highly complex and uncertain contexts of the environmental and social challenge, on monitoring and reviewing, on evaluation methods allowing for this integration and flexibility, describes well the approach shared by all the contributors to the section.

References

Barbanente, A., Borri, D. and Pace, F. (1993), Micro-problems and Micro-decisions in Planning by Artificial Reasoners, in Klosterman, R. E., and S. P. French, Eds., *Procedings of the 3rd International Conference on Computers in Urban Planning and Urban Management*, **2**, City Planning Program Georgia Tech, Atlanta, pp. 17-34.

Barbanente, A., Borri, D., Concilio, G., Macchi, S. and Scandurra, E. (1996), Dealing with Environmental Conflicts in Evaluation: Cognitive Complexity and Scale Problems, *Proceedings of the 3rd International Workshop Evaluation in Planning*, September, University College, London.

Borri, D. (1993), Formal Vs. Informal Evaluation Procedures in Planning Expert Systems, in Khakee, A., and K. Eckerberg, Eds., *Process & Policy Evaluation in Structure Planning*, Swedish Council for Building Research, Stockholm, pp. 91-102.

Borri, D., Khakee, A. and Lacirignola, C. (1997, Eds.), *Evaluating Theory-Practice and Urban-Rural Interplay in Planning*, Kluwer Acadelmic Publishers, Dordrecht.

Faludi, A. (1973), *Planning Theory*, Pergamon, Oxford.

Forester, J. (1989), *Planning in the Face of Power*, University of California Press, Berkeley.

Friedmann, J. (1987), *Planning in the Public Domain. From Knowledge to Action*, Princeton University Press, Princeton.

Healey, P. (1997), *Collaborative Planning. Shaping Places in Fragmented Societies*, Macmillan, London.

Heiner, R. A. (1983), The Origin of Predictable Behavior, *American Economic Review*, **73**, pp. 560-95.

Khakee, A. (1996), The Communicative Turn, *Proceedings of the 3rd International Workshop Evaluation in Planning*, September, University College, London.

Schön, D. A. (1983), *The Reflective Practitioner. How Professional Think in Action*, Basic Books, New York.

Soderbaum, P. (1996), Economics and Ecological Sustainability: An Actor-Network Approach to Evaluation, *Proceedings of the 3rd International Workshop Evaluation in Planning*, September, University College, London.

LINKING *EX ANTE* AND *EX POST* EVALUATION IN BRITISH TOWN PLANNING

N. LICHFIELD & A. PRAT

1. Introduction

Statutory *town planning* in Britain can be seen as an elaborate system of development planning and implementation set in place in order to directly shape and, more often, indirectly affect the process of change and conservation of the built and natural environment. This traditional term is still used, despite being somewhat outdated. Amongst current synonyms are: urban and regional, town and country, land use and community planning.

In simple terms, this institutional and legal system produces, at the local authority level: 1. processes, i.e. the cyclical sequences of plan making/implementation/review, where many actors, interests and decisions are involved; 2. products, i.e. the plans, comprising both maps and supporting documents.

This paper focuses on how evaluation is conceived and used in this distinct policy system, both in relation to its processes and products; on the present shortcomings; on the possible explanations for the current situation; and, finally, on suggestions for improvement, referring to the need for:

1. better integrating of evaluation in the planning system
2. greater strengthening of the link between *ex ante*, *in itinere* and *ex post* evaluation

In undertaking this analysis, it has to be recognised that the town planning field overlaps more and more with other policy fields. The means, objectives and institutions involved in this system embrace those of other planning systems dealing with other aspects of the urban and rural change (such as transport, urban renewal, urban and rural conservation and environmental protection) or with other activities that, while having a physical component, are more focused on economic and social objectives (urban programmes, economic development policies, housing). Thus, it would also worth investigating if this proximity with other policy fields has some impacts on evaluation in town planning, and if some learning between different evaluation traditions can be

N. Lichfield et al. (eds.), Evaluation in Planning, 283–298.
© 1998 *Kluwer Academic Publishers. Printed in the Netherlands.*

initiated. This apart, evaluation in the British planning system is, or could be, affected by three major developments:

- recent changes in the town planning system itself and their shortcomings as far as evaluation is concerned;
- recent changes in evaluation in the environmental field;
- evaluation research tradition beginning to be applied in the evaluation of urban programmes.

2.   Evaluation in the British Town Planning System

2.1.   A BRIEF HISTORY OF THE STATUTORY PLANNING SYSTEM

It is well known that the British planning system had an early origin. The original planning framework was designed by the 1909 Housing and Town Planning Act, followed by the 1932 Town and Country Planning Act, which gave Local Authorities the power to control development by means of planning schemes. These were local acts setting out regulations for the use and development of land. The schemes had to be approved by the Minister and both Houses of Parliament: this process gave them the force of law and consequently allowed permission to develop in conformity with zoning prescriptions.

The 1947 Town and Country Planning Act, which is often considered a turning point not only in British planning, introduced a new paradigm in planning, through two major changes: (1) a new development planning system; (2) financial provisions for the compensation and betterment problem. The core of the system was the development plan, which was conceived as a "... framework or pattern of proposed land use, in the form of coherent set of proposals for the use of land, against which day-to-day development can be considered" (MoLGP 1951, p. 23) It essentially consisted of a survey, a statement of the main policies and proposals, and maps. Consideration of day-to-day development against the plan meant the use of discretion in the planning process: each permission would be granted or refused on the basis of its merits against the policies indicated in the plan binding and "other material considerations". Conformity to the plan did not immediately meant permission to develop would be granted. The plan itself did not convey development rights. Certainty for the developer had been reduced in favour of flexibility of the system.

The 1947 style of planning proved to be extremely slow in the process of approval and review and demanding in terms of analytical studies. Moreover, the Act was built on the logic of a positive development planning, which was rejected during the 50s, for ideological and practical reasons. Following the widespread criticisms and the report prepared by the Planning Advisory Group (1965), a new framework for the planning system was conceived. The 1968 Town and Country Planning Act, which was then incorporated in the 1971 Act, introduced the hierarchical distinction between strategic and tactical planning issues. Both levels had their own planning instrument: the Structure Plan, prepared by the County, and the Local Plan, prepared by the District and the Borough. Since then, the skeleton of the system has remained the same. The Structure Plan provides the strategic framework for policies: it is in practice mostly focused on land use control. Within the context of the Structure Plan, the Local Plan provides detail guidance for smaller areas. Furthermore, two additional levels were envisaged to co-ordinate the policy: national policy statements (since 1988, called Planning Policy Guidance; PPG) and regional policies from central Government.

During the 80s the system was more or less explicitly undermined by the reduction in importance of the County and its Structure Plan, a series of informal plans, the elimination of the metropolitan planning level, the spread of exceptional policy instruments, such as the central Government urban programmes. Another modification to this framework has fused the two levels of planning in a "unitary" planning Authorities in London boroughs and the metropolitan districts. These Authorities are now requested to provide a Unitary Development Plan, which comprises in it the two "structure" and "local" components. The purpose of the changes introduced by the Conservative Government has been to "lift the burden" of planning constraints in favour of private development, to limit public expenditure, and to increasingly centralise the control of planning decisions.

In recent times, this trend has been curbed. Following the 1990 Town and Country Act, the 1991 Planning and Compensation Act, reflects this change. The general aim of the Act is to make the planning system become "simpler and more responsive, reducing costs for both the private sector and local authorities and making it easier for people to be involved in the planning process" (DoE, 1992a, 1.10). The Act introduced the so-called *plan-led system*. The four significant changes are: (1) the adoption of local plans has become mandatory (2); the plan has become the main instrument for planning decisions; (3) the requirement for central approval of structure plan is removed; (4) Counties are requested to produce mineral and waste plans. In practical terms, although it is too early to appreciate the consequences of this change, the additional emphasis on the plan and the reduction of emphasis on "other material considerations" means that the whole planning framework is given a new strength. Furthermore, the balance between discretion and rigidity has apparently shifted back to the latter (Grant, 1992), giving Local Authorities a strong means for development control and forward planning.

2.2.   THE PLANNING PROCESS WITHIN THE SYSTEM

In the classical view of planning, which corresponds to the one depicted by the legislation, the planning process is oriented towards the production and management of its main product - the plan itself – as a framework for the control of development aimed at contributing towards its implementation. Accordingly, the formal local authority planning process includes three elements:

- plan making;
- plan implementation;
- plan review.

2.2.1. *Plan Making*

Plan making is by tradition the most formalised phase of the whole planning process. The relevance of this phase can be understood by reference to the very nature of the planning system itself. It essentially offers an anticipatory system of control on development (Ball and Bell, 1991), i.e. a system that provides a framework for addressing future private and public development decisions. Hence, preparing this framework is of central relevance.

In this view, the plan must offer a balance between certainty and flexibility in order to be able to anticipate, or at least react, to the foreseeable and unexpected decisions made by other players in the urban arena and to the changes occurring in the urban system.

It is generally held that the British town planning system performs its function in a more policy-oriented, flexible and discretionary way than other European planning systems, that are mainly founded on zoning approaches (Grant, 1992). It tends to be prepared taking into account realistically predictable changes (such as in terms of demographic changes and demand for development) and existing constraints. The policies included in the plan are supposed to outline the local authority framework to assess future applications and not an ideal, visionary end-state.

The plan making traditionally implies phases, with different mixes of technical and political components: survey and analysis of the planning areas, preparation of a draft plan following consultation, check of the conformity of plans, intervention by the Secretary of State and adoption.

Two potentially contradictory trends in plan making are emerging. On one side, the trend is to make the system "simpler and more responsive, reducing costs for both the private sector and local authorities and making it easier for people to be involved in the planning process" (DoE, 1992a, 1.10); in a cost-cutting efficiency interpretation, this could mean simplifying and expediting the plan making phase. On the other side, the 1991 Planning and Compensation Act has increased the complexity of plans, the size of the areas covered and the participation of interested groups and individuals during the phase of plan preparation (DoE, 1991, 1997). It has also introduced a significant shift of emphasis in flexibility and discretion, compared with the situation between 1947 and 1990 Town and Country Planning Acts, thereby introducing a "plan-led system". This so-called "plan-led" orientation introduced by this Act has meant an increase in the weight of the plan itself, by reducing the emphasis on so called more discretionary "other material considerations" (any other matter of planning interest). The extent to which this is actually happening is arguable, and where the balance between discretion and rigidity will actually reposition itself is still an open question.

2.2.2. *Plan Implementation*

The plan implementation phase relies upon a series of measures, that can be presented in terms of ascending order of social control over the non-planned situation (Lichfield, 1996, p. 17):

- general influence;
- intelligence and information;
- organisation and co-ordination;
- financial inducement and penalties;
- development control;
- carrying out development projects.

2.2.3. *Plan Review*

Plan review is the phase linked with the subsequent plan amendment stage. Formal review is a statutory requirement (at least every five years). From this perspective, review is the decision-making act of changing the existing policies in the light of experience. Examples are to amend them in term of implementation measures (means), objectives (policies and proposals), unclear wording and conformity with other plans and national statements. Plan review is fostered and dependent on the existence of formal requirements, by the changes in national guidance and European policies, by the changes in local conditions and development pressure, and for local and national political reasons.

2.3. EVALUATION IN THE PLANNING SYSTEM IN PRINCIPLE

In the preceding section we have described the planning process under the statutory planning system. Here we describe the evaluation process within this planning process.

For the purpose of this analysis, we can define evaluation as the systematic procedure of observing, analysing and formulating a judgement on the operation *and* the results, *both* foreseen or effected, of a plan/programme/policy process. The kind of evaluation we wish to investigate concerns the formal evaluative requirements within the local planning system. The aim is to see how evaluation is supposed to be undertaken and contribute to the planning activities. In order to proceed, we need first to define: (1) the concept of evaluation as a process; (2) the evaluative functions.

2.3.1.   *The evaluation process*

As seen above, a plan (or similarly a policy or a programme) undergoes different phases: making, implementation and review. In practice, identifying the beginning, the end and all intermediate phases of a plan/programme/policy can be very difficult, especially in the case of long-term policy commitment such as education policies, where incremental changes are the rule. However, in general terms, there is always a moment in which some kind of financial or political commitments are undertaken (policy making), a period in which no major policy changes arise and the policy live its "normal" life (policy implementation) and a period in which political and financial commitments are questioned or interrupted (policy review). In the case of planning, it is easier to identify these phases because the planing system itself is structured very much in this way.

Evaluation can be applied in all the different stages of the life of a plan/programme/policy. Three evaluative perspectives can therefore be identified, each of them implying a different angle in relation to the planning and management process.

Ex ante evaluation. *Ex ante* evaluation methods and approaches have been created in order to aid choice by decision-takers and stakeholders in the decision making process. *Ex ante* evaluation has to deal with two decision-making problems; (a) uncertainty in respect of the future; (b) balancing efficiency, effectiveness and equity of the decision.

In itinere evaluation. *In itinere* (or intermediate, *interim*) evaluation concerns two aspects: (1) the collection and analysis of data (monitoring) of both policies and socio-economic changes; (2) the evaluation of on-going policies.

Ex post evaluation. *Ex post* evaluation concerns: (a) understanding the (anticipated and unanticipated) changes produced by the policy, and the process that has induced them; (b) judging the degree of achievement; (c) concluding on any needed review of the plan and the planning process.

2.3.2.   *Evaluative functions*

In general terms, evaluation can have six functions, to which correspond six alternative or complementary ways of providing a *raison d'être* to this activity (Prat, 1995). These functions, which can overlap in certain respects, generate the motives for undertaking an evaluation study in order to assist in:

1.   Choice/design function: framing the decision-making process and in choosing the most satisfying decision. Traditionally, this process includes the analysis and

evaluation of various costs (inputs), and benefits (outputs and outcomes) on different people, that are generated by the policy options. Most of established *ex ante* evaluation methods refer principally to this function (Lichfield, 1996).

2. Knowledge function: throwing light on complex policy processes and on their costs and benefits. The judgmental component of evaluation is ideally reduced to the minimum. In practice, many evaluative studies have given a strong contribution to the understanding of social processes and are excellent examples of social research.

3. Feed-back function: producing technical knowledge (data) aimed at improving and fine tuning on-going policies and affecting decisions, through the detection of implementation deficits. This function, very much based on a top-down approach to policy analysis, is often claimed to be the main justification for monitoring activities.

4. Social/policy learning function: providing policy practitioners, researchers and society with information and interpretation of a variety of components: impacts, values, operation of political and administrative phenomena, etc. This contributes to the long-term process of learning within institutions. This is the function that is often claimed by supporters of *in itinere* and *ex post* evaluation, who take into consideration the difficulty of a mechanical, negative feed-back in public affairs, but believe in the capacity to contribute in improving the whole policy process.

5. Accountability function: making specific public actions (and actors) answerable in the public arena, i.e. to other institutional levels, citizens and taxpayers. This function focuses on helping to legitimate (or delegitimate) current policy practice, by exposing it to external judgement. Critics argue that evaluation is often used as an *ex post* legitimisation of the effects of a public intervention in order to praise public intervention, even in the case of unforeseen or useless effects ("post hoc ergo propter hoc"). A second criticism refers to the risks of the inquisitive bias that evaluation can acquire in this perspective, in order to combat "waste, fraud and abuse", as it is stated in many American evaluation statements, or to control and readdress local authorities and public agencies activities.

6. Communication function: providing all stakeholders in the policy process with information on facts and values and help their dissemination. This function is fundamental in increasing real and effective participation in the policy process itself and improving the negotiation process. In addition, as highlighted in other papers in this collection (see papers by Voogd, Khakee), if planning is conceived as a communicative practice, evaluation itself can be modelled in order to give its contribution in this sense.

2.4.   TRENDS IN EVALUATION PRACTICE IN THE PLANNING SYSTEM

The recent shift of emphasis to a plan-led system (2.1.1) can be expected to affect evaluation. Planning guidance is more focused on the degree of implementation, which should become the criteria for judging the planning success (DoE, 1992b, p. 101). It seems that the plan-led system would ask for more attention to evaluation of plans, in order to be more up to date, responsive and accountable. This means, in terms of evaluation, that the monitoring/*ex post* evaluation process could acquire a central role. The system is supposed to become more reactive to external changes, to its failures or inconsistencies and to political, stakeholder and public opinion demands; in this perspective, evaluation is favourably considered as a formal way to collect and integrate part of this information into the planning process.

2.4.1.   *The evaluation process in practice*

Since there are different activities going on at different levels within the planning process *ex ante* evaluation also portrays this variety. It can relate to the plan as a whole (including policies, proposals, projects and programmes of projects) or get into greater depth on selected projects. In this latter case, it is often linked to feasibility studies. This type of evaluation has a recent but solid tradition. In the planning field, it is now more theoretically consolidated and institutionally accepted than *in itinere* and *ex post* evaluation. The introduction of an appraisal phase within the planning process is aimed at searching for the most socially advantageous proposal. It, thus, concerns the process of analysis of a number of plans or projects with a view of searching out their comparative social and economic advantages and the act of setting down the findings in an analytical framework (Lichfield *et al.*, 1975, p. 4).

An exploration of 1970 showed that it is only in the sixties that evaluation of options was incorporated into planning practice and that there existed some 23 methods of *ex ante* evaluation (Lichfield, 1970). In a later review of methods, Lichfield (1996) has identified some that have escaped the early analysis, such as multi-attribute evaluation (Winterfeldt and Edwards, 1986) and strategic choice (Friend and Jessop, 1969) and some later ones, such as multi-criteria evaluation (Voogd, 1983). In this it can be seen that the different evaluative methods emphasise differences in measuring problems, choice of evaluation criteria, relationship with the planning and political process, an so on.

The British town planning requirements used to make no explicit references to a systematic use of evaluation during the plan preparation. The earlier indications regarded the need to take into consideration the effects and impacts of the proposals, and to make a selection between different options, but freedom was left in interpreting how to apply these indications. In practice, this implies that *ex ante* evaluation exercises have been conducted in a *ad hoc* way on specific projects for the purpose of preparing plans, choosing amongst different projects, giving evidence for appeals, inquiries.

Ex ante evaluation received an independent boost in 1988 from the introduction in the system of the Environmental Impact Assessment (EIA) for significant development projects (EC Directive 85/337). The number of Environmental Impact Statements (EISs) has increased dramatically; this has contributed significantly to a more positive attitude to *ex ante* evaluation. Very often EISs have affected the decision-making process, by means of leading to important modifications of the proposals (Glasson, Therivel and Chadwick, 1994, p. 198). In addition, thanks to the awareness on the strategic and multifaceted dimension of environmental problems, there is a trend to promote the introduction of evaluation procedures also for plans and policies (Strategic Environmental Assessment (Therivel *et al.*, 1992; DoE, 1991) and to extend the consideration of impacts to the socio-economic dimensions (Carley and Bustelo, 1984). But unfortunately for the town planning process, the EIA system is operated alongside but not necessarily integrated within the planning system (Lichfield, 1992)

In itinere evaluation is very strongly emphasised in the latest planning guidance, in terms of monitoring requirements. Planning Policy Guidance Note 12 (DoE, 1992a, 4.21) states that "local planning authorities are required to keep under review the matters which may be expected to affect the development of their area or the planning of that area".

Plan monitoring has two functions. The first concerns the regular observation of the non-planned area changes; it is identified by the national guidance as regarding the collection of data about the principal physical characteristics of the area; the size, composition and distribution of the population; the communications, transport system and traffic. The second regards the proper monitoring of the effects of planning decisions, in order to evaluate and review the plan. There are no clear statements form the DoE on how to define and measure the effects and impacts.

The monitoring exercise is supposed to be dependent on the formulation of policies and proposals. The emphasis on monitoring is consequently flanked with a recall to the need for clear policy objectives and the need to consider the monitoring operation during the plan preparation. The system for reaching judgement on the effectiveness of the plan needs to be straightforward, in order to allow local authorities to have a framework for measuring progress in implementation (DoE 1992a, 4.23). Just because implementation is continuing it follows that – ideally - so will monitoring of experience under the plan.

In practice, plan monitoring documents usually include a series of data, which seems not to be related to a specific investigation purpose. They are usually organised around themes that follow a sectoral view of the fields involved in planning (transport, housing, land use, etc.). They often pay some attention to the planning activity itself, usually looking at basic statistics on development control activities. However, in this case, it is often not clear what could be the criteria for judging the success of the planning authority. It would be hard to call this a proper process evaluation.

Ex post evaluation concerns the assessment of the success and effectiveness of policy. This assessment is aimed at assisting a review of policies and plans, based on regular monitoring. of existing policies, in order to alter or replace them. Government guidance on methods and approaches to be used is vague; DoE (1992b, 5.19) only indicate the existence of a distinction between quantitative and qualitative assessments. Moreover the distinction between outputs and outcomes is almost ignored.

In practice, plan review is very poorly linked to monitoring and *ex post* evaluation, the latter being also almost non-existent. The influence of monitoring on review is very weak. Research shows (Winter, 1989) that local plan review is precipitated at first by the modification to the Structure Plan and, secondly, by major changes in local issues. Monitoring reports lead to review only in 20% of the cases. Review is more seen as the quickest way to prepare a new plan rather than a reaction to systematic evaluation.

Finally, it can be noted that *ex post* evaluation is much more developed in the close field of urban programmes, such as City Challenge and the Single Regeneration Budget (DoE, 1996; DoE, 1995) For this reason, some cross-fertilisation between urban planning and urban policy, in terms of evaluation methods and management of the evaluation process, could be valuable, as also could some knowledge of evaluation research applied to social programmes (see e.g. Rossi and Freeman, 1982)

Conclusion

Although there is an apparently growing interest in the systematic introduction of evaluation within the planning process, this has hardly had major practical consequences until now. *Ex ante* evaluation appears to be strong in methods and theoretically accepted by the professional and political community. The legal introduction of environmental appraisal for significant development projects has also proved that there is a growing institutional acceptance of the need for *ex ante* considerations of the effects and impacts of planning decisions. Monitoring is a patchwork of methods and approaches; this reflects the administrative character of this activity, which is strongly related to contingent and contextual aspects and the difficulty in spreading good practice. The difficulty in making monitoring both scientifically consistent and contributive to the planning process relates to a very general difficulty in the relationship policy/research, i.e. transferring research findings and observations into a policy process, and, vice versa, transferring back inputs from policy practice to research. Lastly, *ex post* evaluation in planning appears to be extremely weak in terms of practice, legitimacy and link with both the evaluation process and planning process.

2.4.2.   *The evaluative functions in practice*

In terms of evaluative functions, for a long time the major motivation for evaluation in planning has come from the relevance in this policy field of the choice/design function, in the phase of plan preparation. This explains why *ex ante* evaluation methods are more developed. As far as concerned to feedback, knowledge and social/policy learning functions, they also look poorly developed.

The accountability orientation of many recent requirements and discourses on the planning system and more in general "the rise of the evaluative state" (Neave, 1988 quoted in Henkel, 1991, 9), in order to make it more responsive and politically and market reactive, has not yet promoted the development of an evaluation approach based on exploiting this function. In other words, the trend to evaluation as a mean for setting targets, and measuring performance, that is intensively used in other fields of public management in Britain (Pollitt, 1990; Power, 1994), has not pervaded the planning system.

2.4.3. *Main findings*

The main findings from this analysis on the use of evaluation in British town planning can be summarised as follows:

1. evaluation is not coherently integrated into the actual planning process;
2. *ex ante* evaluation is strong in terms of methods and acceptance, *In itinere* is inconsistent and *ex post* evaluation is almost non-existent.

Attempting to explain why

Before exploring how these shortcomings could be addressed, it is worth presenting some possible explanations for this distinct development of evaluation in this field. This investigation has to refer back to some characteristics of planning in relation to other policy fields.

As the British planning system allows for some discretion, in gemral terms, more *in itinere* and *ex post* evaluation could be expected, because the "success" of a British plan can not simply be evaluated in terms of conformity to a definitive end-state or set of rules. To a certain extent, the process orientation and the flexibility of the system would therefore naturally require some form of process and ex post evaluation. Why is this not happening?

One possible explanation is that, despite the existence of relevant provisions for adaptive implementation, the old concept of the plan as an end state to be achieved, a blueprint of the future, may be harder to die than it is usually thought. This means in practice that attention is still focused on plan-making. If this view still prevails, consequently, attention is concentrated on *ex ante* evaluation. A second - and symmetric - explanation refers to the fact that the planning community may have actually become very well aware of this adaptive nature of the plan, but is not confident that existing evaluation methods in the field could effectively deal with it, also as frames for *in itinere* and *ex post* evaluation.

3.   Tackling the Findings

3.1.   EVALUATION IN TOWN PLANNING AS A PLANNING AND MANAGEMENT LEARNING SYSTEM

As just seen, the first shortcoming of current evaluation practice is in terms of its link with the overall town planning and management practice. There is therefore a need to reappraise the relationship between evaluation, planning and management and to modify evaluation in order to make it a useful tool for all actors involved in the process. In this it must be recognised that planning and management are interlinked: planning *is* management for urban change, and planning implies management for implementation while management implies have a policy, plan or strategy (Lichfield, 1988). Second, evaluation is not a discrete element in planning and must be involved throughout the process (Lichfield *et al.*, 1975). The interaction could be achieved by providing within the planning department with a Planning and Management Learning System (PMLS). It would be based on:

1.   an evaluation framework, which constitutes a firm methodological base for evaluation over the whole time of the planning process;
2.   organisational aspects, such as staff, resources, expertise, information systems and so on, that allow it to work properly and effectively in the political and administrative environment.

The introduction of the PMLS would help to overcome the existing gap between a need to introduce evaluation in the town planning process and the evaluation practice itslef. A PMLS could help in providing useful and timely information and interpretation on the changes that are happening in the system, induced by the policies or other factors and on the values that are underlying political decisions and conflicts. Evaluation can be seen as a contribution to the planning and management process, in all its phases. It should be a flexible framework, that should react to the discovery of its defects and to external changes.

The kind of contribution is not purely in terms of feed-back, but more in terms of learning. Since it is just in the middle ground between basic research and management, evaluation can help to bridge the gap between these two fields, by transferring, to one side, inputs from practice and, to the other side, explanations and applied intelligence. Furthermore, the system should ideally help to increase and enrich the dialogue and reciprocal understanding between politicians, planners, developers and other stakeholders and public opinion.

3.2. STRENGTHENING THE LINK BETWEEN *EX ANTE* AND *EX POST* EVALUATION

The second shortcoming, relates to the need to fortify the link between all components of the evaluation process, namely *ex ante*, *in itinere* and *ex post* evaluation. Therefore, the same methodological framework used *ex ante* should be applied *in itinere* and *ex post*. This would help to apply all the questions identified above on efficiency, effectiveness, process and so on, throughout the whole life of the plan and minimise *ex post* problems due to the lack of useful data collection on baseline conditions and implementation. A PMLS is not only an evaluation division within the planning department, but should also provide methodological consistency to the evaluation activity.

In practical terms, methodological coherence in the evaluation activity is an help to learning. Policy actors will not be forced to master ever changing evaluative methods and new types and styles of analyses; they could focus on the substance of the policy problems, more than on understanding methodological aspects. The process of learning between the policy process and the evaluation process can hence become reciprocal and this would limit the perennial problem of the low interest in the evaluation findings from policy-actors.

3.3. THE APPLICATION OF COMMUNITY IMPACT EVALUATION AS AN EVALUATION FRAMEWORK

The link just explained is important for any method of *ex ante*, *in itinere* and *ex post* evaluation. We demonstrate here in relation to one particular method, namely Community Impact Evaluation (CIE) (Lichfield, 1996). This has been primarily devised for *ex ante* evaluation but proposals have been advanced on how it can be used for *ex post* evaluation.

CIE is a method that results from the adaptation of Cost-Benefit Analysis (CBA) to urban and regional planning. The first variation from traditional CBA was introduced in the Planning Balance Sheet (Lichfield, 1960), in that in planning there is a need to consider not only financial and direct costs and benefits to the promoter, but also indirect and external ones. The second adaptation came in the 70s in conjunction with the spread of Impact Assessment theory and practice. These methods were more elaborate than PBS in investigating the characteristics of the processes leading to the impacts of projects on the natural environment and, few years later, on society. CIE integrates the PBS approach with some formal means of opening the black box of the policy process and take into account the chain of effects and impacts (on physical, social , economic, cultural dimensions, etc.). One fundamental feature of the CIE concept is that it provides not only a measure of the total costs and benefits, but also their distribution between impacted sectors of the community, to enabling the equity and social justice implications of the decisions to e taken into account.

The rationale of CIE can be described with a generic algorithm, which constitutes the framework to be adapted to diverse circumstances (Lichfield, 1996, 110-111). The fundamental steps of the generic algorithm are the following (adapted from Lichfield, 1996, 109-111):

1.   *Project description*
     a. planning process in project
     b. projects in the system
     c. options and specification by plan variable
     d. system change

2.   *Analysis and Evaluation*
     e. framework for decision
     f. effect assessment
     g. impact evaluation

3.   *Conclusion*
     h. decision analysis
     i. evaluation report
     j. decision and communication

Our concern here is not to describe the method, but to show how it can be used as a PMLS throughout the planning process. From a methodological point of view, the essence of the *ex post* evaluation is to use the report of the *ex ante* in order to trace through from monitoring the new situation in a systematic way, following the same steps, so that differences can be brought out and the *ex ante* approach can be improved. Here we present the process, by referring to the six evaluative functions

1.   CIE and the choice/design function
     From an *in itinere* perspective, the role of the choice/design function is limited. This means that there is no need to apply the steps that involve comparison between different options, unless there is a real opportunity to affect the rationale of the on-going policy. From an *ex post* perspective, the choice has been made but consideration can be given as to whether the choice was the correct one in the light of the later situation. The *ex post* comparison with the *ex ante* of the different options can allow reflection on the advantages and disadvantages of the actual chosen one. Since ex post evaluation is often done for the purpose of review, this a useful contribution to the political debate on subsequent policies.

2. CIE and the knowledge function

    From an on-going and ex post perspective, the CIE can be considered as applied research on the policy and on case studies. This means creating knowledge on the processes, which could be useful not only for management aims but also for basic research. Put another way, new theories (i.e. new explanations of the possible chains of effect and impacts) can be integrated in the model.

3. CIE and the feed-back function

    The ex ante CIE framework is already a framework for data collection throughout the whole policy process. It specifies what kind of information is needed. If the CIE evaluation has been built within the actual policy environment that is implementing it, the kind of information provided by monitoring activities could be timely and easily understood. In the case of town planning this is particularly relevant, inasmuch as a PMLS can make the system more reactive. Since the British system allows for discretion, if ex ante forecasts or assumptions prove to be inexact, or major socio-economic changes occur, there is a way to modify the attitude in granting permissions.

4. CIE and the social/policy learning function

    The contribution of CIE to this is similar to that of the knowledge function. From the very nature of this evaluative approach, it helps to build a common understanding among all actors involved since the beginning of the policy process. The theoretical assumptions on the prediction of the chain of effects and impacts can be verified and amended throughout the evaluation process.

5. CIE and the accountability function

    Obviously, a PMLS is a good means for accountability. CIE, applied throughout the policy life, can be adapted in order to provide information to external bodies, stakeholders and citizens, via regular reporting and communication. Since CIE is focused on the chain of effects and impacts, the accountability function can shift from a function imposed from above to a more learning-oriented and critical one.

6. CIE and the communication function

    The main purpose of the communication function is to facilitate the democratic participation of relevant stakeholders in the process. CIE does this by introducing the decisions to be taken in terms of the impact on the sectors of the community. In the following *in itinere* and *ex post* stages, stakeholders can be informed via regular dissemination of information and contribute to the improvement of the framework itself. The initial classification of the sectors can be modified throughout the process, in order to better portray the changes in the social environment, in the distribution of impacts and in the perception of problems.

References

Ball, S. and Bell, S. (1991) *Environmental Law*, Blackstone Press, London.

Carley, M.J. and Bustelo, E.S. (1984) *Social Impact Assessment and Monitoring. A guide to literature*, Westview, London.

DoE (1991) *Policy Appraisal and the Environment*, HMSO, London.

DoE (1992a) *Development Plans and Regional Guidance. PPG12*, HMSO, London.

DoE (1992b) *Development Plans. A good Practice Guide*, HMSO, London.

DoE (1995) *Single Regeneration Budget Challenge Fund. Guidance Note no. 2. Monitoring and Periodic Review*, HMSO, London..

DoE (1996) *City Challenge Interim National Evaluation*. Regeneration Research Report, HMSO, London.

DoE (1997) *General Policy and Principles*, PPG1 (revised), HMSO, London.

Friend, J.E. and Jessop, W.M. (1969) *Local Government and Strategic Choice: an Operational Research Approach to the Process of Public Policy*, Pergamon, Oxford.

Glasson, J., Therivel, R. and Chadwick, A.(1994) *Introduction to Environmental Impact Assessment*, UCL Press, London.

Grant, M. (1992) Planning law and the British planning system, *Town Planning Review* **63** (1), pp. 3-12.

Henkel, M. (1991) *Government, Evaluation and Change*, Jessica Kingsley, London.

Housing and Town Planning Act 1909

Lichfield N. (1988) *Economics in urban conservation*, Cambrifge University Press, Cambridge.

Lichfield, N. (1960) Cost-benefit analysis in city planning, *Journal of the American Institute of Planners* **26**(4), pp. 159-169

Lichfield, N. (1970) Evaluation methodology of urban and regional plans: A review, *Regional Studies* **4**(2), pp. 151-165.

Lichfield, N. (1992) The integration of environmental assessment into development planning: Part 1, some principles, *Project Appraisal* **7**(2), June, pp. 58-66.

Lichfield, N. (1996) *Community Impact Evaluation*, UCL Press, London.

Lichfield, N. P. Kettle and Whitbread M. (1975) *Evaluation in the planning process*, Pergamon, Oxford.

MoLGP (1951) Progress report: town and country planning, 1943-51 [Cmd 8024]. London: HMSO

O'Riordan, T. and Turner, R.K. (1983) *An annotated reader in environmental planning and management*, Pergamon Press, Oxford.

Planning Advisory Group (1965) The Future of Development Plans, HMSO, London.

Planning and Compensation Act 1991

Pollitt, C. (1990) *Managerialism and the Public Services. Cuts or cultural change in the 1990?*, Blackwell, Oxford.

Power, M. (1994) *The Audit Explosion*, Demos, London.

Prat, A.(1995) *Il problema dell'efficacia nel dibattito e nella pratica recenti dell'urbanistica*, unpublished thesis, Facoltà di Architettura, Politecnico di Torino,Torino.

Rossi, P.H. and Freeman, H.E. (1982) *Evaluation. A systematic Approach*, 2nd ed., Sage, Beverly Hills.

Therivel, R. *et al.* (1992) *Strategic Environmental Assessment*, London: Earthscan.

Town and Country Planning Act 1932

Town and Country Planning Act 1947

Town and Country Planning Act 1968

Town and Country Planning Act 1971

Voogd, H. (1983) *Multi-criteria evaluation for urban and regional planning*, Pion, London.

Winter, J. (1989), Is local plan review now underway?, *The Planner* **75**(1), pp. 22-27.

Winterfeldt, D.V. and Edwards, W. (1986) *Decision analysis and behavioural research*, Cambridge University Press, Cambridge.

EVALUATION IN ISRAELI SPATIAL PLANNING

Theory vs. Practice

E.R.ALEXANDER

1. Introduction

Spatial planning in Israel takes place in a well defined hierarchical statutory planning system (Alexander, Alterman and Law-Yone, 1983; Torgovnik, 1989). The planning process referred to here covers statutory land use plans, and the development policy plans which often precede them at the national, regional, district and local levels. It also includes sectoral planning (e.g for transportation, energy, solid waste disposal, and tourism) and strategic infrastructure elements and projects, such as harbors, airports, and major highway construction. In this sense, it is fair to describe Israel as highly planned.

The planning process used in all these aspires to conform to the normative rational model. This is obvious from review of the relevant plans and planning documents, which present the stages of this planning process and their products. Even when the rational planning process is not in fact followed, plans and reports make every effort to present the proposals as if they had been arrived at through using the rational planning process[1].

A salient feature of the normative rational planning model, of course, is the development and evaluation of alternatives. This element, therefore, is an integral part of the planning process which produces all the Israeli plans and policies referred to above. In Israel, rationality in the form of the requirement for alternatives' development and evaluation has even become institutionalised as the basis for legal accountability of planning and admininstrative decisions.

One form this has taken is as the basis for judicial review in administrative law, as exemplified in the precedent-setting decision in Berger vs. Minister of the Interior (High Court of Justice No.297/82, 1983). Another more recent manifestation of this requirement is in the guidelines for the Environmental Impact Statements that are required for most major projects, which demand consideration and evaluation of alternatives (Ministry of the Environment, 1992). The development and evaluation of

N. Lichfield et al. (eds.), Evaluation in Planning, 299–310.
© 1998 *Kluwer Academic Publishers. Printed in the Netherlands.*

alternatives, then are integral parts of Israeli planning. Systematic evaluation is an almost universal element of the types of plans referred to above.

2.   Plan Evaluation in Israel

Evaluating plan alternatives invokes both instrumental and substantive rationality. Instrumental rationality assumes or adopts a given goal, which is reflected in the chosen evaluation methods and criteria. Investment analysis and its criteria (such as benefit cost analysis and net present value) are a common expression of instrumental rationality in plan and project evaluation in Israel, to identify the most economically rewarding alternative. However, for the purpose of this paper this aspect of evaluation is relatively trivial, because these evaluations are usually only one element in a more comprehensive assessment of the plan or project alternatives which is the subject of our discussion.

Most of the critiques of rationality and its derived methods (such as benefit cost analysis) refer to formal instrumental rationality. In response to these critiques, evaluation methods have been developed which aim to reflect substantive rationality. In other words, these methods include the articulation and prioritization of the goals and objectives themselves which the subject plan or project is expected to achieve. These evaluation approaches are what has been called multi objective or multi-criteria decision analysis[2]

Some form of multi-objective decision analysis is applied in the systematic evaluation of alternatives in most planning in Israel above the scale of local project plans. This includes national statutory land use and sectoral plans, district statutory land use and development plans, metropolitan and some local development plans, and most sectoral planning for major strategic projects. Citation of cases would produce an almost exhaustive inventory of the major Israeli plans of the last twenty years. An illustrative set of recent examples includes the national development policy plan recently completed, "Israel 2020", the latest national statutory land use plan (TMA-31), the development and statutory plan for the Central District (TMA-3), the development policy for the Jerusalem Tel Aviv axis, the development and statutory plan for the Ben Gurion international airport (TMA-4/2), the development plan for the central area of the Beersheba metropolis, the development plan for Haifa port (TMA-13), and the plan for the Trans Israel Highway #6 (DPW, 1991).

On its face, this extensive use of multi-criteria decision analysis for plan and project alternatives' evaluation reflects a praiseworthy aspiration towards substantive rationality in planning. However, a more critical and detailed review of some of these applications reveals a yawning gap between the prescriptions of normative theory and the experience of real life practice. Often, it becomes obvious to the educated reader of planning documents that the development and evaluation of alternatives is a token exercise in rationality designed to support a predetermined decision or course of action.

One classic symptom of this syndrome are the three options with the preferred alternative bracketed between two almost unrealistically extreme variations. Another is a limited set of objectives obviously weighted to favor a particular alternative, without any rationale for the selected priorities or reasons for omitting other, no less important goals[3]. The absence of systematic sensitivity analysis (Alexander, 1992) or its very limited use is also symptomatic of an evaluation approach which is clearly more symbolic than substantive.

In view of the widespread availability of well developed and quite systematic multi-objective evaluation methods, and extensive literature presenting appropriate methodology, the low methodological quality of most of these applications is surprising. Methododological flaws include information loss due to failure to systematically transform available quantitative data into standardized scores, unintended weighting of criteria due to the absence of standardized scores, duplication of criteria or use of several highly correlated criteria, and even the use of different criteria for each alternative[4]. Sometimes the evaluation is sensitive to these errors, and their correction produces different outcomes. A detailed analysis of one such case follows, which is presented here not so much for its intrinsic interest, but as a typical example of an Israeli application of multi-criteria decision analysis[5].

3.   Evaluating Organisational Alternatives for Highway #6

The "Trans-Israel Highway", HW #6, is designed to serve as a the second major North-South axis for interurban traffic, linking Beersheba in the south and the Galilee in the north with the central Tel-Aviv metropolitan area. Among relevant considerations were the possible ways in which its planning, construction, and operation would be organized and funded. Three alternative forms of organisation were:

1. *Government:* The currently prevailing process, in which government plans and constructs the highway. Funding for planning, construction and maintenance is allocated from the government budget.

2. *Authority:* Creation of an independent public agency (the Trans-Israel Highway Authority) to plan, construct, and operate the highway. The Authority would borrow funds for construction, and debt service and operation costs would be met from revenues raised by operating the highway as a toll road.

3. *Private:* Privatisation of the highway by franchising its construction and operation to a private investor, whose costs and profit would be recouped from toll revenues.

The planners evaluated these three organizational alternatives using a multi criteria evaluation approach, shown in Table 1 below. The alternatives were evaluated on each criterion, from negative performance (scored 0) through neutral (1) to positive (2). Several features are noteworthy:

- Criteria are not explicitly weighted;

- Scores (0-2) are not standardised (i.e different criteria exhibit different ranges);

- Irrelevant criteria (i.e. factors on which there is no difference between the alternatives, and which therefore cannot affect the choice between them) are included.

In Tables 1.a and 1.b some of these errors are corrected: scores are standardized and the scores for capital funding are adjusted to reflect the actual estimated cost differences between the alternatives. Table 1.b consolidates multiple cost related criteria under a single one: cost as reflected in net present value. This corrects the information loss of the original evaluation. However, the final ranking of the alternatives is not sensitive to these errors and remains unchanged, though the intervals between their total scores have narrowed considerably.

Organiztnl. Alt.	Govt.	Auth.	Priv.	a			b		
				Govt.	Auth.	Priv.	Govt.	Auth.	Priv.
Legal Measures	2	1	0	1.00	0.50	0.00	1.00	0.50	0.00
Cost (NPV)							0.00	1.00	0.82
Construction									
Duration	0	1	2	0.00	0.50	1.00	0.00	0.50	1.00
Effiency	0	1	2	0.00	0.50	1.00	0.00	0.50	1.00
Cost	0	1	2	0.00	0.50	1.00			
Operations									
Efficiency	1	1	2	0.50	0.50	1.00	0.50	0.50	1.00
Maintenance	0	1	2	0.00	0.50	1.00			
Funding									
Capital	0	0	1	0.00	1.00	0.82			
Proj. Contin.	0	0	1	0.00	0.00	0.50	0.00	0.00	0.50
Govt. Allocs.	0	0	2	0.00	0.00	1.00			
Off-Budget Fin.									
Domestic	1	1	0	0.50	0.50	0.00	0.50	0.50	0.00
Foreign	0	0	2	0.00	0.00	1.00	0.00	0.00	1.00
Invst. Attract.	0	0	2	0.00	0.00	1.00	0.00	0.00	1.00
Users									
Level Serv.	1	1	2	0.50	0.50	1.00	0.50	0.50	1.00
Costs	1	1	0	1.00	0.14	0.00	1.00	0.14	0.00
Covtl. Consids.									
Risk	2	1	1	1.00	0.50	0.50	1.00	0.50	0.50
Control	2	2	2	1.00	1.00	1.00	1.00	1.00	1.00
Natl. Econ.	2	2	2	1.00	1.00	1.00	1.00	1.00	1.00
Sett. Pol.	2	2	2	1.00	1.00	1.00	1.00	1.00	1.00
Total	14	16	27	7.50	8.64	13.82	7.50	6.64	10.00

TABLE 1. Evaluation of organisational alternatives HW#6

Another significant error is the absence of explicit weighting of the criteria. Naturally, in this case each criterion implicitly receives equal weight. Here, since there are 18 criteria, each criterion's weight is 5.55%. This produces some results that raise questions whether they really reflect decision makers' values. If they do, why were these priorities not expressed in explicit weights?

Analysis shows that cost and funding considerations had the highest priority, which eventually led to the recommendation of privatization. These factors amounted to over 40% of the total "weight". By contrast, user considerations, and the impact on the national economy, had relatively low priorities (11% and 5.5% respectively). This is not necessarily to suggest that these priorities are flawed: if they really express decision makers' preferences they are quite legitimate. In fact, it is quite likely that they do reflect the values of the (then ruling) Likud government, whose intention to privatise the Highway 6 project was no secret.

It is not the relative priorities of the respective criteria that is open to criticism, but the absence of transparency in this evaluation of the organizational alternatives. The evaluation under discussion seems more designed to camouflage the essentially political considerations that led to selection of the preferred alternative, under the guise of an objective and "scientific" evaluation method.

To test this hypothesis, let us check whether the privatization alternative is the robust result of "objective" intrinsic characteristics that make it the optimal choice, or whether it is sensitive to changes in the relative priorities of criteria which express different values and preferences. To answer these questions, the evaluation framework was "cleaned up"[6], and sensitivity analysis was done assigning different weights to the various criteria (Tables 2-5 below). The results of this sensitivity analysis are revealing.

Tables 2 and 3, which essentially reflect the same preferences as the original evaluation, and only contain some methodological corrections, already show a significant narrowing of the gap between the total weighted scores of the Authority and Private alternatives. Tables 4 and 5 explore the results of changes in the weighting of some criteria, and result in a reversal of the rankings of the Private and Authority forms of organization.

In Table 4 there are some relatively marginal redistributions between weights. The coefficient of the cost revenue criterion is raised from 0.4 to 0.5, with half of the increase coming from a 5% reduction in the weight of construction considerations, and the other 5% from Investment Attraction. In Table 5 Legal Considerations and Operational Efficiency are reduced to 2.5% each, the priority of Financing considerations is cut from 20% to 10%, and instead the importance of User costs and services is raised from 5% to 20%.

Criterion	Weight	Government		Authority		Private	
		Score	W*S	Score	W*S	Score	W*S
Revenue/Cost (NPV)	40.0	0.00	0.00	1.00	40.00	0.82	32.80
Legal	5.0	1.00	5.00	0.50	2.50	0.00	0.00
Construction (10%)							
Duration	5.0	0.00	0.00	0.50	2.50	1.00	5.00
Effic.	5.0	0.00	0.00	0.50	2.50	1.00	5.00
Operation							
Effic.	5.0	0.50	2.50	0.50	2.50	1.00	5.00
Funding							
Proj. Cont	5.0	0.00	0.00	0.00	0.00	0.50	2.50
Financing (20%)							
Domestic	5.0	0.50	2.50	0.50	2.50	0.00	0.00
Foreign	5.0	0.00	0.00	0.00	0.00	1.00	5.0
Invest. At	10.0	0.00	0.00	0.00	0.00	1.00	10.00
Users (5%)							
Cost	2.5	1.00	2.50	0.14	0.35	0.00	0.00
Service	2.5	0.50	1.25	0.50	1.25	1.00	2.50
Govtl. Consids. (10%)							
Risk	2.5	1.00	2.50	0.50	1.25	0.50	1.25
Control	2.5	1.00	2.50	1.00	2.50	1.00	2.50
Natl. Econ	5.0	1.00	5.00	1.00	5.00	1.00	5.00
Total	100.0		23.75		62.85		76.55

TABLE 2. Weighted criteria

Criterion	Weight	Government		Authority		Private	
		Score	W*S	Score	W*S	Score	W*S
Revenue/Cost (NPV)	40.0	0.00	0.00	1.00	40.00	0.82	32.80
Legal	5.0	1.00	5.00	0.50	2.50	0.00	0.00
Construction (10%)							
Duration	5.0	0.00	0.00	0.50	2.50	1.00	5.00
Effic.	5.0	0.00	0.00	0.50	2.50	1.00	5.00
Operation							
Effic.	5.0	0.50	2.50	0.50	2.50	1.00	5.00
Funding							
Proj. Cont	5.0	0.00	0.00	0.00	0.00	0.50	2.50
Financing (20%)							
*Domestic	5.0	0.50	2.50	1.00	5.00	0.00	0.00
Foreign	5.0	0.00	0.00	0.00	0.00	1.00	5.00
Invest. At	10.0	0.00	0.00	0.00	0.00	1.00	10.00
Users (5%)							
Cost	2.5	1.00	2.50	0.14	0.35	0.00	0.00
Service	2.5	0.50	1.25	0.50	1.25	1.00	2.50
Govtl. Consids. (10%)							
Risk	2.5	1.00	2.50	0.50	1.25	0.50	1.25
*Control	2.5	1.00	2.50	0.50	1.25	0.00	0.00
*Natl. Econ	5.0	0.50	2.50	1.00	5.00	0.00	0.00
Total	100.0		21.25		64.10		69.05

TABLE 3. Weighted criteria

Criterion	Weight	Government		Authority		Private	
		Score	W*S	Score	W*S	Score	W*S
Revenue/Cost (NPV)	50.0	0.00	0.00	1.00	50.00	0.82	41.00
Legal	5.0	1.00	5.00	0.50	2.50	0.00	0.00
Construction (10%)							
Duration	2.5	0.00	0.00	0.50	1.25	1.00	2.50
Effic.	2.5	0.00	0.00	0.50	1.25	1.00	2.50
Operation							
Effic.	5.0	0.50	2.50	0.50	2.50	1.00	5.00
Funding							
Proj. Cont	5.0	0.00	0.00	0.00	0.00	0.50	2.50
Financing (20%)							
*Domestic	5.0	0.50	2.50	1.00	5.00	0.00	0.00
Foreign	5.0	0.00	0.00	0.00	0.00	1.00	5.00
Invest. At	5.0	0.00	0.00	0.00	0.00	1.00	5.00
Users (5%)							
Cost	2.5	1.00	2.50	0.14	0.35	0.00	0.00
Service	2.5	0.50	1.25	0.50	1.25	1.00	2.50
Govtl. Consids. (10%)							
Risk	2.5	1.00	2.50	0.50	1.25	0.50	1.25
*Control	2.5	1.00	2.50	0.50	1.25	0.00	0.00
*Natl. Econ	5.0	0.50	2.50	1.00	5.00	0.00	0.00
Total	100.0		21.25		71.60		67.25

TABLE 4. Weighted criteria

Criterion	Weight	Government		Authority		Private	
		Score	W*S	Score	W*S	Score	W*S
Revenue/Cost (NPV)	40.0	0.00	0.00	1.00	40.00	0.82	32.80
Legal	2.5	1.00	2.50	0.50	1.25	0.00	0.00
Construction (10%)							
Duration	2.5	0.00	0.00	0.50	1.25	1.00	2.50
Effic.	2.5	0.00	0.00	0.50	1.25	1.00	2.50
Operation							
Effic.	2.5	0.50	1.25	0.50	1.25	1.00	2.50
Funding							
Proj. Cont	5.0	0.00	0.00	0.00	0.00	0.50	2.50
Financing (20%)							
*Domestic	2.5	0.50	1.25	1.00	2.50	0.00	0.00
Foreign	2.5	0.00	0.00	0.00	0.00	1.00	2.50
Invest. At	5.0	0.00	0.00	0.00	0.00	1.00	5.00
Users (5%)							
Cost	20.0	1.00	20.00	0.14	2.80	0.00	0.00
Service	5.0	0.50	2.50	0.50	2.50	1.00	5.00
Govtl. Consids. (10%)							
Risk	2.5	1.00	2.50	0.50	1.25	0.50	1.25
*Control	2.5	1.00	2.50	0.50	1.25	0.00	0.00
*Natl. Econ	5.0	0.50	2.50	1.00	5.00	0.00	0.00
Total	100.0		35.00		60.30		56.55

TABLE 5. Weighted criteria

This sensitivity analysis reveals that the preference of the Private alternative over the public Authority is a direct result of the relative importance assigned to financing considerations, and is not related to any priority of cost revenue criteria. Such sensitivity analysis could be extended ad infinitum, but even this limited exploration is enough to show that the choice among two of the alternatives in this evaluation (the government Authority and franchising to a Private operator) is not the result of significant intrinsic and objective differences between them. Rather, it is an artefact of the evaluation itself, reflecting the decision makers' priorities and values.

In itself, there is nothing problematic about this conclusion. After all, that is exactly what multi-criterion decision analysis is designed to do. The problem is the way in which this evaluation was communicated in the context of the planning process. Rather than identifying and highlighting the value related aspects of this decision, the plan presented it as the result of objective evaluation of the expected outcomes of the three organizational alternatives. The main way in which this was acheived was by suppressing the weighting of the criteria, so that only a relatively expert observer could deduce (as this author did) that this choice was based more on the decision makers' priorities than on intrinsic differences between the options.

4.   Discussion

Though its methodological and substantive errors are specific (though not unique), the multi criteria evaluation application used to choose between organisational alternatives for constructing and operating Highway #6 is typical of such applications in its orientation and presentation. Like many other evaluations in the context of Israeli spatial and project plans, this evaluation was not really undertaken to aid in making a decision. The objective of such evaluations is not to decide on a choice between the options that are presented, but to present an ostensibly objective and "scientific" rationale for a choice that has already been made.

In itself this is not necessarily bad. The link between rationality and accountability is recognized, expressed in the positive associations with "rationale", or providing reasons for a decision[7]. But the way in which these evaluations are carried out and presented (as suggested by the example analysed above) implies an alternative explanation. The kinds of applications of multi-criteria decision analysis which have been reviewed all suffer from an absolute lack of transparency and suggest a technical-professional expertise and scientific "objectivity" that is completely incompatible with the essentially value based reasons for their choices, which they reveal under closer analysis. Here evaluation does not provide the rationale for a decision, but a rationalization.

These applications of multi-criteria decision analysis do not use their evaluation of alternatives as part of an attempt to apply substantive rationality to solve their problems, as prescribed by the "classic" model of rational planning. Rather, they are rhetorical devices (Throgmorton, 1993) used in a context of communicative practice (Sager, 1993; Innes, 1995). But, contrary to the assumptions and prescriptions of some of the proponents of communicative practice (e.g. Forester, 1989, 1993; Healey, 1995), the planners here are not engaged in communicative action, but in strategic action[8]: they are using the plan, and the evaluation that rationalizes the choice between its supposed alternatives, to accomplish their (or their political masters') strategic goals: adopting the plan and ensuring its implementation[9].

In any normative perspective (whether in the ethical sense, or even just as sound professional practice) these applications of multi-criteria evaluations are a misuse of a method which has potentials that are either unrealized or wilfully ignored. Multi criterion decision analysis offers the unique possibility of combining objective empirical information about alternatives' consequences and more subjective or experience based assessments of possible outcomes with systematic articulation of goals and priorities in an integrated decision framework. This can be used in two ways, neither of which is the kind of application we have seen in the Israeli cases.

One way is as a decision making aid, essentially applying substantive rationality in making a choice between alternatives when decision makers are uncertain about their own preferences. This is the conventional role of evaluation of alternatives in the rational planning process. The other way is as a communicative tool, in what the communicative practice model of planning sees as an interactive planning process. It is this potential that is perverted in the prevailing use in Israel of multi-criteria decision analysis as a rhetorical tool for rationalizing preconceived plans.

What creates the potential of multi criteria evaluation methods as a communicative tool is their capability to combine the assessment of alternatives' performance with the assignment of relative priorities to goals and criteria, and observe their respective effects on the resulting decision. The analysis shown above of the evaluation of organizational alternatives for HW#6 demonstrates this potential.   Performing sensitivity analysis of alternative goal priorities showed that the choice between the options as presented was essentially a matter of relative priorities between objectives, and revealed the value base for the policy that was actually chosen.   In a planning system that was structured and worked in conformity to communicative practice norms, such insight would be valuable. Formal and systematic evaluation should be the point of departure for what would become an essentially political discourse leading to consensus on the appropriate priorities for selecting between these alternatives.

Multi-criterion evaluation, then, has the potential for identifying the essential nature of a decision. Is it a technical choice between options which are sufficiently different in their objective performance for the choice between them to be robust in relation to any reasonable combination of goal or criterion priorities? This is sometimes the case, but such situations are increasingly rare with expanding scale and growing complexity of the planning object. If evaluation shows that the choice between alternatives is essentially a matter of the relative priorities assigned to criteria, then the evaluation itself can and should become a communicative tool for interactive discourse designed to generate consensus on a mutually agreed upon set of priorities, and with it the selection of the preferred alternative[10].

Unfortunately, planning in Israel is institutionalized in a way that is rarely conducive to communicative practice. Consequently, unless there is a radical restructuring of conventional planning processes in Israel, it is unlikely that these suggestions concerning the use and misuse of evaluation methods will be heeded. This is because the preparation of development and statutory plans and plans for major strategic projects is entrusted to a small and relatively closed community of professionals (planners, social-science academics and researchers, architect-town planners, engineers in relevant sectors and other sectoral experts), and bureaucrats representing the interested and affected units of government and public agencies.

Major plans are developed by (sometimes quite large, and often multidisciplinary) planning teams combining planners and experts in the appropriate sectors. Each such planning team is directed by and interacts with a Steering Committee which is made up of officials representing the relevant institutional interests. In spite of some enhanced consciousness of the politics of planning, there is little more than lip service to any more openness of the planning process, and the only stakeholder input is indirectly through the members of the Steering Committee. These could conceivably be participants in an interactive evaluation of alternatives, using a multi criterion decision framework, but planners' professional interests and technical orientation have rarely led them to suggest such a possibility[11].

More open and participative planning, in which, for example, a wider range of political interests was directly involved, and representatives of more varied constituencies and stakeholders could participate, would find an interactive form of evaluation an invaluable tool. Such an evaluation could combine multi criterion decision analysis with sensitivity analysis of alternative sets of goal priorities, to stimulate discussion of the implications of stakeholders'different preferences in the context of the planning process itself[12]. Effectuating such a planning process in Israel, however, will need a major effort in institutional design[13] to restructure the way in which major development and statutory plans and national strategic projects are planned today.

5.   References

Alexander, E.R. (1996) After rationality: Towards a contingency theory for planning, in S.J.Mandelbaum, L.Mazza and R.W.Burchell (eds.) *Explorations in Planning Theory*, Rutgers - State University of New Jersey, Center for Urban Policy Research, New Brunswick, NJ, pp.45-64.

Alexander, E.R. (1995) *How Organizations Act Together: Interorganizational coordination in theory and practice*. Gordon & Breach, Montreaux/Trenton NJ.

Alexander, E.R. (1993) The politics of evaluation; The relationship between method and results in multi-objective plan evaluation. *Environmental Planning* **48-49**, 41-52 (Hebrew)

Alexander, E.R. (1989) Sensitivity analysis in planning: Using sensitivity analysis for complex decision models. *Journal of the American Planning Association* **55**, 323-333.

Alexander, E.R., Alterman, R., and Law-Yone, H. (1983) Evaluating plan implementation: The national statutory planning system in Israel. *Progress in Planning* **20** Part 2

Cohon, J.L. (1978) *Multiobjective Programming and Planning*, Academic Press, New York.

DPW (1991) Highway No.6. Department of Public Works, Jerusalem (Hebrew)

High Court of Justice No.297/82 (1983). Berger et al. vs. Minister of the Interior. *Verdicts 1983* pp.29-57 (Hebrew).

Flyvberg, B. (1996a) The dark side of planning: rationality and realrazionalitΣt. in: S.J.Mandelbaum, L.Mazza and R.W.Burchell (eds.) *Explorations in Planning Theory*. Rutgers - State University of New Jersey, Center for Urban Policy Research, New Brunswick, NJ, pp.383-394.

Flyvberg, B. (1996b) *Rationality and Power*. Avebury, Aldershot, Hants.

Forester, J. (1989) *Planning in the Face of Power*. University of California Press, Berkeley, CA.

Forester, J. (1993) *Critical Theory, Public Policy and Planning Practice*. SUNY Press, Albany, NY.

Innes, J. (1995) "Planning theory's emerging paradigm: communicative action and interactive practice". *Journal of Planning Education & Research* 14 (3): 183-189.

Healey, P. (1995) "The communicative turn in planning theory and its implications for spatial strategy formation". *Environment and Planning B: Planning & Design* 23 (2): 217-234.

Hill, M. (1968) A goals-acheivement matrix for evaluating alternative plans. *Journal of the American Institute of Planners* **34**, 19-28.

Hill, M. (1973) *Planning for Multiple Objectives*. Monograph series 5. Regional Science Institute, Philadelphia, PA.

Keeney, R.L., and Raiffa, II. (1976) *Decisions with Multiple Objectives: Preferences and value tradeoffs*. Wiley, New York.

Lichfield, N. (1996) *Community Impact Evaluation*, UCL Press, London.

Metropolitan Planning Team (1996) Document of Principles - Development Policy for the Tel-Aviv Metropolis - Stage B: Scenarios and Metropolitan Development Alternatives; Interim Report No.3 24.7.1996 pp.6.1- 7.10 (Hebrew)

Ministry of the Environment (1992) Guidelines for Environmental Impact Statements, Jerusalem (Hebrew)

National Council for Planning and Building (1986) TMA 4/2: Report of the Professional Committee in the Matter of Ben Gurion Airport, Jerusalem, October 1986 (Hebrew)

Saaty, T.L. (1990) *Multicriteria Decision Making: The Analytic hierarchy process*, AHP Series Vol.1, Pittsurgh, PA.

Sager, T. (1994) *Communicative Planning Theory*, Avebury, Aldershot, Hants.

Shefer, D., Amir, S., Frenkel, A., and Law-Yone, H. (1997) Generating and evaluating alternative regional development plans. *Environment and Planning B: Planning & Design* **24**, 7-22.

Throgmorton, J. (1992) "Planning as persuasive storytelling about the future: Negotiating an electric power rate settlement in Illinois". *Journal of Planning Education & Research* **12**, 17-31.

Torgovnik, E. (1990) *The Politics of Urban Planning Policy*, University Press of America, Lanham, MD.

Voogd, H. (1983) *Multicriteria Evaluation for Urban and Regional Planning*, Pion, London.

Weistroffer, H.R. (1983) An interactive goal programming method for non-linear multiple crtieria decision making problems. *Computer and Operational Resources* **10**: 311-320.

---

[1] This was the case, for example, in the current plan for developing Ben Gurion International Airport, which a commission report described as the result of rational development and evaluation of alternatives, though in

fact it was the product of an incremental communicative planning process (National Council for Planning and Building, 1986).

[2] For the purposes of this discussion this designation means evaluation methods involving the assignment and aggregation of performance scores of alternatives (Voogd, 1987), such as Goals Acheivement Matrix (Hill, 1968, 1973), Hierarchical Decomposition (Saaty, 1991), and Planning Balance Sheet and Community Impact Evaluation (Lichfield, 1996). These types of methods are the focus of this paper, though multi-objective evaluation methods include approaches involving trade off functions between objectives and criteria (Keeney and Raiffa, 1976), and goal programming (Cohon, 1978; Weistroffer, 1983).

[3] A frequent form this takes is the high priority assigned to economic goals and the omission or extremely low weights assigned to environmental objectives (which, when they are included in such cases, have obviously only been added pro forma).

[4] All these methodological errors have been found in evaluations and plans reviewed by the author; since it is not the purpose of this paper to criticise specific applications, but rather to review the culture of evaluation in Israel as a whole, the relevant citations are omitted.

[5] Based on Alexander (1993).

[6] "Cleaning up" comprised some methodological and some substantive corrections. These included: Consolidating duplicative cost criteria under one: NPV; Omitting redundant criteria (Table 2 a and b); Correcting the scores for some criteria (marked * in Table 3) which did not reflect true objective considerations.

[7] Contrast this with the negative association between rationality and rationalization. This is related to a problem solving orientation, which sees ex-post rationalization of a decision (i.e where rationality was not invoked to make the choice itself) in an unfavorable light. We can associate the positive link between "rationale" and rationality, on the other hand, with a communicative orientation, on which more below.

[8] Habermas makes this distinction; for a discussion of its relevance for the communicative practice model of planning, see Alexander (1996).

[9] The analysis could be carried further in this sense, implying a planning that is far from the Habermasian ideal of communicative action, and much closer to a Foucaultian view of planning as an exercise of societal power (cf. Flyvberg, 1996.a,b).

[10] There are evaluation programs which combine multi criteria evaluation with sensitivity analysis (e.g. "Expert Choice") which are eminently adapted to interactive applications.

[11] One of the rare exceptions is the Development Policy for the Jerusalem-Tel-Aviv Axis, for which members of the Steering Committee were polled to obtain aggregated priority weights for the goals and crtieria which they identifiied, and which were used in the subsequent multi-criteria evaluation of three spatial development options. A recent case of (otherwise technically excellent) multi-criteria evaluation where this opportunity was typically foregone is the Development Plan for the Northern Region (Shefer et al. 1997) where the planning team evaluated the alternatives and the evaluators tested the sensitivity of alternative combinations of weights. This is also the course followed in the multi-criteria evaluation application used in preparing the Principles for Development Policy of the Tel-Aviv Metropolis (in progress; Metropolitan Planning Team, 1996).

[12] The absence of broader participation in planning, and the limitation of public involvement to statutory arenas has had negative effects on the acceptance and implementation of plans. A current case is the national development strategy "Israel 2020" which has just been completed by an interdisciplinary planning team: its publication is receiving negative comments on its "closed" planning approach.

[13] See Alexander (1995: 51-52) for a definition and discussion of institutional design.

DEVELOPMENTS IN TRANSPORT APPRAISAL IN BRITAIN

C. NASH

1. Introduction

The transport sector was an early sector in which social cost-benefit analysis was applied. Whilst the pioneering studies of the M1 motorway and the Victoria line underground railway were essentially ex post studies that were independent of the decisions to proceed with this projects, by the early 1970's the technique was being extensively used by central government to appraise trunk road schemes, railway investment and closure proposals. Local authorities were more resistant to its use, often preferring simple point-scoring assessments against a range of objectives. However, they were increasingly forced to use it when making the case to central government for grant-aid towards the costs of transport projects.

In the meantime, developments within the public transport sector made the use of cost-benefit analysis less relevant there. In 1986 the bus network was deregulated and - over the following year - progressively privatised. On the majority of routes services were provided on a purely commercial basis and local authorities had no right of intervention. Where a local authority perceived the need for a service that the commercial operator was not willing to provide without subsidy, it could secure provision of the service by competitive tendering, but even here use of SCBA to decide on service levels and fares was rare, some form of accessibility standards being the more common approach. In the rail sector, inter-city passenger and freight services had been identified as to be provided on a purely commercial basis; other passenger services outside the major cities were to be provided on the basis of maintaining broadly existing service standards at minimum cost. Passenger Transport Executives in the major cities still had control of local rail service fares and service levels in return for which they paid the subsidies and could use SCBA. Finally London Regional Transport was still able to use cost-benefit analysis systematically in the planning of public transport services as, despite privatisation of the bus network, it still controlled fares and services across the network as a whole, bus services being provided by means of competitive tendering.

Thus it appeared that a highly differentiated and arguably totally inconsistent approach to transport appraisal was developing, with trunk roads being subject to a fairly narrow cost-benefit analysis, local authority schemes to a much wider multi-criterion approach and public transport typically to a purely commercial or accessibility

311

N. Lichfield et al. (eds.), Evaluation in Planning, 311–327.

based approach (see, for instance, the comprehensive review provided by the special issue of Project Appraisal devoted to the Transport Sector, which was editted by Lichfield and appeared in December 1992). However, in a number of respects a more coherent approach has now started to emerge. Firstly, there have been moves towards a more strategic approach to trunk road appraisal. Moreover, as trunk road appraisal adopted a wider framework approach that did not consider solely economic efficiency, the two approaches of central and local government began to come together. Secondly further developments in the public transport sector appear to expand the potential role of cost-benefit analysis there, particularly in the case of rail services. Nevertheless, many problems of inconsistency remain, in particular as a result of the rules applied to public transport schemes, and of the lack of public control over the majority of decisions regarding bus services.

This paper considers these developments in more detail. Firstly, a brief outline of the basic methodological approach to transport appraisal in Great Britain is given. Then developments are considered in turn in the trunk roads, local transport and public transport sectors respectively. (The latter two obviously overlap).

2.   Methodology

As most commonly defined, social cost-benefit analysis is concerned with a single objective - that of economic efficiency. Seeking economic efficiency means undertaking all projects for which the willingness to pay of the beneficiaries exceeds the compensation required by the losers. This section first considers the methodologies used to find the willingness to pay of beneficiaries and the compensation required by losers from the effects of transport projects, and then considers the issue of the inclusion of other goals in transport appraisal.

Consider the appraisal of a typical road scheme (Table 1). Like any other project, it will involve capital, maintenance and operating costs. Unlike many other projects, the bulk of the operating costs will be incurred by people other than the agency undertaking the project - namely motorists, bus companies and road hauliers. Moreover to the extent that in the absence of the scheme they would have used poorer quality more congested roads, then operating cost savings appear as a benefit of the scheme; it is only in respect of any traffic generated by the road itself that they appear as a cost. There may also be some maintenance cost savings on existing roads as a result of the reduced level of traffic using them.

So far, the cost benefit analysis appears very straightforward. All the above items are readily measured in money terms. The one adjustment to market prices that is undertaken is to remove any element of taxation from these costs.

But many costs and benefits of transport projects - time savings, pain and grief resulting from accidents, environmental effects - do not have a market price. In this case, a variety of methods have been used to try to establish what those affected would be willing to pay for the benefits or would require in compensation for the costs.

TABLE 1. Costs and benefits of a road scheme[2] (Note: Present values, £k in 1979 prices, discounted at 7%)

	Traffic growth (alternative assumptions):	
	High	Low
Costs		
Construction cost	2491	2491
Maintenance cost	72	72
Delays during construction	32	32
Total	2595	2595
Benefits		
Time and operating cost savings	4218	2658
Accident savings	417	304
Total	4635	2962
Net Present Value	2040	367

Source: Institute for Transport Studies, University of Leeds, Economic Evaluation in Transport Planning Case Notes.

In the case of time savings, there is a distinction to be made between time spent travelling during working hours (which includes bus and lorry drivers as well as business travellers), and time spent travelling during one's own time. In the former case, it is usual to value the time at the wage rate of the employee concerned plus a mark-up to allow for overhead costs of employing labour (such as social insurance charges). This assumes that the time saved can be gainfully employed, and that the gross wage represents the value of the marginal product of labour in its alternative use. Doubts, however, may be raised on a number of grounds. Is the time saving large enough to be of use, or will it simply be wasted as idle time? (individual transport projects often yield savings of less than a minute, although these may be aggregated with savings from other schemes to form more useful amounts of time). Will the labour released find alternative work, or add to unemployment? If it does find alternative employment, does the gross wage really reflect the value of its marginal product in the new use?

For non-working time, the problem of valuation is greater. The approach here has been to try to discover what people are willing to pay to save time, either by `revealed preference' or by `stated preference' methods. Revealed preference methods rely on studying people's behaviour in situations in which they reveal an implicit value of time.

The most popular case is that of the choice of travel mode, where people may have a choice between two modes one of which is faster and more expensive than the other. If a model is estimated which forecasts the probability that someone chooses one mode rather than the other as a function of journey time, money cost and any other relevant quality differences, then the relative weight attached to time and money can be used to estimate their *value of time*.

TABLE 2. Example of a Stated Preference Question

Please compare the following alternative combinations of train fare and service level:							

					A	
LONDON dep.		250	320	350	420	450
Stockport...	510	540	610	640	710	
Manchester, arr.		520	550	620	650	720

Fares: One way £12, Return £24
Scheduled Journey Time: 2h 30 min
Reliability: Up to 10 min late

			B		
LONDON, dep	250	.	350	.	450
Stockport...	540	.	640	.	740
Manchester, arr.	550	.	650	.	750

Fares: One way £10, Return £20
Schedule Journey Time: 3h
Reliability: Up to 30 min late

Do you:

Definitely Prefer A	Probably Prefer A	Like A and B Equally	Probably Prefer B	Definitely Prefer B

*Source: Institute for Transport Studies, University of Leeds questionnaire

This approach was used for many years, but it suffered from some problems. One had to find cases where such trade-offs really exist and are perceived by a representative cross section of the population. To estimate the value of time to a reasonable degree of accuracy, samples running into thousands are needed, and the data usually have to be collected specifically for this purpose by means of a questionnaire survey. If *stated preference* methods are used, then respondents to the survey are asked what they would choose given hypothetical alternatives (an example is given in Table 2). This enables the individual trade-offs to be designed to reveal the maximum information about the value of time; moreover, each respondent can be asked about a number of different choices. This allows great economies in sample size (MVA et al, 1987). After piloting and testing to ensure that the results were similar to those produced by revealed preference methods, this approach was used extensively in the studies that determined the values of leisure time currently used by the British Department of Transport.

Turning to accidents, the costs may be divided into those that are readily valued in money terms, and those that are not. The former include damage to property and vehicles, health service, ambulance and police costs, and loss of production due to victims being unable to work (this again is typically valued at the gross wage). What is more difficult is to place a money value on the pain, grief and suffering caused by death or injury in an accident. For many years, in Britain, this value was determined by the political process rather than the preferences of those directly involved. However, it is possible to apply both revealed preference and stated preference techniques to this issue as well. The way to do this is to recognise that transport improvements do not save the lives of specific known individuals; rather they lead to a reduced probability of involvement in an accident for all users. Thus real or hypothetical trade-offs between safety and cost may be used to derive the *value of a life*. Such a stated preference study (Jones-Lee, 1987) is indeed the basis of the value currently used by the British Department of Transport, although there may be doubts as to how well people are able to respond to questions involving changes in very small probabilities.

The British Department of Transport utilises a computer program (COBA) (DoT, 1997b) to calculate the value of its trunk road projects. This programme includes the money values of all the items so far discussed, as indeed do the methods used in all the major countries of Western Europe. But it does not value any of the other effects of road schemes, of which by far the most important and controversial are the environmental effects of such schemes. Although a lot of effort has been put into valuation of environmental costs and benefits of road schemes in recent years, the view of DoT is that these valuations are as yet not sufficiently accurate to be used in practice.

Another problem with SCBA as so far explained is that it takes no account of distributive issues.

If one is concerned with the distribution of income in the economy, then it is necessary to know not just whether a project contributes to economic efficiency but also who gains and who loses from it. This can be undertaken by identifying groups in terms of function (motorists, consumers, bus users, government, transport operators, residents) and income level, and disaggregating costs and benefits to these groups. Then it is possible, explicitly or implicitly, to attach higher weight to poorer sectors of society.

However, this makes the analysis much more complicated, as the repercussions of projects have to be traced through all those affected. For instance, suppose that a project reduces the costs of freight transport to a city centre. The results obtained may be a mixture of: higher profits for freight operators, higher profits for retailers, higher rents for property owners, higher tax revenues for central and local government, and lower prices for consumers. As can be seen, it is much easier to measure the transport cost saving than attempt to trace through who ultimately gains from it.

As soon as it is accepted that distribution is important as well as efficiency, then logically we are involved in some form of multi-criterion decision-taking; this is also true if environmental costs and benefits are seen to be relevant but are not valued in money terms. As used in practice, multi-criterion approaches require three stages: firstly definition of a set of objectives, which may for instance relate to accessibility, the environment, safety, economy and equity; next, measurement of the extent to which each project contributes towards the desired objective; finally, weighting of the measures in order to aggregate them and produce a ranking of projects.

As it stands, this method would be quite consistent with the principles of cost-benefit analysis if the following conditions held:

1. all the objectives related to impacts on the welfare of the population concerned, and
2. the measures of achievement and weighting of them are based on the preferences of the people affected by the projects, subject possibly to some form of equity weighting.

In practice, the first condition probably generally holds but the second does not. Measurement of the degree of contribution to objectives is often based not on detailed measurement, and valuation but on the judgement of the professional staff planning the projects. This might be defended either on grounds of convenience (it is easier to ask professional staff then the public at large) or on the grounds that professional staff know better what matters than does the public at large.

Whether these weightings are expressed in money terms or not, they are essentially performing the same function as money values in expressing relative valuations. Moreover, to the extent that at least one of the performance measures - cost or economy -is expressed in money terms, they can readily be transformed into money values. There is therefore less difference between this approach and traditional cost-benefit analysis than might at first sight be supposed.

It appears then that, as currently practised, multi-criteria decision-making techniques are essentially concerned with aiding and ensuring consistency in this latter stage of weighting by the decision-taker. This is a separate role from that played by the cost-benefit analysis, and may be complementary rather than competing. To the extent that the information provided by a cost-benefit analysis is seen as relevant to the decision-taker, it still needs to be provided. But what is clear is that it must be provided in a sufficiently disaggregate form for the decision-taker to, explicitly or implicitly, apply his or her own weights.

A consistent approach to the specification, measurement and reporting of effects on incidence groups at a disaggregate level has been developed by Lichfield (1996). Initially developed as the Planning Balance Sheet and later extended and renamed to form what is now known as a Community Impact Evaluation, this approach clearly specifies all the relevant groups and consistently identifies costs and benefits to them, either in whatever the natural units of measurement of the effect in question are, or in money units where valuations are deemed to be sufficiently reliable to be of value. The Lichfield approach may essentially be seen as an amalgamation of the best features of both cost-benefit analysis and multi-criterion techniques.

One final point may be made on the distributional issue. When evaluating methods of valuing time and accident savings, there is - not surprisingly -clear evidence that these are related to ability to pay. Thus if one were applying cost benefit analysis purely as an efficiency test, one would need to disaggregate benefits by income group and apply higher values of these benefits to the better off. This would systematically bias decisions towards improving roads used more by the affluent, for instance those in wealthier parts of the country.

In practice, this has never been seen as politically acceptable in Britain. Thus it is usual simply to apply average values to all road users. This in itself could introduce some curious biases to decisions however. For instance, it may lead authorities to spend money on securing time savings for travellers in poor areas on the basis of the average value of time, when in fact those travellers value the time savings at less than the cost of the scheme, and would rather have received the cash as a tax reduction. This illustrates the problem that effectively uprating *one* item of benefit for the poor whilst not applying similar weights to all others distorts the relative values of different types of cost and benefit. It is more consistent to value all costs and benefits at people's own willingness to pay or to accept compensation, and then to weight the sum total of costs and benefits.

3. Roads

Trunk road projects have been subject to CBA for more than 20 years. Until recently the economics involved was very simple. Projects were defined as small sections, typically only a few miles, of new or upgraded road. Forecast traffic levels were loaded on to a computerised network to simulate traffic flows and times on individual links with and without the project. From this, changes in operating costs, journey times and accident levels were estimated, to be set against the increased capital and net increase in maintenance costs resulting from the new road.

This approach to appraisal has been heavily criticised on many occasions over the years. The first key criticism is its narrow approach with heavy emphasis on time and accident savings (with time savings playing the leading role - table 3) as the justification for new roads.

TABLE 3. Typical Breakdown of Benefits of Road Scheme

WORKING TIME SAVINGS	43%
NON-WORKING TIME SAVINGS	46%
ACCIDENT SAVINGS	14%
OPERATING COST SAVINGS	-3%

Source: DoT (1987a) Values of Journey Time Savings and Accident Prevention

This criticism came to a head in the road protests of the 1970's, and culminated in the appointment of an Advisory Committee on Trunk Road assessment (the Leitch Committee) to consider the issues. This committee reported in favour of supplementing the existing COBA analysis with a 'framework' showing effects on a wide group of interested parties, and measured in a mixture of monetary , quantitative and qualitative units (Leitch, 1977). This approach has survived, with some degree of modification, as chapter 11 of the current Design Manual for Roads and Bridges (DoT, 1993). A summary of the elements included in the original Leitch framework is shown in Table 4. From this it will be seen that there is a wide variety of measures, in different units, viz. physical measures, numbers of houses, rankings, and verbal descriptions, as well as the financial ones. At the same time, no measures are included of non-local environmental effects of schemes. This is because, with traffic assumed constant regardless of what road schemes are built, the level of these pollutants hardly varies.

TABLE 4. Summary of the measures used in the Leitch framework for assessing the costs and benefits of road schemes.

Incidence group	Nature of effect	No of measures financial	Other
Road users	Accidents	1	3
	Comfort/conveni	6	
	ence	5	
	Operating costs		2
	Amenity		
Non-road users directly affected	Demolition disamenity (houses, shops, offices, factories, schools, churches, public open space)		37
	Land take, severance, disamenity to farmers		7
Those concerned	Landscape, scientific, historical value, land-use, other transport operators		9 (+ verbal description)
Financing Authority	Costs and benefits in money terms	7	
TOTAL		19	59

Source: Leitch, 1977

Thus the standard approach to trunk road decision taking is a form of informal multi-criteria analysis bearing some resemblance to the Lichfield Community Impact Evaluation, but with less coherence in the way incidence groups are identified and effects set out (For instance, accident savings are shown in physical and monetary units, whereas time savings are shown in monetary units alone). Nevertheless when the committee (by this time the Standing Advisory Committee on Trunk Road assessment - SACTRA) looked at the issues again in 1992 (Wood, 1992), criticism still existed that the monetised factors were given systematically more weight in decision taking than the non-monetised. Moreover the DoT has persisted in quoting CBA ratios in support of the case for road investment, even though these miss out many relevant costs and benefits. SACTRA considered that there was no objection in principle to monetary value of environmental costs and benefits, and considered that more work should be done to bring these into the monetary measures. However, it saw this as inappropriate in two circumstances; firstly, where unique environmental assets were involved which were not susceptible to routine valuation and where a direct political decision was inevitable, and environmental effects which were a matter of critical importance to survival. Whilst the first of these exceptions appears reasonable (one of the big merits of cost-benefit analysis - the use of consistent valuations across projects - clearly does not apply in this case) the second is more puzzling. It is regarding emissions such as greenhouse gases that consistent valuation across all projects - not just transport ones - would seem to have a major role to play in terms of ensuring that environmental constraints are met at minimum cost.

However, by this time, an even more fundamental problem with the current appraisal method was apparent and this was the lack of a strategic approach. Essentially what the current appraisal did was to ask "given that traffic is going to grow by a certain percentage over the next 20-30 years (under the 1989 National Road Traffic Forecasts, roughly 100% by 2025), would you rather it was on the existing road network, or on the network embodying specified improvements". This has been likened to the familiar question as to whether you would prefer to be hanged or guillotined - presumably the answer is usually neither. Yet at the level of detail of a small section of road, no alternative strategy of traffic restraint, land-use planning or public transport improvement was likely to make much difference. All alternative policies needed to be pursued at a much more strategic level if they were to have any significant effect on traffic levels. Although SACTRA strongly recommended such strategic studies, progress has been slow. Nevertheless the recent decision to bring the trunk road programme within the framework of regional and structure planning may reflect a significant move in this direction. One of the strongest criticisms of the previous regime (voiced particularly in a minority report by a member of the SACTRA committee, Audrey Lees) was its divorce from land-use planning. Trunk road projects were often seen as opening the way to significant land-use changes, both by attracting particular activities to certain locations (e.g. retail and distribution activities to motorway interchanges) and less directly by influencing future planning decisions. Yet all such effects were ignored in appraisal and trunk roads were taken as given in the land-use planning process.

Two other significant criticisms can also be related to the lack of a strategic approach. As explained above, the appraisal approach generally assumed that overall traffic levels would be identical whether or not a particular scheme were to be built. In the case of a major bypass or radial road into a congested urban area this was in itself implausible (Wood, 1994), but much less plausible was the suggestion that overall the growth of traffic in an area would be independent of the transport strategy followed. Together with a shift to a more strategic level of appraisal would come a clear need to model the full range of responses to an investment programme as a whole- namely frequency, destination, mode, time of day and route.

Secondly came the long running debate on the significance of small time savings. At the level of the individual scheme many of the time savings involved were estimated to be less than a minute. Some studies had suggested that such time savings were typically of much less unit value than larger savings (MVA et al, 1987). Certainly the notion of undertaking serious destruction of property or natural resources in order to save road users less than a minute each was not generally seen as supportable by many lay commentators. However, DoT had always countered this argument by saying that these savings would be aggregated with savings from other transport projects as well as savings from developments in other areas of time allocation. Attaching a smaller value to these savings would favour individual large schemes over sets of smaller ones. A move to a strategic level of appraisal in which groups of schemes would be considered together would go a long way to overcoming this objection to adopting a lower value for small time savings than large.

Thus we have slowly moved away from a narrow cost-benefit analysis of individual schemes towards a strategic multi-criteria appraisal of transport policies and projects as a whole. There is still a long way to go regarding inter-urban roads. As will be seen progress has been faster at the urban level.

4.   Local Transport

As commented in the introduction, local authorities in Britain have generally been less keen on espousing simple cost-benefit analysis than has central government. In many cases, a main motivation behind transport projects is the encouragement of economic development and the promotion of particular patterns of land use. Thus for instance better roads to remote areas may be built to reduce their disadvantage in terms of transport cost; improved public transport to a city centre may be used to try to reduce decentralisation of jobs.

It is clear from the above that the approach of Central government in Britain has generally been to concentrate on the direct transport benefits of projects, on the assumption that these are overwhelmingly the most important factors. Part of the reason for this is that, in a small country with an already well developed transport system, even a major transport project will only have a small effect on the total costs of production and distribution of most industries in a particular location. Typically, in Britain, transport costs amount to some 5-10% of total production and distribution costs, and even major projects will change total cost by less than 1 per cent (Parkinson, 1981).

Nevertheless, there clearly are cases where transport improvements do affect land-use and economic development. A major estuary crossing, for example, may enable firms to concentrate their distribution facilities (or even production) on one side of the estuary, with consequent exploitation of economies of scale (Mackie and Simon, 1986). A major motorway development close to a major conurbation will tend to attract distribution and retailing activities, particularly at junctions with other motorways (MacKinnon, 1988). Improved rail services to the city centre may well trigger house building for commuter purchase (Harman, 1980). It should also be noted that these developments are not always beneficial. In the case of the M25 motorway around London, much new development has been attracted to a green belt area, at considerable environmental cost. Improved roads to remote areas may promote tourist travel, but they also enable firms to serve those areas from major centres, leading to the closure of local facilities such as bakers and distribution depots. Improved rail services may lead to the growth of long distance commuting and urban sprawl.

It is perhaps not surprising that, with their responsibilities for planning, local authorities are far more concerned about such effects that is the Department of Transport. Moreover, to the extent that transport improvements do attract jobs from a neighbouring locality, the authority involved may see this as a benefit, whereas from the national viewpoint it is not. The development of a common appraisal framework (ITS and MVA, 1991) (see Table 5 for an overview of factors considered in this analysis) has brought together some of these considerations in a framework approach similar to that of Lichfield discussed above.

5. Public Transport

In Britain it is now the case that most public transport operators (bus companies, Train Operating Companies) and even the main provider of rail infrastructure (Railtrack) are purely commercial organisations, and will thus be primarily interested in financial appraisals.

Debate over whether using financial appraisal in some sectors and social cost-benefit analysis in others biases investment decisions has gone on for decades. Following the Leitch report, the Department of Transport commissioned a study to compare the effects of using financial appraisal and social cost-benefit analysis for rail schemes, and concluded that they gave similar results. However, this is only true of investments where the majority of benefits are cost savings; it would not hold true of investments to improve the quality of service or to introduce new services.

In the bus industry, the use of cost-benefit analysis now is primarily by local authorities considering what services to support, and at what level of fares and frequencies. However, relatively few authorities actually use CBA for this purpose; most apply some form of accessibility analysis in comparison with predetermined standards (Bristow, Mackie and Nash, 1992). However, there now seems to be widespread agreement that there is a need for local authorities to become more involved in issues of bus industry investment, not just in terms of infrastructure and information but also in the quality of the vehicles themselves. Whether these can be achieved by voluntary 'quality partnerships', as the government hopes, or will require legislation, remains to be seen.

In the rail sector, with all former British Rail passenger services now being put out to franchise, there is much more scope for the use of cost-benefit analysis by the Office of Passenger Rail Franchising (OPRAF) in determining minimum service levels, maximum permitted fares and whether particular routes should be supported at all. Moreover OPRAF is in a position to influence investment in rolling stock (by requiring this as part of the franchise agreement) and in infrastructure (by underwriting higher access charges or charges continuing beyond the current franchise period), and thus needs an appraisal method for dealing with these issues. OPRAF has now published the criteria by which it will undertake such appraisals (OPRAF,1996) and these are based on the use of social cost-benefit analysis together with wider appraisals of accessibility and of environmental impact. Institutionally, however, the fact that OPRAF has no direct local government or even Department of Transport input to its decision means that integration of its decisions with those on other modes as part of an integrated strategy may be difficult (except in major cities covered by Passenger Transport Executives).

One complication of the new arrangements for the rail sector is the number of bodies who need to reach agreement on the investment and its funding; a major infrastructure project may affect several train operators, rolling stock leasing companies and OPRAF and the charges are subject to approval by the Rail Regulator. These tends to make the voluntary agreement of investment programmes complex, and suggests the necessity for OPRAF to play a key strategic role if major investment projects are to go ahead.

For local public transport schemes, local authorities or private operators may apply for capital grants under Section 56 of the 1968 Transport Act. However, the conditions are stringent, and benefits to public transport users are disallowed from the case for grant. Such benefits include not just time and cost savings, but also that part of accident savings experienced by those diverting to public transport from car as a result of a scheme. The difficulty this poses in justifying schemes is well illustrated by the case of the Manchester Metrolink scheme, where a scheme that looked very healthy in terms of a full social cost-benefit analysis, became very marginal on Section 56 criteria. (Nash et al, 1991).

TABLE 5... Common Appraisal Framework

Impact groups	Users of Different Modes					Occupiers of Property				Users of facilities		Operators & Government				
	Car	Bus	Rail	Cycle	Walk	Residents	Industry	Commerce	Social	Shoppers	Community centre	Bus operator	Rail operator	Freight operators	Local Authority	Central Government
Disaggregate groups	Journey purpose/Spatial/Mobility Group/Socio-economic					Spatial/Socio economic				Spatial/Socio-economic						
Supplementary tables																
IMPACTS																
Planning blight		•		•		•	•	•	•	•	•	•	•	•	•	
Disruption during construction		•		•		•	•	•	•	•	•			•	•	
Land take			•	•	•	•	•	•	•	•	•					
Noise and vibration				•	•	•	•	•	•	•	•					
Air pollution				•	•	•	•	•	•	•	•					
Community severance			•	•	•	•	•	•	•	•	•					
Threats and intimidation				•	•	•			•	•	•					
Visual intrusion			•	•	•	•	•	•	•	•	•					
Accidents	•			•	•	•	•	•	•	•						
Energy	•	•	•	•		•			•	•						
Accessibility	•	•	•	•		•	•	•	•	•						
Existing businesses																
Regeneration									•	•					•	
Job creation						•	•	•	•	•	•				•	
Capital cost												•	•			•
Time changes	•	•	•	•	•							•	•	•		
Operating cost changes	•	•	•	•	•							•	•	•		

TABLE 6. Sources of new rail traffic (%)

	Bus	Car	Rail	Other	Generated
Birmingham Cross City	36	11	27	0	26
Glasgow (Argyle line)	54	4	0	7	25
Liverpool (Loop & Link)	46	20	0	10	24
West Yorks (New stations)	56	16	13	2	13
Nottingham (Robin Hood) Phase I	37	43	0	2	18
W.Yorks (Wakefield-Pontefract)	61	9	3	1	26
Tyne & Wear*	79	8	7	6	-
Mean	**53**	**16**	**7**	**4**	**17**

*excludes generated trips

The reasoning behind this approach to investment appraisal lies in the current government approach to subsidies. In the early 1980's, several studies indicated that very large subsidies to reduce fares and improve services could be justified (DoT, 1982). However, there were fears that subsidies leak into inefficiency and have a high opportunity cost (see Glaister, 1987). Thus current government policy is to rely on market forces wherever possible, and to accept cases for grant or subsidy only in the presence of externalities as opposed to user benefits.

The result is not only that grants for public transport schemes are more difficult to justify than for roads, but also that because of the small budget allocated to Section 56, even when justified a scheme may have to wait for years for funding.

There is also strong interdependence between public transport investments and policies and projects in the road sector. The principal issue that arises in estimating the effect on road users and the environment of public transport investment is that of the degree to which extra public transport traffic has actually diverted from road. Examination of a number of urban rail investments suggested that this may typically be 20% or less (Table 6) (Chartered Institute of Transport, 1996), although this may understate the long term effects; rail investment may influence not just the choice of travel mode for existing workers but also the home and workplace locations of those newly locating. This must be the explanation for the very large increases in commuter traffic recorded following electrification of many London commuter routes. Similar conclusions arise from studies of the impact of new high speed rail services, where some 50% of traffic was generated rather than diverted (Bonnafous, 1987) and the potential for diverting freight from road to rail also appears relatively small except in particular markets (bulk commodities, and general merchandise travelling long distances (Fowkes, Nash and Tweddle, 1991).

What this seems to imply is that even though there is evidence that the volume of rail traffic can be significantly influenced by providing more attractive fares and service level packages, this course of action will only make a modest contribution to solving the problems of road and air traffic growth. If we do take the congestion and environmental problems posed by the transport sector seriously, then we have to discourage the growth of car and air traffic more directly. The sort of measures which might be considered are higher prices (especially for motoring in cities and for air travel), widespread traffic calming and parking controls. These measures all sound very negative. But if they are seen in that way, there is a danger that all that will happen is the further decentralisation of the population away from cities into smaller towns where the use of the car is not so restricted. If cities are to be places where people actually want to live and work, then we need to create developments in which the attractive environment these controls bring will make them welcome. Such cities would encourage the use of walking and cycling for shorter journeys, and of bus and train for longer journeys. In other words the benefits of public transport investment in an integrated transport policy may be much greater than the evidence of diversion from isolated investments implies (May, Guest and Gardner, 1990).

6. Conclusions

For reasons that have been explained above, there is little consistency between methods of project appraisal applied in different parts of the transport sector in Great Britain. Central Government tends to place heavy weight on a relatively narrow cost-benefit analysis, whereas local government is concerned with a greater range of objectives. The current approach in Britain regards public transport as a largely separate issue from roads, with different appraisal methods and different approaches to funding. The justification for this is that it promotes x-efficiency by giving public transport managers clear commercial objectives, but it undoubtedly has costs in terms of allocative efficiency.

What this paper suggests is that a narrow approach to the benefits of transport investment will be inappropriate. Transport investment has to be seen as part of the overall transport strategy of a city or region, and plans for investment integrated with land-use and economic development plans. In many countries this takes place to a far greater extent than in Britain (SDG, 1992), although the growing acceptance by DoT of a 'package approach' to investment decisions, and the introduction of trunk road planning into local and structural planning procedures is encouraging.

Nevertheless, putting together appropriate packages outside the major cities covered by Passenger Transport Executives is institutionally complex (local authorities, the Department of Transport and OPRAF all being involved), and the package approach remains flawed as a result of differing funding rules for different modes, and lack of control over commercial bus services. Since the date of the conference there has been a change of government in Britain, and the new government has committed itself to development of an integrated transport policy. These problems will have to be tackled before truly integrated transport strategies can be pursued.

7. References

Bonnafous, A. (1987) The Regional Impact of the TGV. *Transportation* **14** (2), pp.127-137.

Bristow, A.L., Mackie, P.J. and Nash, C.A. (1992) *Evaluation Criteria in the Allocation of Subsidies to Bus Operations*, Proceedings of the PTRC Summer Annual Meeting, PTRC, London.

Chartered Institute of Transport (1996) *Better Public Transport for Cities* Chartered Institute of Transport, London.

Department of Transport (1982) *Urban Public Transport Subsidies: An Economic Assessment of Value for Money*, **1** Summary Report; **2** Technical Report, DoT, London.

Department of Transport (1984) *Economic Evaluation Comparability Study - Final Report*, DoT, London.

Department of Transport (1987a) *Values of Journey Time savings and Accident Prevention*, the Department of Transport, London, DoT, London.

Department of Transport (1987b) *Coba 9 Manual: A Method of Economic Appraisal of Highway Schemes*, London, The Department of Transport, May 1987, DoT, London.

Department of Transport (1993) *Design Manual for Roads and Bridges*, The Department of Transport, London, DoT, London.

Fowkes, A.S., Nash, C.A., And Tweddle, G. (1991) Investigating the Market for Inter-Modal Freight Technologies, *Transportation Research* **25a** (4), pp.161-172.

Glaister, S. (Ed.) (1987) *Transport Subsidy*, Policy Journals, Newbury.

Harman, R. (1980) *Great Northern Electrics in Hertfordshire*, Hertfordshire County Council, Hertford.

ITS and MVA (1991) *The Development of a Common Investment Appraisal for Urban Transport Projects*, Working Paper 348, Institute for Transport Studies, University of Leeds, Leeds.

Jones-Lee, M. (1987) *The value of transport safety*, Policy Journals, Newbury.

Leitch, Sir George, et al. (1977) *Report of the Advisory Committee on Trunk Road Assessment*, HMSO, London.

Lichfield, N. (1996) *Community Impact Evaluation*, UCL Press, London.

Lichfield, N. (ed.) (1992) Special issue on comparability in transport evaluation, *Project Appraisal* **7**(4), pp. 193-264.

Mackie, P.J. and Simon, D. (1986) Do road projects benefit industry? A case study of the Humber Bridge, *Journal of Transport Economics and Policy*, **2** (.3), pp. 377-384.

May, A.D., Guest, P.W., and Gardner, K. (1990) Can Rail-Based Policies Relieve Urban Traffic Congestion?, *Traffic Engineering and Control* **31** (7/8), pp. 406-407

McKinnon, A.C. (1988) Recent trends in warehouse location, in Cooper,J. (ed.) *Logistics and Distribution Planning. Strategies for Management*, Kogan Page, London.

MVA Consultants, Institute for Transport Studies, University of Leeds and Transport Studies Unit, University of Oxford (1987) *The Value of Travel Time Savings*, Policy Journals, Newbury.

Nash, C.A. et al (1991) *The Future of Railways and Roads*, Institute of Public Policy Research, London.

Office of Passenger Rail Franchising (1996) *Appraisal of Support for Passenger Rail Services. A Consultation Paper*, report published by OPRAF, London.

Parkinson, M. (1981) *The Effect of Road Investment on Economic Development in the UK.* Government Economic Service Working Paper 43, Department of Transport, London.

Steer Davies and Gleave (SDG) (1992) *Financing Public Transport: How Does Britain Compare?* Transport 2000 et al, London.

Wood, D A (Chairman) (1992) *Assessing the Environmental Impact of Road Schemes.* The Standing Advisory Committee on Trunk Road Assessment. Department of Transport, London.

Wood, D A (Chairman) (1994) *Trunk Roads and the Generation of Traffic.* The Standing Advisory committee on Trunk Road Assessment. Department of Transport, London.

MANAGING UNCERTAINTY IN THE EVALUATION PROCESS: A LEGAL PERSPECTIVE

D. MILLICHAP

1. Introduction

The *ex ante* evaluation of projects attempts something that is fundamentally impossible - predicting the future. The future is uncertain and those undertaking *ex ante* evaluation of development projects have to address that uncertainty. The following examines the challenges of managing uncertainty. The main focus of this discussion is on the role and responsibilities of the public sector decision-maker: in examining this role a deliberately "legalistic" perspective adopted. This is justified on the grounds that the current "official" justification for planning in the UK employs language that does not adequately focus on the duties of the decision-maker and so fails to provide an appropriate structure for illuminating the challenge of managing uncertainty.

The approach also harks back to concepts found in the UK planning system of the inter-war era: they supply, it is argued, a better conceptual framework for improving *ex ante* evaluation. This approach provides criteria that impels decision-makers to take account of the problem of uncertainty at a systemic level. Such an approach (as it involves concepts such as "community" and "rights") therefore provides a better basis for addressing uncertainty and taking account of the interests of those at stake. Such an approach would improve the quality of decision-making. More importantly, perhaps, it would also increase the legitimacy of decisions - particularly those that might put the future community at risk. However, it is important to note that basic elements of this rediscovered approach mean that such benefits can be achieved without causing disproportionate harm to the functioning of the planning system, the present community or place excessive economic burdens on those affected by the decision-making process.

2. The Fallibility of the Predictive Technique

2.1. THE UNCERTAINTY OF THE FUTURE

The predictive techniques employed by *ex ante* evaluation appear to be very sophisticated and numerous. Some of these techniques can focus on economic aspects - the traditional focus of cost-benefit analysis; some techniques can focus on social issues

N. Lichfield et al. (eds.), Evaluation in Planning, 329–341.

- such as community impact analysis/evaluation. Yet such tools suffer from one fundamental weakness. None of them can predict with 100% accuracy all the significant impacts of a detrimental nature that will arise from the development project that is being analysed. For all their sophistication the tools of prediction have this one fundamental flaw. Decision-makers thus rely on imperfect instruments when they undertake their public duties. If the decision-maker had the ability to see into the future and see not only the full impacts that would arise but also how those impacts affected the environment (perhaps for many years into the future) then the decision-maker would be able to make a full *ex ante* evaluation of the development proposals and fully meet his political duty to those who might be affected by his decision. Such a crystal ball is not part of the tool kit available to a decision-maker. However, this does not mean that the decision-maker can disclaim responsibility for not taking account of the risks thereby created. Nor can those who designed the various decision-making tools and procedures avoid their obligations to community so put at risk. The problems affecting communities (that may crystallise in the future) are problems for which the planning system and its actors must bear some responsibility. These problems are "political", in the broadest sense of that term, and addressing them is a matter of political responsibility The focus of this paper is on the heavy burden and responsibility of the decision-maker - a burden that requires a fundamentally different approach to the activity decision-making where that activity can put at risk (because of the lack of a crystal ball) both present and future communities.

2.2.   MANAGING UNCERTAINTY IN THE ECONOMIC SPHERE

The fallibility of the predictive technique is a problem inherent (we might say "systemic" as it is so fundamentally a part of the planning process) in *ex ante* evaluation. If the planning is supposed to operate for the wider benefit of the community then the problem of fallibility is one that raises issues of (*inter alia*) legitimacy. If he fails to take steps to address the uncertainty of the future then the decision-maker, as "guardian of the public interest", might fairly be criticised for not protecting those communities that, later on, are detrimentally affected by unforeseen impacts. Development projects (either on their own or cumulatively) have very significant effects on communities - present, local, future or global: those impacts can be complex (social and economic). Such impacts can be long term, secondary etc. The impacts can be spread over time and space - and with large development projects such dynamic and "temporal" perspectives are clearly of relevance to the decision-making process. Such complexity is not novel. A similar complexity (in the economic field) has long been an issue which investors etc. have had to address. Thus problems of predicting future demand etc. are, at base, founded on the same limitation of the predictive technique - no-one has the perfect crystal ball. But with such economic uncertainties the investment community is protected (to some degree) by having at hand a range of mechanisms. Some economic risks can be countered by taking out insurance. Foreign investors can "hedge" currency market risks. Investors can sell options in the development - allowing them to pull out if they consider the risks are moving against them. The economic mechanisms for discounting the risks posed by the fallibility of the predictive technique are numerous and sophisticated. The investment community has the ability to manage the risks of uncertainty - to some degree. The question then arises as to whether public-sector decision-makers involved in the assessment of development projects can also, by parallel mechanisms, protect the interests of the various communities that may be put at risk by uncertainty.

2.3.   TAMING TIME BY ADDRESSING SYSTEMIC ISSUES

A key issue is that of time. Time drives uncertainty - whether the issue is one of economics or the environmental impact of development. The decision-maker, if he is to take the interests of all communities into account, needs to tame time. In doing so he will protect those communities in ways parallel to those employed in the economic sphere. The failure to use parallel mechanisms in public-sector, regulatory decision-making allows uncertainty to pose a significant risk to the future community in particular. This failure is not, however, solely the fault of the individual decision maker. The approach generally taken by the planning system as a whole must be held to account: decision-makers operate within the institutional framework of law, policy and practice established by a variety of actors. If we take a "systemic" viewpoint - comparing the management of uncertainty in the environmental sphere with that of the economic - then it is clear that a "systemic" failure is largely to blame. If investment decisions can be structured so as to manage uncertainty why cannot decisions supposedly taken to protect communities achieve the same? If there are parallel mechanisms available to the public-sector decision-maker then the failure to implement them is a matter of political responsibility. From a legal perspective such a debate must involve considerations of legitimacy, fundamental issues of democracy and the rights of those affected by decisions made in the public realm to have some meaningful influence on such decisions. This broader perspective on decision-making and the need to address the uncertainties of the future cannot be ignored.

2.4.   EXISTING PROTECTIONS IN LAND-USE PLANNING

So is the planning system able to manage the uncertainties of the future? The UK planning regime does allow the decision-maker (often the local authority) to "buy out" a project whose implementation is found to be damaging to planning interests. Such a remedy is found in the "discontinuance order" power under the Town and Country Planning Act 1990, section 102. Such an order can be made "in the interests of the proper planning... [and] amenity" of the area. It enables the local authority to stop the activity because of its detrimental impact. However, since compensation is payable to the owner the economic implications (for the local authority) of taking action under this provision mean that it is rarely used. In practice therefore the future community that is harmed by unforeseen detrimental impacts has no real power to protect its interests. The discretion vested in the decision-maker provides the degree of power to the authority that allows it to avoid the expense of putting right an error that was due to the systemic limitations of *ex ante* evaluation. Overcoming the disincentive arising from compensation liability is central to addressing such problem sites. So although this one example of taming uncertainty it appears to be a rather impractical mechanism. This example might also suggest that *ex post* factor "sticking plasters" are unlikely to be the answer - they will tend to be subject to the same limitations that currently prevent the discontinuance order provisions working. (The discontinuance notice does not keep pace with the dynamic of time - it is attempting to put right past mistakes but puts the blame for those mistakes squarely on the shoulders of the community: consequently the landowner is able to ask for compensation.) What is needed is a mechanism that avoids the economic disincentive of *ex post facto* reaction. The mechanism must bridge the gap between the *ex ante* evaluation process and the *ex post facto* stage - the stage at which systemic flaws in the decision-making process become apparent. This suggests that the mechanism has to be part of the *ex ante* evaluation process and yet apply in an ongoing fashion to the development project after completion. In this way (as with the various mechanisms for addressing economic risk) the solution shares a temporal dimension with the problem that is seeks to address. The solution has to be rooted in time. An obvious candidate is the planning condition. It can be mechanism by which the decision-maker reaches forward in time and, where appropriate, fine tunes the controls applied to a development project. (Environmental assessment is also another example of the current system having some regard, for large projects that fall within the ambit of the procedures, to the problems of predicting the future. Baseline assessments and monitoring may be included in the mechanisms employed - but current UK practice is not particularly strong at present. This may be due to a lack of emphasis on addressing the fundamental problem of uncertainty and the risks it poses for various communities.)

3.   Monitoring, Planning Conditions and "Public Interest"

3.1.   THE MONITORING ELEMENT

A "temporal" mechanism that effectively tames uncertainty must, as an initial requirement, be able to compare predictions about the future with what actually happens. Achieving such a comparison requires that impacts be described. This in turn requires an information-gathering procedure - in other words, monitoring. A monitoring mechanism is thus the first instrument that the decision-maker needs to address. However, the use of monitoring techniques is, currently, not encouraged by the policy framework governing the planning system in the UK. The ability of planning conditions to fulfil such a role is plainly possible: but policy advice central government strongly dissuades decision-makers from employing conditions for this purpose. Planning conditions:"should not be imposed to require the provision of information to enable the impact of the development to be monitored, either on-site or in adjacent areas unless that information is required to ensure that a planning condition can be properly monitored and planning enforcement procedures could be effectively carried out, if required."(1)

Two points arise. First, such policy advice does not have any bearing on the underlying legal position. Just because policy opposes the use of certain powers in certain ways does mean that such powers cannot, in law, be used. There is no legal bar to using conditions to set up a monitoring mechanism. Second, the policy advice is flawed by a "systemic" blind-spot. It seems to ignore the systemic challenge facing all decision-makers of managing uncertainty in an effective way. Policy from the central government thus (without really addressing the real issue of "systemic" fallibility of the planning system in predicting the future) dissuades planning authorities from using a mechanism that could play a central role in managing uncertainty. The policy does not explicitly consider the importance of this issue - the importance of protecting communities put at risk by a planning decision whose basis is weakened by systemic fallibility. (This raises the question of whether such a "material consideration" has therefore, unlawfully, been disregarded: it is not a question that can be answered here for reasons of space.) The communities for whom such issues are of key concern are thus ignored. This can hardly be said to bode well for a system that depends on legitimacy for its continued smooth running. Such concerns become more apparent when questions are raised about the underlying rationale that seems to encourage a less than transparent discussion of the interests of communities in being protected from uncertainty. The current "public interest" rationale may help explain why there is a fundamentally flawed approach to the challenge of uncertainty.

3.2.   THE "PUBLIC INTEREST" RATIONALE - PART OF THE PROBLEM

One reason why the policy in PPG23 may have such a "systemic" blind spot is the lack of a demanding standard that explains what the decision-maker is supposed to do when operating the planning system. The orthodox "public interest" explanation of planning (expressed for example in paragraph 39 of PPG1) is not one which places much of a burden on the decision-maker (2). "The planning system regulates the development and use of land in the public interest" - that is the message of the policy. Such a notion has become more of a closed circle justification of policy and decisions. Any policy aim is, almost by definition, "in the public interest" and any planning decision that is referable to such a policy is also "in the public interest". This deference to the decision-maker (evident in judicial decisions which accept such a characterisation of the rationale of planning) can all too easily lead to the decision-making process being inadequately policed. A lack of procedural rigour will encourage substantive decisions that are of dubious legitimacy. When planning decisions affect communities that are already disenfranchised (i.e. the future community) then this is worrying. Yet the "public interest" rationale does not really help us appreciate why there are problems of legitimacy. A more demanding explanation of what planning should aim to achieve could, however, provide a more illuminating set of criteria - helping decision-makers to approach their responsibilities in a more effective (and transparent) manner. A more demanding standard for decision-making would avoid the opaque language of "public interest" and require decision-makers (and policy-makers) to spell out the underlying reasoning for decisions. A rationale that, for example, put "communities" and their interests at the centre of the decision-making process would force issues such as legitimacy, accountability etc. into the open. This would not necessarily mean that the courts would then jump in and interfere with the "political" aspects of decision-making. An effective rationale for planning should ensure that these are issues are reserved for the decision-makers. Such a rationale would, however, require that such political issues be addressed in a transparent and open fashion - so that those communities affected by decisions can see how their interests have been taken into account. A new rationale for planning would thus demand openness on the political aspects of decision-making. It would force the various actors in the planning process to meet head on the challenge of addressing the interests of those communities (particularly the future) that are not immediately in view. Such a rationale would bring into focus the political obligations owed to communities other than the present and local. The "public interest" rationale can all too easily mask these issues - allowing decision-makers to avoid being open and transparent about the important issues that they must address. Such sleight of hand eats away at legitimacy - when decisions appear to ignore the interests of communities. A rationale that demands a more open approach to decision-making would help improve the quality of decisions and also the legitimacy of the system.

3.3.   THE COMMUNITY RIGHTS RATIONALE

One approach that impels the decision-maker to consider and justify his approach to the systemic risk posed by uncertainty is the "community rights" rationale. This appears to be a novel concept - especially given the dominance of the "public interest" justification for planning that is so loudly and repeatedly voiced. Yet the concept of "community rights" has its roots in the earliest years of planning in the UK. Those roots can be discerned in the case of Re Ellis and the Ruislip-Northwood Urban Council(3). This case was concerned with the 1909 planning legislation - the Housing, Town Planning, etc. Act 1909. One of the first planning schemes (similar in concept to zoning ordinances and comprehensive development plans of US planning practice) to be proposed under this legislation was challenged by a landowner - Mr. Ellis. In the Court of Appeal Lord Justice Scrutton voiced his views on the fundamental impact that the new regime had had on the rights and obligations of landowners. He commented that:

"I can quite understand that Parliament may have taken a view that a landowner in a community has duties as well as rights, and cannot claim compensation for refraining from using his land where they think that it is his duty so to refrain."

To find such a statement made in respect of planning legislation in its infancy is surprising: the idea that, as result of planning control, private property interests are constrained by "landowner duties" is a significant judicial statement of the underlying impact of the 1909 planning legislation. (It equally applies, of course, to the post-war planning system in the UK.) The correlative notion derived from "landowner duties" is "community rights". So this case provides us with a succinct expression of planning's rationale. Planning is concerned with delineating and protecting "community rights". However, this recognition of the radical impact of the planning regime was somewhat lost as the years progressed. Although references to "communities" etc. are to be found in cases and central government policy statements in the post-war planning regime the debate soon came to be dominated by the "public interest" rationale. (See D. Millichap, Planning Perspectives, 10 (1995) 279-293, "Law, Myth and Community: A Reinterpretation of Planning's Justification and Rationale" for a fuller discussion of the "community rights" rationale and its displacement by the "public interest" rationale.) "Community rights" never developed into the sophisticated rationale that could have served planning from its earliest years to the present. The "public interest" rationale was adopted instead. Although such a concept has its merits it also has significant weaknesses. In the context of uncertainty, risk and the future community such a rationale fails to provide any illumination as to why current approaches to decision-making are, fundamentally, flawed. Such a concept also fails to suggest remedies for such problems. "Community rights", however, is a much more fecund notion of what land-use planning should achieve and how it should go about resolving the fundamental challenges of decision-making - such as addressing the risks posed by uncertainty. It actively requires the decision-maker to answer questions such as "which communities" and "what rights". This is the more demanding standard that impels the decision-maker (and the policy-maker - see the discussion above on PPG23) to be more transparent, more comprehensive in analysing all relevant issues and so address the problems of uncertainty in a more effective way. Communities have the right to be protected from the adverse impacts of uncertainty: the decision-maker who fails to address this when undertaking his role fails to meet the high standards that can properly be demanded of him. The public interest rationale does not highlight so easily the importance of addressing such issues so thoroughly and in such an open manner.

4.   The Costs of Monitoring and the BAPNEEC Principle

4.1.   THE POLITICAL ISSUE OF COSTS AND BENEFITS

A "new" rationale should not, however, impose unreasonable costs on the present
community. To do that would mean that the rights of the present community are being
ignored. One obvious cost, that will fall firstly on all proponents of development - but
then indirectly on the wider community - will be the monitoring costs required by
decision-makers. The economic impacts of such costs may be significant - both for the
owner/developer and the (present) community. However, the guidance that stands
wholly against such imposition (PPG23) reminds us of a legal principle that might help
the decision-maker to balance, in a transparent manner, the two sides of the equation.
Under the integrated pollution control regime polluters are required to take appropriate
action to limit the impacts of their activities. However, the statutory regime does not
require the impossible nor the uneconomic. The duties placed on polluters to control
their activities are applied by reference to the overarching concept of "Best Available
Techniques not Entailing Excessive Cost" – BATNEEC (4). The polluter does not,
therefore, write a blank cheque when submitting to regulation. His duties to use the best
in pollution-control technology are limited by reference to an "excessive" economic
burden standard - the "NEEC" element in the BATNEEC concept. Since the planning
system is also regulatory and cannot, for practical purposes, pursue environmental goals
without regard for economic costs, a similar standard should also be relevant in
designing an appropriate "systemic" response to managing uncertainty. As the primary
issue is the fallibility of prediction then improving prediction (and providing a safety
mechanism in cases of failure) is a central element of the response. So perhaps we
should apply (with modification) the concept from air pollution control to this new
mechanism for dealing, at a systemic level, with uncertainty. BAPNEEC ("Best
Available Prediction Not Entailing Excessive Cost") is perhaps the principal concept
which could guide us. This would encapsulate the balancing of costs between
landowners and the "community" when designing a mechanism for managing
uncertainty in the decision-making process. (I shall return the "cost" problem when
examining the second key element in the mechanism that can be used to address more
fully the interests of all communities potentially at risk from uncertainty.)

4.2.   HOW LONG SHOULD MONITORING BE CARRIED ON?

The costs of monitoring have an economic impact on the landowner (and, indirectly)
the present community. So one of the first issues in managing such an impact is this -
"what is the appropriate period over which monitoring should continue?" The answer
can be an arbitrary one (perhaps set out in primary legislation): we might choose two
years, four year or ten years from the date the project is completed and fully operational.
The answer might be more sophisticated. It might depend on the types of impacts that

could reasonably be predicted: traffic impacts of the local road network might be monitored for a 2 year period while "health" impacts on the local population might be monitored for 10 years of more - the period depends on the time scale during which impacts can be expected to be manifested. There might therefore be categories of impact which merit different periods: a form of "scoping" might therefore be required - to determine the possible categories of impact and make provision for different monitoring exercises carried out for different periods. (The underlying residual risk of unforeseeable detrimental impacts is a particularly difficult issue: one clear line would argue that such issues are best left to the wider community to address by publicly-funded research etc. - rather than project-funded monitoring. It could be argued that such costs would breach the BAPNEEC principle - and should be left to the public purse.)

4.3.   WHAT SHOULD BE MONITORED?

This issue is intimately linked to the foregoing. Different potential impacts manifest at different times. So choosing the types of impacts to monitor will affect decisions about the length of the monitoring obligation. One solution to this problem might be to categorise the adverse effects in terms of how (and not just when) they might be manifested. Monitoring might focus on impacts on human health of a direct nature (e.g. air pollution, radiation in its various forms); impacts on human groups of an indirect nature (e.g. decline in social interaction of a specified population - reflecting "barriers" such as those created by patterns of activity newly-created by development); economic impacts on established areas of economic activity - impacts arising from a new form of distribution or exchange of goods and services. Monitoring can therefore take a number of forms - from traffic counts, surveys of local populations (to ascertain changes in health) to social and economic surveys. They can be qualitative, quantitative or a mixture of both. Again the BAPNEEC principle would be relevant: very minor development proposals (e.g. the construction of an extra room in a dwellinghouse) might merit very little in the way of monitoring. Such proposals might be so common and well-understood that there is little uncertainty to be managed - and any residual risk would be minimal. The variety of monitoring approaches and the sectoral effects they could be designed to catch indicate that there is some argument for establishing a standard list of such obligations with the possibility of an additional category for larger development   proposals whose impacts might be more wide-ranging, cumulative or long-term. (Parallels with environmental assessment are pertinent here). However, as noted above, the costs must be proportionate - to comply with BAPNEEC. It should also be constantly borne in mind (and here the emphasis on the community plays a crucial role) that the effects should primarily be measured in terms of their impacts on communities - not individuals).

5.   The Mitigate Element: Bridging the *Ex Ante* and *Post Facto* Evaluation Divide

5.1.   THE MITIGATION ELEMENT

The above has focused on the "monitoring" process. However, that is clearly not the whole story if we are to take full account of the future community and its interests. Monitoring is merely a stage - providing the information base for an informed choice as to appropriate action when detrimental impacts (unforeseen at the time of project approval) do in fact arise. This is where the second key element comes in - mitigation. The obligation to mitigate is the real meat of the "insurance policy" that, by way of he decision-making process, affords the future community the protection it deserves. It is an obligation that flows from the decision-maker meeting his duties to all communities affected by the development. The concomitant obligation on the developer to set up economic instruments that can fund mitigation frees the decision-maker from the temporal restraints that would otherwise allow uncertainty to put communities at risk. Naturally, the BAPNEEC concept also comes into play at this second stage. It enables the costs of mitigation to be constrained - and a balance to be struck between all those communities whose interests are affected by the decision-making process.

5.2.   SIGNIFICANCE, CRITICAL CAPITAL AND THE COSTS OF MITIGATION

The mitigation element depends on other subsidiary concepts and standards for its implementation. One such criterion is to be found in the concept of "significance". An impact on the community might be discovered by monitoring but it might not necessarily be "significant". So a key is to address, by way of the mitigation mechanism, those effects that are "significant" to the community. (A failure to establish "significance" criteria would weaken the political case for imposing costs on the present and local community). The clearest candidate for applying "significance" to any fact-situation is to look for threats to the continued existence of the community in question. The destruction of physical or social capital on which a community depends ("critical capital" is sometimes the term used in this respect) would clearly fall with the category of effects that would merit mitigation - whether by way of avoidance or restitution action. (The term "mitigate" should in this context be extended to cover more fundamental protective action rather than merely reducing detrimental impacts). Clearly, the costs of mitigation can be significant (especially if critical environmental capital or the continued existence of the future community is at stake). However, the economic impacts (in so far as they affect the proponent of the project) of having to call on the "insurance policy" should have been discounted by the economic mechanisms that he has employed. Further, the triggering of the mitigation element should be a rare event. This is because the expense of mitigation (an expense that would be borne by the investment community) would constitute an economic risk which would be carefully factored into the investment decision made by that community. Thus where such a mechanism becomes the standard (i.e. the systemic response) to uncertainty the market will become sensitised to the (economic) risks and uncertainties to which the promoter of a project will be subject. Self-interest will force the investment community to require developers to be much more rigorous in their analyses of impacts and effects. It will force them to look at the limitations of scientific knowledge about potential impacts - an important aspect of uncertainty. So the very existence of a reasonably-stringent mitigate obligation will of itself decrease the number and cost of those mitigation obligations that are in fact triggered by adverse unforeseen impacts that are significant. Uncertainty will never be tamed - we would need a crystal ball to achieve that. However, the very fact that this (environmental) insurance policy could be called on will encourage a more objective and thorough analysis of those risks and effects that appear, at the moment, to be examined in a cursory way. In this way the solution proposed for taming uncertainty problem becomes a positive virtue - by improving decision-making and the information base upon which predictions are made. Dealing with uncertainty would thus not necessarily increase costs disproportionately: BAPNEEC, the increased sensitivity of the market to the economic risk posed by taming uncertainty and a more objective professionalism among experts should ensure improved decision-making without destroying the market in development sites. Protecting "community rights can thus be achieved without unduly prejudicing the interests of the landowner, developer or investor - or, ultimately, the present community.

6.   Planning Obligations

The final element is the precise mechanism to be used to implement a "monitor and mitigate" mechanism. It was noted above that planning conditions could provide the basis for such a requirement. In fact current examples of "monitor and mitigate" (to be found in cases where highway improvements are required to address increased traffic flows in the local network surrounding a new office development) will invariably use the machinery of a planning obligation to set up the appropriate structures. The use of this contractual (invariably) instrument under section 106 of the Town and Country Planning Act 1990 provides a better basis for the routine use of the "monitor and mitigate" mechanism. The flexibility of agreements to set up appropriate criteria and procedures for both limbs of the mechanism is clearly more attractive to all concerned than the rather restrictive format provided by planning conditions. Planning conditions that do address complex issues (such as the need to design and implement appropriate schemes for remediation etc.) invariably fall back on planning obligations in order to put such controls into practice. Planning obligations have a broad legal basis in the primary legislation: monetary payments to the local authority are permissible - relevant of course when it comes to the economic impacts of the monitoring and mitigation phases of the mechanism. Planning obligations are also favoured because of the more effective means at hand to enforce the duties placed on the landowner/developer. A planning obligation can also be the basis of imposing control on future occupiers of the project site - to ensure that detrimental impacts that area significant are effectively mitigated. A planning obligation can provide for supplementary mechanisms (such as creating "community liaison for a" that enable communities to be kept informed of developments and allow them to comment on monitoring data) that further emphasise the (procedural) "rights" of those communities in the matters at hand. Planning obligations (rather than conditions) thus provide the flexibility that is important in instituting appropriate procedural protections for the communities at risk. They are the best way of protecting those "community rights" that are at stake when uncertainty is an issue.

7.   Conclusion

"Monitor and mitigate" is the basis of a mechanism that addresses the systemic fallibility of decision-making. If the development project is implemented subject to such a mechanism will almost invariably mean that monitoring brings to light significant detrimental effects on communities. This then allows for effective mitigation measures to be designed and implemented. Such activities will take place in the light of criteria that allow for the efficient and fair allocation of costs and benefits. Instead of the static perspective normally involved in using *ex ante* evaluation techniques we have a process that faces up to the systemic limitations facing decision-makers. The decision-making procedure is not artificially frozen in time - nor is it rendered incapable of dealing with problems that were unforeseen. The "monitor and mitigate" approach crosses the temporal boundary that tends to separate *ex ante* from *post facto* evaluation. Such an approach will, in principle, protect at a more fundamentally-effective level the interests of those communities whose voice cannot be heard by the decision-maker. Such a mechanism will, in principle, reflect the legitimate demands of those who can be affected detrimentally by development and yet have no ability in practice to control such development. Such an obligation develops the vague notions (expressed often as "sustainability") that the present community has obligations to the future: from these vague notions we can produce concrete protections for that future community.

Existing planning tools can be employed to implement this mechanism - the contractual form of control under section 106 of the Town and Country Planning Act 1990 is the obvious candidate. The costs of implementing such protections will be minimised in two principal ways. First, the use of BAPNEEC. Second, the readjustment by investors etc. to the (economic) risks posed by the "mitigation" obligations. (Market disciplines will thus impinge on owners and developers and they will develop more sophisticated mechanisms to manage the extra costs of "monitor and mitigate". They are already skilled at dealing with other economic risks of a temporal nature.) The costs of such a systemic response to the problem of fallibility can thus be kept proportionate.

Such procedural changes will also result in substantive improvements. These will not only be discerned in the character of development. Improvements in terms of legitimacy and respect for the planning process will also be felt. Such are the benefits of a legalistic approach, with its "rights" talk: it demands that we think in terms of "whose" rights, "what" rights etc. It requires that such issues be addressed explicitly and not hidden from view. It is a language that is more likely, if the solutions thereby suggested are implemented, to produce mechanisms for decision-making that promote legitimacy - because it is more likely to address the problem of identifying communities likely to be affected by detrimental impacts. It forces decision-makers and policy-makers to heed the interests of those communities and address their concerns. Transparency, accountability and other similar attributes are more likely to be found when procedures are developed in response to a series of questions that are "rights"-orientated. Communities are less likely to feel excluded and marginalised. By pairing "community" with "rights" (a focus suggested by the courts in 1920) the activities of decision-makers etc. will be more likely to give proper consideration to the interests of all those communities that are affected by development. "Monitor and mitigate" planning obligations (reflecting "BAPNEEC" criteria) offer an effective mechanism for managing uncertainty. The planning system can thus, with minor adaptation, ensure that the fallibility of the predictive technique does not continue to ignore "community rights" and put them, and the system as a whole, at risk.

References

1. Planning Policy Guidance Note 23 "Planning and Pollution Control", HMSO 1994, para 3.26.
2. Planning Policy Guidance Note 1 "General Policy and Principles", HMSO 1997. [1920] 1 KB 343.
3. Environmental Protection Act 1990, section 7. The pollution control regime introduced by this legislation uses "BATNEEC" as a central concept.

TOWARDS THE ECONOMIC EVALUATION OF BRITISH LAND USE PLANNING

J.T. CORKINDALE

Abstract

The system of land use planning in Great Britain derives from the 1947 Town and Country Planning Acts. The effect of this legislation has been that the existing land use, and the value of it, is the owner's but that there is no entitlement on his part to develop the land through changing its use. The key questions for the economic evaluation of the British land use planning system are how far this approach can be justified and whether alternative approaches might not be more beneficial. Economic research is gradually developing ways of measuring the costs associated with the system, in particular, the mechanism of development control acts to restrict the supply of land for development and these restrictions have been found to exert an upward influence on land and property prices. Less progress has been made in measuring the benefits of the system. One reason is that, on the whole, planning policy objectives have not been specified in terms which make it possible readily to measure progress towards them. This paper explores the problems of redefining and measuring the benefits of British land use planning and draws some tentative conclusions about the possibilities for reform involving the privatisation of development rights.

1. Introduction

The use of the term 'evaluation' in land use planning is generally taken to refer to the ex ante or ex post evaluation of particular decisions taken within the land use planning process. In this paper, however, the purpose is to try to evaluate the framework of land use planning within which individual planning decisions are made.

N. Lichfield et al. (eds.), Evaluation in Planning, 343–354.

2.   The British Land Use Planning System

The system of land use planning in Great Britain derives from the 1947 Town and Country Planning Act and the Town and Country Planning (Scotland) Act of the same year. The general principles under which the planning system operates in England and Wales are set out in Planning Policy Guidance Note 1 (Department of the Environment, 1992) published jointly by the Department of the Environment and the Welsh Office. Although the system is governed by different legislation in Scotland, it is, in its essentials, the same throughout Great Britain.

The general position is that any person who wishes to develop land by carrying out a substantial physical operation or by making any significant change to the use of land or building must obtain planning permission from the local planning authority (LPA). If the LPA refuses planning permission or imposes conditions which are not acceptable to the developer, the latter may appeal to the appropriate government minister, in England the Secretary of State for the Environment. Although changes of national government since 1947 have led to differing approaches to the apportionment of the development value of the land between the developer and the community, the effect of the planning legislation that has actually been in force has been that the existing land use, and the value of it, is the owner's but that there is no entitlement on his part to develop the land through changing its use (Stephen, 1988).

The fundamental concern of planning since the 1947 legislation has been to intervene in the land development market in the public interest. Through the development control system, the public authorities seek to ensure that the public interest is accounted for in decisions regarding the use and development of land. Decisions taken by LPA's, with rights of appeal to the relevant minister, aim to achieve a balance between individual and public interests. The 1991 Planning and Compensation Act requires planning decisions to accord with the development plan for the area unless 'material considerations' indicate otherwise. The courts may quash a planning decision if there has been a failure to consider a material consideration or if some irrelevant factor has been taken into account. Planning Policy Guidance Note 1 specifies that, in principle, any consideration which relates to the use and development of land is capable of being a material consideration for planning purposes. Perhaps not surprisingly therefore, whilst many decisions have been quashed by the courts for failure to take account of a material consideration, there have been very few occasions on which courts have said that a particular consideration is not relevant to planning. Equally, it is a weakness of the British legal system that it provides no satisfactory mechanism for establishing the proper relationship between the conditions for planning permission and the projected impact of the proposed development. Indeed, in the House of Lords Judgement on Tesco Stores plc v Secretary of State for the Environment (House of Lords, May 1995), all responsibility for the task was abdicated (Grant, 1995).

Nevertheless, what is a relevant or material consideration is of crucial importance for the system of development control, since it is through this concept that the limits to public intervention in the land use planning field are defined (Stephen, 1988).

3.   Assessing the Costs and Benefits of the Planning System

Despite a claim in Planning Policy Guidance Note 1 that the system has served the country well, the effects and effectiveness of the British land use planning system has not in fact been the subject of sustained economic evaluation, despite the fact that it is, arguably, one of the most comprehensive, complete and sophisticated in the world, and has been in place for half a century (Pearce, B. 1992). One of the reasons for this has been the failure of planners and economists to speak the same language. The core of the problem is the differing criteria being used for determining the success or otherwise of land use planning. The first is the sense in which the Apollo programme to land a man on the moon was a success: the cost is largely irrelevant, success is achieved if the objective is attained. The second is the sense in which a business is successful: its product is in demand at a price which allows its business to continue to be profitable (Evans, 1985). It is the second criterion with which the economist is concerned. Of course, the economist engaged in the evaluation of land use planning is not seeking to answer questions about the profitability of land use as such. In essence, he is asking the same kind of question as the accountant asks of a private firm, but the question is being asked about a wider group of people - who comprise society as a whole - and is being asked more searchingly. Instead of asking whether the enterprise will become better off by the firms engaging in one activity rather than another, the economist asks whether society as a whole will become better off by taking a particular land use planning decision than by not taking it, or by taking instead any of a number of alternative decisions (Mishan, 1988). The methodology used by the economist for this purpose is cost-benefit analysis (Lichfield, 1996; Schofield, 1987).

4.   The Costs of the Planning System

In the financial year 1995/96, the cost of running the planning system in England totalled some £650 million. Most of this consisted of recurrent expenditure by LPA's. A further £26 million was spent on the running costs of the Planning Inspectorate. These sums do not include capital expenditure, nor do they include comparable recurrent and capital expenditure in Scotland, Wales and Northern Ireland. All told, the public expenditure costs of the planning system are probably of the order of £1 billion per annum. Yet it is not these sums, but the economic effects of the way the planning system operates that has attracted most interest from economic analysts. This interest has focused on two issues. First, there is concern that the planning system is imposing a burden on business, through controls on the development of green field sites, particularly in the Green Belt, which leads to high land costs in some of the areas preferred by business for development, and which may result in the diversion of investment to locations less favoured by business and, in some cases the loss of prospective investment altogether. There is also concern that excessive bureaucracy associated with the planning system may impose delays, and therefore costs, on business, either directly through the time spent in processing planning applications or indirectly because of the delays to new public infrastructure investment, particularly in the field of road transport. Second, there is the related concern about the implications of the way the planning system operates for land house prices.

It is the latter issue on which research effort has concentrated. Research carried out for the Department of the Environment sought to identify the extent to which the operation of the planning system has affected the supply of land and house prices in different parts of England. Land supply constraints, including those imposed by the planning system, were found to have increased house prices in the South East Region by 35-45 per cent. Restricted land supply had also affected the density of new house building, with an increase in the proportion of flats and terraced housing and a reduction in the proportion of semi-detached houses and bungalows (Eve, 1992). Using comparative data from Reading and Darlington, two towns chosen for their differing degrees of planning restrictiveness but which were 'similar' in many other respects, Cheshire and Sheppard (1989, 1996) examined how far house prices in Reading might have differed had the town adopted the more relaxed planning regime found in Darlington. The results suggested that plot sizes would have been 65 per cent bigger, and that the area of the town would have been 50 per cent bigger, not because of any increase in population but because people would have bought bigger houses with larger gardens. Such changes would have been facilitated by a significant drop in land prices, the impact of which could be expressed as equivalent to a change in household income in Reading ranging from £640 per annum or 6 per cent of household income at the urban periphery to £775 per annum or 8 per cent of household income at the urban core.

Perhaps the most ambitious attempt to model the workings of the planning system in England to date is that by Bramley and Watkins (1996). Using data for the ten year period 1987-96, they simulated a series of policy changes to try to trace and quantify their effects. For example, one scenario entailed a requirement that all structure plan local authorities immediately increase their housing provision levels by up to a half. This was predicted to lead to an average increase in housing output by 9.3 per cent and an average fall in house prices of 4.3 per cent. Another scenario involved new settlements in seven districts in the south of England. For each of these districts, the structure plan housing provision was trebled and land allocated for a ten year period. The impact, nationwide, was an increase in the rate of house building of 5.6 per cent and an average reduction in house prices of about 3.5 per cent.

5.   The Benefits of the Planning System

Whilst the above information suggests that the planning system is costly, both to run and also in its impacts, the important question for policy purposes is how these costs compare with the benefits of the system. If the costs were clearly outweighed by the benefits, rather than vice versa, the policy implications would obviously be very different. Unfortunately however, whereas estimating the costs of planning has proved to be a difficult research area in which to make progress, very little progress at all has been made on the even more difficult task of estimating the benefits. The starting point for this is to examine the objectives of the planning system in order to determine what it is designed to achieve. This is easier said than done. In their study of 'The Containment of Rural England', Peter Hall et al (1973) were only able to spell out 'a very short and simple list of criteria (which) had only the most indirect relevance to any index of welfare' that might be familiar to a social scientist. This problem has recently been encountered again in research by the Department of Land Economy, Cambridge (1995) for the Department of the Environment. This has found that, although many of the objectives of land use planning policy sought to achieve some standard for the policy object, the gap between the existing and the desired state of the policy object was rarely made explicit. Neither the extent of the improvement nor the timetable by which it was to be achieved were stated. A number of policy objectives for the land use planning system were stated in terms of achieving a balance of objectives, but only rarely were the trade-offs between the objectives concerned specified in a way which quantified the 'exchange rate' between one objective and another.

The problem of trade-offs in planning policy has recently surfaced in relation to sustainable development, defined by the World Commission on Environment and Development (1987) as that form of development which meets the needs of the present whilst not compromising the ability of future generations to provide for their own needs. Analytically, this concept is difficult to interpret, though recent discussion of the meaning of 'strong sustainability' and 'weak sustainability' as advanced by Pearce, D et al (1989) has led one author to suggest that the pursuit of sustainable development should be interpreted as the pursuit of 'the old fashioned economist's concept' of economic optimality (Beckerman, 1995). In essence, the economics of sustainable development are very similar to the economics of conservation; thus, investments in conservation, such as soil conservation and land reclamation, should be governed by considerations of how discounted returns from such investments compare with those from alternative kinds of investments. Objections to this procedure generally focus on the supposed unfairness of discounting to the interests of future generations. To the extent that the rate of discount is governed by considerations of pure time preference - that is impatience to consume now rather than later - there may be something in this, and there may be a case for trying to address the problem through political exhortation, etc. From an analytical point of view, however, it needs to be recognised that the optimum level of conservation is what each succeeding generation thinks it to be. Today's generation may choose to be spendthrift and to squander resources at the expense of future generations or, alternatively, to be miserly, merely subsist, and hoard resources which are then used for luxury consumption by future generations (Heady, 1956). The nature and extent of the bequest to future generations will ultimately depend on these choices made by millions of different individuals all over the world.

The concept of economic optimality is also of more general relevance to land use planning. The rules for optimum resource allocation constitute an important part of the theoretical framework of welfare economics which underpins cost benefit analysis, and they have proved very helpful in answering questions about the allocation of resources for production and consumption (Baumol, 1965). The details of the rules need not be gone into here. Suffice it to say that there are two well known difficulties concerning their application for public policy purposes, including land use planning. The first is that the application of the rules calls for data on costs and benefits which do not always show up in financial accounts, and the second is that there is nothing in the rules themselves which allows any judgement to be made about the appropriate distribution of income and wealth.

The first of these difficulties relates to the divergence between private and social costs and benefits. It is central to the underlying purpose of land use planning which, essentially, is to secure an efficient means of internalising external costs where the development of land and property can adversely affect the interests of other members of the community. It is usual to distinguish between two categories of external cost (or 'externality' for short): pecuniary externalities and technological externalities (Viner, 1931). The former relates to the distributional consequences of land and property development. It has been a subject of concern in the UK at least since the time of the Uthwatt Expert Committee on Compensation and Betterment of 1942. Although pecuniary externalities are important, they are not discussed further here. The focus of the remainder of the paper will be on the problem of internalising technological externalities. An example quoted by Knetsch (1983) is of a block of flats so sited that it interferes with the water regime in the area and induces flooding of the ground floors of existing flats. Technological externalities generated by the development of land and property can however vary widely in their nature and extent. They can range from the case of an extension to my house which causes my neighbour's view to be obstructed to the case of a new fossil-fuelled power station which generates carbon dioxide and thus contributes, albeit marginally, to global warming. In economic terms, the central purpose of land use planning is, or should be, to secure the efficient internalisation of technological externalities. The problem of economic evaluation is to determine how efficiently this is currently being done.

6.   The Efficient Internalisation of Technological Externalities

If the task of the planning system can be characterised as being to arrive at a judgement between competing interests in order to secure the efficient internalisation of technological externalities, this task is, in the UK, discharged mainly by local planning authorities because the majority of proposed developments affect essentially local interests. Where wider interests are likely to be affected, the Secretary of State for the Environment can 'call in' a planning application. The Secretary of State for the Environment also exercises a quasi-judicial function in determining the outcome of appeals against decisions made by local planning authorities, though such appeals can only be made by the applicant for planning permission and not by third parties who may be affected by the proposed development. In discharging his responsibilities, the Secretary of State will generally appoint a planning inspector to look into the details of the case, and he will take evidence from interested parties. Under current arrangements, a charge is made by local planning authorities for handling planning applications. No charge is made for dealing with planning appeals. As has already been indicated, an appeal may be made to the courts in the event of dissatisfaction with the Secretary of State's decision, though, as we have seen in the final paragraph of 2 (The British Land Use Planning System) above, the courts have interpreted their role in a rather limited way.

The approach used in the UK can be compared with possible alternative approaches. The original source for one of these is the classic article by Ronald Coase (1960) entitled 'The Problem of Social Cost'. Coase recognised that the law had an important role to play in the settlement of disputes over land use and development. Voluntary bargaining was often costly because discovering an agreed solution might require extensive negotiation, whilst enforcing it might require monitoring and policing. Negotiation, in particular, involves communication, and the costs of communication depend, in large part, upon the number of parties to the dispute and their geographical dispersion. Property law facilitates private agreements, and the idea that it ought to do so was argued powerfully by Coase (Cooter and Ulen, 1988). The relevance of this normative principle is that, where a technological externality has arisen, the courts should choose between compensatory damages (or a liability rule) and an injunction (or a property rule) on the basis of the parties choosing to cooperate in resolving the dispute. Where there are obstacles to cooperation, the preferred remedy is a liability rule involving the award of compensatory monetary damages. Where there are few obstacles to cooperation, the preferred remedy is a property rule involving the award of an injunction against the plaintiff's property (Calabresi and Melamed, 1972).

When this standard is actually applied, the preferred legal remedy depends in large part on how many parties must participate in a settlement. Where the dispute involves a small number of contiguous property owners, the costs of private bargaining are likely to be low, bargaining is likely to be successful and, therefore, the most efficient remedy for resolving these property disputes is injunctive relief. In contrast, where disputes involve a large number of geographically dispersed individuals. the costs of private bargaining would be high, bargaining will not work, and the efficient legal remedy is for the courts to determine compensating damages.

The importance of private property rights in the resolution of disputes over land and property is illustrated by Littlechild (1978) regarding a common law judgement relating to the River Spey in Scotland. The judgement held that the owners of salmon fishing rights on the river did not have the right to prevent public use of the waters for canoeing and sailing. The problem with judgements of this kind is that, although the decision establishes a property right where the situation was not previously well defined, the resultant right, being held 'by the public', is not transferable. Even if the value of (uninterrupted) salmon fishing were higher than the value of canoeing, it is difficult to see how potential fishermen could buy the right to fish from potential canoeists. Had the decision gone the other way, it would have been straightforward for potential canoeists to negotiate with easily identified owners of fishing rights. Far from protecting the rights of the public at large, this legal decision may have prevented the use of resources the public would prefer.

7.   The Privatisation of Development Rights

The case for the privatisation of development rights is that, in the absence of arrangements to facilitate the voluntary exchange of development rights, efficient land use will be impaired (Fischel, 1985). In the UK, this would of course entail reversing the effect of the 1947 Town and Country Planning Acts. The principal objection to doing this has been articulated by Malcolm Grant (1988). If one were redesigning property rights to substitute for the existing system of regulation, the maintenance of existing environmental quality would require a redefinition in order to restrict lawful externalities beyond the tolerances currently actionable in nuisance or trespass. This is no function to be left to the courts, to hammer out the necessary extensions to existing liability rules. The gradual evolution of the common law to meet changing circumstances is one thing; the need to fill a vacuum previously filled by regulation is quite another. Yet some progress has already been made in this direction through the introduction of the General Development Order and through Enterprise Zones and Simplified Planning Zones. These policy developments can be seen as limited steps in the direction of returning development rights to the landowner. The purpose of Simplified Planning Zones, for example, was to enable planning authorities to pursue a positive approach to development control by specifying in advance the type of development which would be acceptable within an area, thus removing the uncertainty and delay perceived as inherent in the existing system of planning control. Conditions attached to development were to be kept to a minimum. At the same time, however, it was considered that there might be a need for upper or lower limits on site size. It was also argued that the SPZ concept would not be appropriate in sensitive areas such as National Parks, Areas of Outstanding Natural Beauty, the Green Belt or conservation areas. Certain types of development which might be hazardous or polluting were also to be excluded (Arup Economic Consultants, 1991).

This confusion about the function of SPZs is symptomatic of a desire on the one hand to 'lift the burden' from business and on the other hand not to lose control over what happens to the environment. In areas of high unemployment, low incomes and social deprivation, it would be quite logical to emphasise the former objective at the expense of the latter. In areas without social and economic problems of this kind, the opposite might very well be the case. In principle, both kinds of objective could be pursued, but with differing degrees of relative importance in different geographical locations. One means of doing this would be through the use of tradeable development rights (TDRs), a species of market-based instrument, the purpose of which is to enable the market to deliver, in a cost effective way, the appropriate combination of economic and conservation objectives. Landowners, in determining whether or not, and how, to develop their land, would have an incentive to take account of the public good value from not developing or developing differently. Two conditions would have to be met for this to happen. First, public conservation objectives would need to be rather precisely defined (as for example in the national biodiversity strategy (UK Biodiversity Steering Group, 1995). Second, the environmental gains and losses generated by particular kinds of land use change would need to be identified and quantified in advance.

One of the best known examples of TDRs used for land use planning purposes is in the New Jersey Pinelands in the USA. Following creation of the Pinelands reserve in 1978, the Pinelands were divided into different land use zones. The most ecologically sensitive areas were classified as Preservation Areas. Most residential, commercial and industrial development is prohibited in Preservation Areas, although some activities, such as forestry, recreation, etc, are allowed if done in conformity with environmental standards. Growth is encouraged in Protection Areas, subject to environmental and zoning standards. These are in turn sub-divided into Forest Areas, Agricultural Production Areas, Regional Growth Areas, and Rural Development Areas, each of which has a predetermined housing density allowance per unit area. Development is regulated through the Pinelands Development Bank, which incorporates a system of TDRs with transfer facilitated by a Credit Bank. The TDRs are allocated to landowners in the Preservation Areas and Agricultural Production Areas; they can then be purchased by developers in Regional Growth Areas in order to increase construction density. (Rural Development Areas are treated as transition zones in which modest development is allowed in order to reduce pressure on Regional Growth Areas.) Landowners seeking development credits retain title and may continue using the land for authorised, non-residential use. Prior to a sale of credits, they must record a deed restriction binding all subsequent owners of that property to the same authorised uses (Clark and Downes, 1995).

8.   Regulation versus Tradeable Development Rights

The current regulatory system in the UK and TDRs can be seen as being at opposite ends of a spectrum of alternative approaches to land use planning. In his comparison of land use planning in the UK and the USA, Wakeford (1990) commented that the UK starts with discretion and imports some certainty, but no guarantees, by ensuring that the development plan influences decisions, and by providing an appeal process to help achieve consistency of decisions. In the USA, on the other hand, zoning started by designating precisely what could be done on land - certainty if ever there was any. Administrative discretion was seen as a significant advantage of the UK system. Other writers have been more critical of the British approach. For example, Stephen (1987) concludes that it is not conducive to clarity in the ultimate objectives of public policy, and, in particular, whether the regulation of land use is to be a means of redistributing income from the owners of land with development potential to other members of the community. He suggests that, if this is an objective, it might be better effected by an explicit legislation intent and definition of property rights than by ad hoc decisions and negotiations in 'smoke-filled rooms'.

   In fact there appear to be three principal reasons for believing that a move to something closer to the American approach might constitute an improvement. First, there are considerations of openness in government of the kind that Stephen referred to. Second, there would be more scope for the introduction of potentially more efficient ways of delivering economic and environmental objectives by using market-based instruments like TDRs. Third, as has been discussed above (see paras 1-2, 5 The Benefits of the Planning System) the objectives of land use planning as currently defined are both vague and numerous. One reason for wanting to have a large number of policy instruments derives from the theory of economic policy developed by Tinbergen (1952). This formulated the decision rule that, in order to avoid confusion in policy and to achieve policy objectives satisfactorily, one needs at least as many independent instruments of policy as one has independent policy objectives. The discipline of the market, embraced in the privatisation of nationalised industry, has been salutary in focusing attention on the need for more precision in the formulation of government policy objectives. This has most obviously been so in relation to the privatisation of the electricity industry where essentially market-based approaches have had to be found to environmental externality problems including sulphur dioxide emissions from fossil-fuelled power station and the decommissioning of nuclear power stations.

Advocates of the privatisation of industry sometimes concede that, in an ideally functioning state, the efficiency gains they associate with privatisation could be achieved by the imposition of a suitably tight budget constraint and the active promotion of competition where required. However, these advocates reject this vision as Utopian, on the grounds that such a state does not exist in reality and that any actual state will not remotely behave in this optimal fashion. On the contrary, it will shield public enterprises from competition and subsidise their inefficiency. Given the supposed inadequacies of any actual state, they conclude that privatisation is the second best solution (Rowthorn and Chang, 1993). The argument in favour of privatising development rights can be couched in similar second best terms. Essentially, the state cannot be relied upon to make economically efficient decisions about development proposals because it will not be able, or will not wish, to make satisfactory judgements about the balance between economic and environmental gains and losses.

9.   Conclusion

It is a weakness of the British land use planning system that its objectives have always been rather vague and ill-defined. It is proposed here then the principal purpose of planning should be to internalise technological externalities. If this were accepted then the planning research agenda would naturally focus on matters such as how to measure externalities, the economic basis of land use zoning, and the application of economic instruments, such as TDRs, in land use planning.

References

Arup Economic Consultants (1991) Simplified Planning Zones: Progress and Procedures, London HMSO
Baumol, W J (1965) Economic Theory and Operations Analysis, 2nd edition, Prentice-Hall Inc, New Jersey
Beckerman, W (1995) Small is Stupid: Blowing the Whistle on the Green, Duckworth
Bramley, G and Watkins, C (1996) Steering the Housing Market: New Building and the Changing Planning System, The Policy Press

Calabresi, G and Melamed, A D (1972) Property Rules, Liability Rules and Inalienability: One View of the Cathedral, Harvard Law Review, Vol 85, No 6

Cheshire, P and Sheppard, S (1989) British Planning Policy and Access to Housing: Some Empirical Estimates, Urban Studies, No 26

Cheshire, P and Sheppard, S (1996) Some Economic Consequences of Land Use Planning, paper presented at the AREUEA International Real Estate Conference, Orlando, Florida, May 1996

Clarke, D and Downes, D (1995) What Price Biodiversity? Economic Incentives and Biodiversity Conservation in the United States, Center for International Environmental Law, Washington

Coase, R (1960) The Problem of Social Cost, Journal of Law and Economics, Vol 3, October

Cooter, R and Ulen, T (1988) Law and Economics, Harper Collins

Department of Land Economy, University of Cambridge (1995) Developing Indicators and Measures for Evaluating the Effectiveness of Land Use Planning, Stage 1 Interim Report, unpublished

Department of the Environment/Welsh Office (1992) Planning Policy Guidance: General Policy and Principles, PPG1, London HMSO

Evans, A (1985) Urban Economics: an Introduction, Blackwell

Eve, G (1992) The Relationship between House Prices and Land Supply, London HMSO

Expert Committee on Compensation and Betterment (1942) Uthwatt Report, London HMSO

Grant, M (1988) Forty Years of Planning Control: the Case for the Defence, The Denman Lecture, Department of Land Economy, Cambridge

Grant, M (1995) If Tigard were an English City: Exactions Law in England following the Tesco Case, unpublished paper

Hall, P et al (1973) The Containment of Urban England, London: George Allen and Unwin

Heady, E O (1956) Economics of Agricultural Production and Resource Use, Prentice-Hall Inc, New Jersey

Knetsch, J L (1983) Property Rights and Compensation: Compulsory Aquisition and Other Losses, Butterworth & Co (Canada) Ltd

Lichfield, N (1996) Community Impact Evaluation, UCL Press

Littlechild, S C (1978) The Fallacy of the Mixed Economy: an Austrian Critique of Economic Thinking and Policy, Institute of Economic Affairs, Hobart Paper 80

Mishan, E J (1988) Cost Benefit Analysis: an Informal Introduction, 4th edition, London: Unwin Hyman

Pearce, D et al (1989) Blueprint for a Green Economy, Earthscan

Rowthorn, B and Chang, H (1993) Public Ownership and the Theory of the State in Clarke, T and Pitalis, C (Eds) The Political Economy of Privatisation, Routledge

Schofield, J A (1987) Cost Benefit Analysis in Urban and Regional Planning, London: Unwin Hyman

Stephen, F H (1987) Property Rules and Liability Rules in the Regulation of Land Development: an Analysis of Development Control in Great Britain and Ontario, International Review of Law and Economics, Vol 3

Stephen, F H (1988) The Economics of the Law, Wheatsheaf Books

The UK Biodiversity Steering Group (1995) Report Vol 1, Meeting the Rio Challenge, London HMSO

Tinbergen, J (1952) The Theory of Economic Policy, North Holland, Amsterdam

Viner, J (1931) Cost Curves and Supply Curves in Stigler, G J and Boulding, K E (Eds) (1952), Readings in Price Theory, American Economic Association, Richard D Irwin Inc

Wakeford, R (1990) American Development Control: Parallels and Paradoxes from an English Perspective, London: HMSO

World Commission on Environment and Development (1987) Our Common Future (The Brundtland Report), Oxford University Press

CONCLUSIONS: WHERE DO WE GO FROM HERE?

Evaluation in Spatial Planning in the Post-postmodern Future

E.R.ALEXANDER

1. Introduction

Evaluation in spatial planning, like spatial planning itself, and planning in general, is in transition. Traditionally, evaluation has been an integral part of the rational planning process (Alexander, 1992.a: 74-75, 82-85). But, with the demise of the "classic" rational model as the dominant planning paradigm, what will be the role of evaluation in spatial planning in the future? What can we learn from our discussion about the new kinds of evaluations and evaluation methods that are emerging to replace the systematic and rational evaluations of the past?

Forty years of critiques have undermined the "classic" model of rational decision making which was the foundation paradigm for planning and related "decision sciences". Envisaging its imminent collapse, suggested replacements ranged from versions of bounded rationality such as "satisficing" to situational decision-making: "theory-in-practice" or the "phenomenology of the episode". But, though agreement was widespread that rationality is passè, none of these alternatives superseded it as the dominant normative planning model (Alexander, 1984).

Another alternative, communicative practice[1] has emerged in the last fifteen years to claim the role of the dominant planning paradigm (Beauregard, 1996: 106-109). But this model, too, has its shortcomings, which are, in a way, the mirror image of the the defects of "classic" rationality. The rational model, focusing on the decision and problem solving, is limited to individual deliberation, and ignores the interactive aspect of planning. Communicative action focuses on just what deliberative rationality neglects: consensus through social interaction and communication. But it fails to account for individual decisons and strategic (self-interested) action. In their normative form, both paradigms include simplifying assumptions which are divorced from reality.

These observations suggest that communicative action, with all its answers to the flaws of the rational paradigm, cannot supersede it. Rather than being mutually exclusive, both models are complementary, each reflecting a different aspect of planning in the real world (Alexander, 1996). Indeed, Mandelbaum (1979) asked whether any one paradigm can answer all our normative questions, given the complexity of planning and the value-laden character of most planning problems and

355

N. Lichfield et al. (eds.), Evaluation in Planning, 355-374.
© 1998 *Kluwer Academic Publishers. Printed in the Netherlands.*

policy issues, and answered that "A Complete General Theory of Planning is Impossible".

But perhaps a contingent framework could provide the comprehensive overview of the real dynamics of planning that no single paradigm can. I have suggested such a framework to integrate four planning models (some of them have been called paradigms) which are prevailing or emergent today. Just as this framework answers the question (asked about planning paradigms): After rationality, what?, we can use it to structure the answers to our question: What will post-rational evaluation be like?

2.   Planning: A "Four-Fold Way"

The contingent framework sees planning as a "four-fold way", integrating four different views of planning: rational planning, communicative practice, coordinative planning, and "frame-setting". This framework reveals that the four planning models are complementary, not conflicting: each model involves different kinds of actors or roles, doing different kinds of planning at different stages or levels in the planning process.

Planning as deliberative: The first is the "classic" rational planning paradigm. This implies a view of planning as a deliberative activity of problem solving, involving rational choice by self-interested individuals, or homogenous social units (organizations, agencies, governments) acting as if they were individuals (Alexander, 1996). The objective of rational planning is for the actor to decide to what ends action should be undertaken, and what course of action would be most effective[2]. This view of planning, of course, embraces quite a wide range of planning models, from "ideal" rationality through various forms of bounded rationality such as satisficing and incrementalism. It also provides the conceptual base for most analytical and planning models and methods in use today, from benefit-cost analysis to strategic games.

Planning as interactive: The new paradigm of communicative practice sees planning as a social interactive process. Planning is not an activity of an individual (as envisaged in the rational model) but happens in the process of their interaction. The focus of this view of planning is communication between the actors (again, individuals or quasi-individuals), which is the subject of positive analysis and normative prescription (e.g. Healey, 1995).   Rather than the individual decision, the material of communicative practice is planners' statements (Forester, 1996), the narrative of plans (Mandelbaum, 1991; Healey, 1993), and the rhetoric of planning research and analysis (Throgmorton, 1992). Interactive approaches ranging from facilitation (e.g in transactive planning - Friedmann, 1973, 1994) through conflict resolution, mediation and bargaining are methods reflecting the ideas of communicative practice.

Planning as coordinative: Another view sees planning as anticipatory coordination (March and Simon, 1958; 158-169). From this perspective, planning is not only about where to go, but how to get there (Alexander, 1994.a: 196). Here the relevant units are not individuals or quasi-individuals, but organizations[3], and the focus of coordinative planning is how to deploy them to undertake the necessary actions at the appropriate time to accomplish mutually agreed upon outcomes (Alexander, 1993). Strategic planning (Bryson and Cosby, 1992; Bryson, 1988) offers some of the methodological tools of coordinative planning, while its structural aspect is institutional design: specifying and creating the processes and organizational frameworks that enable coordinated action (Bolan, 1996; Alexander, 1995.a: 51-52).

Planning as frame-setting: The "frame-setting" aspect of planning (Alexander, 1992: 51), though one of its traditional functions[4], has recently enjoyed renewed recognition. "Framing" or frame-setting describes the social process of constructing and defining a problem situation and developing appropriate responses (Schön and Rein, 1994). It involves interpretive schemes and metaphorical expression (Faludi, 1996), and reflects the structuring power of ideas through a process of policy discourse (Hajer, 1995). Here the relevant arena is the community: a policy network in a particular arena of concern (Laumann and Knoke, 1987; Marin and Maynz, 1992; Sabatier and Jenkins-Smith, 1993) or a planning community involved with a specific spatial area (Alexander and Faludi, 1996).

What is striking about these four "paradigms" of planning is how different they are. Their protagonists are playing different roles, enacting different processes, and have different purposes. In the traditional rational model and in the new paradigm of communicative practice, the actor is the individual (or quasi-individual) planner or decision maker. In coordinative planning it is a heterogenous social unit - an organizational unit, organization, or interorganizational network; and in the "framing" model of planning it is the relevant policy or planning community.

The planning process in the rational model is deliberative, internal to the "individual" protagonist. Here is the critical distinction between this model and communicative planning, which is interactive *between* individuals. Coordinative planning and "framing" are also interactive, but the actors and arena are different: in coordinative planning the interaction is between organizations, while in frame-setting the process is essentially a discourse (involving individuals, organizations and institutions) in a policy or planning community.

We can also distinguish between these models by their purpose and products. Deliberative rational planning is oriented toward solving problems, and the result of the planning process is a decision committing the "individual" to a particular course of action to address the problem or acheive the objectives she has set himself. Communicative action involves a set of individuals in a process intended to reach a collective consensus. In the ideal case this may indeed be their common purpose; in more realistic circumstances participants may be using the interaction to achieve their particular goals[5]. Coordinative planning involves organizations in the production of actions that are concerted to acheive mutual goals, while in "framing" a policy or planning community is developing a common image, meta-policy or planning doctrine that will serve as a frame-of-reference for future decisions and actions.

Each of these paradigms is an "ideal type", but there is ample evidence that they exist: each of them has been identified in observed cases of planning. In the dynamic, messy process of planning in real life, they often overlap and appear at different stages of the process[6]. Or, with shifts in observers' cognitive approach[7], the same case can often be analyzed using different planning paradigms[8]. What can we learn about evaluation in spatial planning, using this contingent framework to structure the review of our discussions?

3.   The "Four-fold Way" in Evaluation

If we look at our discussion of evaluation in spatial planning in this light, we find an excellent fit between this contingent framework and the universe of evaluation approaches, methods, and issues that is represented here. We can easily relate each contribution to one or several planning paradigms, making manifest the link between evaluation and its planning contexts.

Deliberative-rational planning: Evaluation in the context of deliberative rational planning (i.e where individual or quasi-individual decision makers are seeking the best means to accomplish their objectives or to solve a problem) is still widespread and useful. Successive generations of evaluation methods, however, themselves reflect the shift from instrumental rationality (formal rationality identifying the optimal means to achieve a defined goal) to substantive rationality, which includes defining and prioritising the goals themselves.

Financial appraisal (in the private sector), evaluates alternative investments on the basis of their potential profitability, and investment analysis in the public sector (such as benefit-cost analysis) asessing projects' economic efficiency are classic expressions of *instrumental rationality*, and the known limits of the latter reflect the flaws of the rational paradigm (see e.g. Barbanente et al,. above). Nevertheless, benefit-cost analysis is still widely used, e.g. in strategic project evaluation in Israel (see Alexander above), in British transportation appraisal (see Nash above), and even (as Community Benefits Analysis) for European Trans-European Networks (TEN) projects (Roy, 1994: 56). So research aimed at limiting its defects (even if they are endemic) is well warranted.

Nash's review of project appraisal approaches in the British transportation sector gives an excellent account of some of these efforts. It also describes the transformation of simpler (instrumentally rational) benefit-cost based evaluations into broader, more complex evaluation frameworks involving substantive rationality. An example is the SACTRA reports' demand to incorporate a wider range of measures into transportation project appraisal, going beyond economic efficiency indicators and including environmental, land use and developmental impacts.

Our discussion did not include many references to, or applications of conventional investment analysis, except as a point of departure for other evaluation approaches. Corkindale's economic evaluation of British land use planning, which opens with a benefit-cost evaluation, is an example. Assessing the costs of British planning is simple, but identifying its benefits is not, let alone estimating their monetary value. In spite of his patent reluctance, Corkindale cannot avoid reviewing some alternative institutional arrangements (privatisation, tradable development rights)[9] in the process of appraising the marginal impacts of the current planning sytem. Whether he knows it or not, Corkindale cannot do a benefit-cost analysis of such a complex project, without being dragged into institutional design (which is addressed below).

Evaluation approaches which relate the ranking of alternatives to the relative priorities of criteria that reflect multiple goals or objectives are aspiring to *substantive rationality*. If we compare attitudes to investment appraisal and benefit-cost analysis to developments and applications of various forms of multi-ojective decision analysis, it is clear that a retreat from instrumental rationality is in progress. Yet that hardly implies an abandonment of rationality: the quest for substantive rationality is replacing the confidence in instrumental rationality as the holy grail of evaluation research.

This quest is also destined to be unsuccessful: multi-criteria evaluation methods have their own intrinsic flaws ensuring that they will always fall short of the ideal. Notwithstanding, ongoing efforts to improve this methodology, such as Glasser's "Theory of Wicked Problem Resolution", are evidence of continued investment in this aspiration, and confidence (perhaps somewhat naive) in the possibility of its acheivement.

Proposals, analyses, demonstrations and appraisals of various multi-criteria evaluation methods in a broad spectrum of sectors are prominent in our discussion. This type of evaluation features in Nathaniel Lichfield's overview of British planning, and Nash contributes an up-to-date discussion of the evolution of multi-criteria evaluation in the transportation sector. Batty's review of computer-based decision support systems also refers to what are essentially multi-criteria evaluation models, linking GIS with location-allocation models in a goal-optimization framework.

Dalia Lichfield presents a recent application of one of this family of methods, Integrated Planning and Environmental Assessment (IPEA) to guide the choice between alternative by-pass routes in Devon. Interestingly (and not uniquely, as we shall see), the IPEA includes interactive elements that reflect an attempt to blend substantive and communicative rationality. A proposal for what is essentially another multi-criteria decision analysis is Lombardi's set of sustainability criteria for evaluating local plans and projects.

Substantive rationality is also Bizarro and Nijkamp's aim, in their use of yet another form of multi-criteria decision evaluation, "Rough Set Analysis", to assess and aggregate a group of sustainable urban revitalization policy cases. However, as we shall see, their analysis of the conservation problems of cultural built heritage just uses this evaluation as one element in a much richer approach which includes significant elements of coordinative planning and institutional design.

Finally, Alexander's critique of evaluation in Israel invokes substantive rationality too. Though frequently used in Israel's statutory planning system, multi-criteria evaluations often serve more as rationalizations than as the decision supports they are supposed to be. His plea for more transparency, and for using multi-criteria evaluation's interactive potential[10], also implies a blend of substantive and communicative rationality that is still more often found in normative prescriptions than as real-life applications[11].

These contributions are evidence that multi-criteria evaluation is widespread, and that its development as a vehicle for substantive rationality is being actively pursued. For better or for worse, multi-criteria evaluation has evolved into a significant part of relatively advanced planning practice. Practicing planners, in fact, in dire need of such (and other) tools, did not participate in "the retreat from rationality" preached by planning academics, in the absence of usable methods based on competing planning paradigms (Wyatt, 1996: 639-641). In this light, the predictions of rationality's demise as a planning paradigm seem premature.

Communicative practice: Undoubtedly, evaluation as communicative practice was the most prominent theme of our discussion. Two papers address this topic directly, though each offers a slightly different point of view. Khakee presents communicative planning and evaluation from a normative perspective, drawing on the classic authors who developed the rationale for communicative practice, and making the argument for a participative and interactive evaluation approach. He describes a case of structure planning in a mid-sized Swedish community as a communicative discourse, but his analysis of the organizational aspects of this experience clearly (if perhaps unintentionally) illustrates how participative planning and evaluation are structured by institutional design.

Less prescriptive is Voogd's contingent approach, classifying planning arenas that vary on several dimensions. He correctly concludes that different planning paradigms fit various planning arenas, and that appropriate evaluation approaches follow suit[12]. This explains the Dutch rejection of systematic evaluation methods in many planning arenas, and their more widespread use of participative-interactive approaches. The observed change parallels a shift from substantively rational planning in sectoral and more hierarchical arenas to more integrated comprehensive planning in a system that is becoming more complex and less hierarchical. This engrossing analysis again illustrates the mixture of "ideal type" planning paradigms: it is really a discussion of the institutional design of evaluation.

Communicative practice is extensively discussed in the context of several other contributions. Barbanente, Borri and their associates present communicative practice as a response at once to the inadequacies of instrumental rationality and to the complexity of planning and environmental problems. Like Khakee's case, however, their description and analysis of one concrete[13] issue (the Bologna-Firenze autostrada widening project) and their involvement in another case (masterplanning for a diverse community) ultimately provoke them to explore questions of institutional design: organizing a participative system, and designing new instruments for environmental protection that will succeed where legal protection has failed.

Söderbaum's proposed actor-network evaluation approach (illustrated in a Norwegian application) is also a form of communicative practice, aiming to elicit consensus in resolving the conflicting claims of ecological sustainability and economic development. Fusco Girard applies communicative practice in a similar way to address a different conflict: between economic and democratic values in the context of cultural and environmental heritage conservation.

Much of the discussion of evaluation in the context of communicative practice is striking in its almost exhortatory tone: interactive evaluation which involves participation of all the relevant stakeholders is good for you. The relative absence of a methodological component (beyond truisms like "identify all the relevant interests and their representatives") is not so much the contributors' fault: it reflects the weakness of the communicative practice paradigm at its present stage of development. Undoubtedly, this model has considerable potential, but its realization is inhibited by continued confusion between its descriptive-positive and its prescriptive-normative components[14].

The other striking commonality between several of these papers is their association, implicit and unintended, but no less obvious for that, of communicative practice with institutional design. Voogd does this on the abstract and general plane: starting out by discussing communicative practice, he ends up developing what is essentially an institutional design of evaluation contexts. On the more practical level of real-world case experience and application, Barbanente, Borri et al., Khakee, and D.Lichfield all undertake what is essentially coordinative planning to structure their communicative practice by institutional design.

Coordinative planning: In its process aspect coordinative planning is hardly addressed here, but many of the papers explore or apply its structural dimension: institutional design. Some of these, as mentioned above, introduce institutional design considerations in framing their discussion of communicative practice-related evaluation. In others (such as Corkindale's benefit-cost analysis of British planning), institutional design -- whether consciously or not -- is an integral component of the evaluations themselves.

The one contribution that is an exception to the above description is Bizarro and Nijkamp's analysis of urban revitalization policies involving cities' cultural heritage. Here coordination is a recurrent theme. This includes linking the various aspects of urban revitalization planning (plan-making, financial planning, implementation, monitoring and evaluation) in an integrated framework, and recognizing the need for concerting action between institutional levels (vertical coordination) and intersectoral coordination. They suggest that effective action will demand government intervention[15], to implement a revitalization strategy based on policies selected from their proposed "Portfolio". This is coordinative planning par excellence.

One example (perhaps the most complete) of the use of institutional design almost as an evaluation approach in itself is Hull's analysis of housing land allocation decisions in North East England. Her account links observed housing outcomes to how the development plans operate as allocatory mechanisms (not the only one; e.g the important role of the recently formed Urban Development Corporation is described) under changing institutional and governance configurations. Her conclusion highlights the role of institutional design: "...how policy outcomes can be redefined through the restructuring of the political setting to encompass institutional and value change".

Lichfield's and Prat's chapter, too, can be viewed in large part as an exercise in institutional analysis and design. The analysis delineates the institutional design "space", identifying the relevant stages of the planning process, and defining three distinct types of evaluation to review their functions, methods, and performance in the British planning system. Based on their analysis, Lichfield and Prat's design proposes linking ex ante with ex post evaluation by developing Community Impact Evaluation as a common framework for both.

The discussion of evaluation and equity in regional planning, contributed by Clemente et al., has a significant coordinative planning component too. Their proposals for instruments and incentives for valuing and promoting conservation, to counter current institutional processes and structures that encourage expansion and production, and to reinforce communities' ability to resist development pressures, are really an outline institutional design. This is also the case in Batty's review of computerised evaluation and decision support systems, which blends a comprehensive account of current and projected hardware and software resources with ideas on their most effective deployment.

What all these efforts at institutional design - of evaluation processes and contexts, of evaluations themselves, or as parts of or the whole of an evaluation - have in common is their unreflexive quality. This is typical: most of us usually do not know that we are engaged in institutional design when we are doing it. Is that necessarily bad? Perhaps not, but institutional design is a deliberative and intellectual activity, as opposed to an intuitive or instinctive one. Therefore, it is probably enhanced, rather than inhibited, by appropriate reflection and self-consciousness, and a better developed knowledge-base would also do no harm[16].

Planning as "frame-setting": The links between evaluation and the "frame-setting" model of planning, unlike evaluation's association with rationality, are more tenuous and indirect. There are three ways in which evaluation can interact with the "frame-setting" process and affect the emerging "frame". Such a "frame" - the doctrine, meta-policy, planning doctrine, policy or plan - reflects a relatively durable consensus of a relevant territorial polity or community, or issue or policy community, after a period of interactive discourse.

Conflicting evaluation approaches and methods can be part of the contested discourse itself, reflecting the competing knowledge claims, ideologies, and values of the interested parties. The debate about assessment of European T-TEN projects (Richardson, 1997) is a perfect example. The conficting interests included on the one side national transportation planning agencies and the program's "spiritual progenitors", the European industrial concerns who lobbied for its adoption and funding as the European Round Table of Industrialists (ERT). With approval of the T-TEN program their platform was essentially adopted by the Commission, which even created a research institute, the European Centre for Transportation Studies (ECIS) to support its case. Entering the discourse later on the other side was the "environmental lobby" - Green parties, national environmental agencies, and environmental NGOs partly associated in the European Federation for Transportation and the Environment.

The basic paradigm for justifying the T-TEN program was economic rationality, claiming that the infrastructure network was essential for realizing the economic benefits associated with the Single European Market. In appraising particular proposals, this paradigm's evaluation approach is Cost- Benefit Analysis (CBA), already well tried in evaluating national strategic infrastructure projects. However, in the fiscal cisis of the 1990s this rationale came under fire from the European Parliament, which demanded justification of the programs' investments in European terms, as opposed to projects' strictly national benefits.

The response was the development of an improved form of CBA that identified "Community Benefits". This became the evaluation method to confirm the benefits of each project, while macroeconomic research was invoked to publish the expected positive impacts (e.g generating 3.2 m. person-years of work between 1998 and 2007) of the program as a whole.

However, at the same time an environmental movement had begun to campaign against the TENs. Many of its objections were targeted on specific projects (such as the Somport Tunnel through the Pyrenees) that had begun without consideration of alternatives or environmental impacts. But these precedents also generated the demand for a more comprehensive assessment that would integrate environmental considerations and alternatives' review into a strategic evaluation framework. The advocated evaluation method was called Strategic Environmental Assessment (SEA).

The debate in EU institutions continued through 1992 (when a pilot SEA of a High Speed Rail link was carried out) to 1995. It involved DGXI (Environment, Nuclear Safety and Civil Protection), DGXVII (Transport), the EU Commission (which initiated several SEA-related methodological studies), the European Parliament's review of the T-TEN program which concluded in a demand for SEA and stronger environmental conditions, and the European Council whose opposition to SEA reflected national governments' resistance to EU involvement in evaluating their pet projects.

A formal conciliation process had to resolve the stand-off between the Council and Parliament, following which a weak form of SEA was incorporated into the T-TEN evaluation process. This is now structured in a hierarchy which reflects the uneasy combination (hardly integration) of opposing paradigms with conflicting knowlege claims: the economic paradigm based on utilitarian rationality, and the environmental paradigm based on ecological principles. Non-mandatory SEA, at the highest level of this hierarchy, is designed to evaluate TEN networks as a whole. Meanwhile, lower in this hierarchy, project proposals are evaluated using CBA identifying "Community benefits", and "Corridor Analysis" of alternative routes and configurations, incorporating environmental considerations (Richardson, 1997: 3-11).

In the second kind of interaction between evaluation and "frame-setting", alternative paradigmatic "frames" - doctrines, meta-policy assumptions, or knowledge claims can be invoked in evaluation itself. While this is rare, due to the limits on interparadigmatic discourse (see below), Milchap's assessment of ex-ante evaluation in British planning and development control is an interesting example. The focus of his discussion is uncertainty: the difficulty of predicting future outcomes, and the problem of "risk management" involved in any evaluation of a proposal's predicted consequences.

The paradigm framing this issue, Millichap suggests, is essentially economic rationality, articulated in the concept of "the public interest". His analysis is explicitly premised on a completely different paradigm: legal entitlement, expressed as "community rights". While the "public interest" legitimises government intervention in development decisions and excuses fallible predictions, "community rights" demand protection from the results of planners' and evaluators' mistakes. Such protection can take a variety of forms: institutionalizing systematic monitoring of plan outcomes[17], improving the quality of predictions by demanding (analagously to pollution abatement) the "Best Available Prediction Not Entailing Excessive Cost", and incorporating "bounded uncertainty" into the consideration of development proposals.

Finally, we can evaluate a "frame", though if we give it paradigmatic status (as we do by defining it as a "frame") we limit the domain of rational evaluation [18]. Nevertheless, suggested criteria for "good" planning doctrines (Alexander and Faludi, 1996: 28-33) offer a clue to how "frames" can be evaluated. Two of these three criteria are external, the third is internal to the "frame" itself.

The first criterion is "objective" validity: to what extent do the information, causal theories or explanations, and knowledge-bases on which the "frame" is premised, correspond with reality. This is the criterion which is most vulnerable to the limits of interparadigmatic discourse, but often the apparent problems of incommensurability are surmountable.

The second criterion is consensus: a good "frame" must reflect the uninhibited consensus of the relevant community. This can be tested in several ways. One test is procedural: was the process which resulted in the "frame" open, participative, and did it involve all the relevant stakeholders? Another is structural: was the social, political and institutional structure framing the discourse enabling or constraining?

The final test is only retrospective: how durable is the "frame", and to what degree has the consensus on which it rests stood the test of time? Less than complete consensus is often revealed by the resurgence of opposition, or reopening of apparently closed issues, after apparent agreement and adoption of a "frame" - a meta-policy, a policy or a plan. Often, the stability of a durable "frame" is not only the result of consensus, but also attributable to the mobilising effectiveness of its core metaphor: the "Balance of Powers" doctrine in 19th. century international relations, the "containment" doctrine of U.S. cold-war foreign policy, the "New Deal" program of the Roosevelt administration, the British "Green Belt" and Netherlands' "Randstad and Green Heart" planning doctrines, the Copenhagen "Finger Plan", the "privatisation" principle of neo-liberal economic policies. Sometimes a core metaphor is not good enough to generate the needed consensus: the Clinton administration's "managed competition" health care system reform proposals is a case in point.

The third criterion is completeness and consistency. Does the "frame" encompass the appropriate decision space, and address the relevant issues or problems. Though this is an internal criterion, this test may also raise the kinds of knowledge-claim, value and ideological issues involved in problem definition, and run into the obstacles of interparadigmatic discourse. The second part of this criterion: internal consistency between the "frame's" various elements, and between the "frame" itself and its knowledge-claim, information, ideological and value premises, is less difficult to apply.

4. Conclusions: Where do we go from here?

The "four-fold way" of planning has proved its value as an organising framework for this review of developments in evaluation. Alternative evaluation approaches, too, as Khakee (above) describes them, fit this framework. The "multiplist" approach, involving pluralistic and open exchange of knowledge, parallels the communicative practice paradigm, while the "reponsive-constructivist" approach, focusing more on stakeholder negotiation, fits coordinative planning. The "design" approach, invoking multiple frameworks of values and technologies to form the evaluation framework, clearly reflects planning as "frame-setting". Evaluation approaches, then, are discussed and evaluation methods applied under each of the four planning paradigms: deliberative rationality (comprising instrumental and substantive rationality), communicative practice, coordinative planning and institutional design, and "frame-setting".

Where are we going? Instrumental rationality, with all its limitations, is still considered a valid paradigm for specific purposes, and evaluation methods such as benefit-cost analysis that are based on this paradigm are still widely used in appraising sectoral investments. Unquestionably, their use is the subject of growing debate, which reflects the aspiration to replace instrumental by substantive rationality. This is also revealed in the tendency (which began as long ago as Lichfield's "Planning Balance Sheet") to subsume benefit cost analysis under a more comprehensive evaluation scheme, usually some type of multi-criteria decision analysis framework. Developments in British transportation project appraisal and in the EU's evaluation in its TEN program illustrate this trend very well.

An impressive array of multi-objective evaluation approaches has been developed over the last three decades, and they are being more and more widely used at increasing levels of sophistication. Multi-objective evaluation methods, or multi criterion decision analysis, which are clearly identified with substantive rationality, are also the object of ongoing efforts at refinement and improvement. This does not provide evidence to confirm any retreat from rationality. On the contrary, it suggests that the rational paradigm, at least in planning practice (as distinguished from planning theory) is alive and well.

Many advanced applications of multi-objective evaluation methods are also adapted to interactive use, offering an attractive link between substantive rationality and communicative practice. The evaluation literature is replete with pleas to use these methods interactively, as decision support systems in arenas of political discourse. Cases of interactive applications, however, are still relatively few, and evaluation as a form of communicative planning is still more prescribed than practiced.

Communicative practice as the foundation paradigm for plan and project evaluation in the context of collective decision making, in communities, regional planning and development policy, and the appraisal of national and transnational strategic projects, is widely advocated. In the form of interactive multi-objective evaluation frameworks at least, its implementation would seem to be easy and obvious. In this light the scarcity of such applications is surprising. Actual examples of successful communicative practice-related evaluations of other types are also few, though there are more cases of aborted efforts and obvious failures. At the same time, the communicative practice paradigm has unrealised potentials which presage developments and applications of which observed trends only herald the beginning.

Its proponents' presentations of the communicative practice model, both in its normative versions -- e.g. in terms of concrete presciptions for action and clear methodologies -- and in its analytic-descriptive modes -- i.e as an "ideal" analytic model for the analysis of planning behavior, or as the conceptual framework for descriptions and accounts of cases of planning and their outcomes -- offer no explicit explanation for this paradox. Here I will offer three possible reasons.

Two suggest themselves on examining the communicative practice paradigm itself, as articulated by its advocates. The first concerns its normative-methodological aspect, which is still relatively undeveloped, providing the would-be practitioner with little more than truisms and exhortations[19]. As a normative model, the communicative practice paradigm also suffers from an unresolved flaw (the mirror image of "ideal" rationality's): its dependence on Habermas' communicative rationality and his "ideal speech situation" premise. Proponents of communicative practice, progressives sharing democratic values, or radicals seeking social transformation, share implicitly benign assumptions about human nature and social behavior, and have yet to come to grips with the implications of the gap between their ideal and cold reality.

The second concerns communicative practice as an analytic-descriptive model, but interacts with the first. Is communicative practice, assuming interaction aspiring to reach consensus, a better explanation of peoples' and organizations' behavior in planning contexts than what Habermas called "strategic action": communicative interaction that is goal-oriented, self interested, and manipulative? Foucaultian analyses such as Flyvberg's (1996.b) case studies of planning events and outcomes, suggests that it is not: another contradiction that is unresolved.

Finally, the focus of the communicative practice model on interaction and process leaves it incomplete: the structural dimension is absent. This is clearly shown in practice (including all the cases discussed above) by the impossibility of implementing communicative practice without considering institutional design, and often implementing the institutional transformations and procedural changes the design proposals demand. This observation highlights the association between communicative practice and institutional design, perhaps a novel insight. It is not to belittle this model to call it incomplete: no single abstract model or paradigm can be exhaustive[20].

Perhaps the most surprising discovery is how much evaluation involves institutional design; this takes three forms. One is designing the evaluation process, which is usually inextricably linked to its institutional context. The demands of communicative practice type evaluations, which cannot be met without significant changes in their governance and organizational contexts, or require the introduction and institutionalization of new instruments and procedures, are one example. This involves changing the given planning context. The contingent use of different evaluation schemes and methods, related to observed relevant characteristics of issues and planning arenas, is another example of institutional design of the evaluation process. But here institutional design is applied to adapt the evaluation to its given contexts.

The second form is designing the evaluation itself: the evaluation scheme, framework or method used to assess a proposal or to compare alternatives. Experience shows that the evaluation and its institutional context are also interdependent, and that adoption of a particular method or approach is contingent on its adaptation to its context, or deliberate changes and removal of institutional constraints.

The third form is perhaps the most widespread: designing the alternatives which are the object of the evaluation. In any but the simplest unisectoral undertakings, specifying the project includes a significant element of institutional design in describing the appropriate delivery system. Assessing the proposed widening of the Bologna-Firenze autostrada and estimating the costs and benefits of the Trans-Israel Highway raise institutional design questions: who is the responsible agency, how will the project be funded and managed, who is accountable for expected and unanticipated impacts, etc.

Such institutional design becomes critical, when the topic of the assessment is the delivery system itself, as is often the case in the context of policy analyses or program evaluations. Evaluating the performance of the UK statutory planning system, appraising a Swedish policy of municipal property privatisation, or assessing a French *commune's* solid-waste collection process, is impossible without institutional design: detailing the proposal to be evaluated, and designing the parameters of the feasible alternatives.

Planning as "frame-setting" involves evaluation too, we have found, though in more indirect ways than its integral role in the rational paradigm. An evaluation approach or method can become the subject of contested discourse between what are essentially competing policy or planning "frames" reflecting different knowledge claims, ideologies, or value sets. Such discourses undoubtedly occur more often than they are documented.

A "frame" -- a doctrine, set of meta-policy assumptions, a planning doctrine, a policy or a plan -- can itself be the subject of evaluation. This may be limited by the constraints of interparadigmatic discourse, but is nevertheless quite feasible. Relevant criteria include a "frame's" objective validity: how does it conform to the real world; finding the answers to this question when there are conflicts between knowlege-claims and contrasting problem definitions may be problematic. Another criterion is the degree of consensus the "frame" commands: this is reflected, ultimately, in its persistence and durability. And "frames" can be assessed on their completeness (which may also be problematic and related to problem definitions) and internal consistency.

Ultimately, a "frame" may be (and often should be) part of an evaluation itself. This means making explicit the paradigmatic or doctrinal premises for the assessment, and, ideally, contrasting them with alternative "frames" as proposed by the "design" approach to evaluation. In this way, perhaps, this succession of planning paradigms comes full circle, with "frame-setting" in interparadigmatic discourse in the context of communicative evaluation acheiving a higher order of substantive rationality.

But in the flow of time all of these are not occuring simultaneously, or to the same degree. If we extrapolate from observed trends, a succession of planning paradigms and their related evaluation approaches is unmistakable. Instrumental rationality, and its associated project appraisal approaches are on the decline. They are being replaced by substantive rationality and its related comprehensive and strategic evaluation frameworks. These will probably still expand before reaching their peak in terms of legitimacy, acceptance, sophistication and extent of application.

Communicative practice, though it has been talked about (mostly by planning academics and theorists) for fifteen years, is still in its infancy. Advocacy has to be followed by better theory, and practice cannot ensue without a better articulated methodology. The interactive potential of multi-criteria evaluations and their use as a framework for political and value-related discourse is gaining increasing recognition, but the scope of applications is limited by institutional constraints and prevailing organizational and political cultures. This enhances the importance of the successful examples we see, as exemplary prototypes for what may be the wave of the future.

The interaction between institutional design and evaluation, though still largely unreflexive, is also growing, as more comprehensive and multisectoral evaluations are replacing simpler types of project appraisal, and as changing forms of evaluation are pushing the "envelope" of existing institutional structures and political-organizational cultures and behavior. The realization that communicative forms of evaluation are often impossible without changes in the planning context, and the need for institutional design to consider what those changes should be and how they could be implemented, is one aspect of this phenomenon.

Deliberate institutional specification of the complex alternatives that are the objects of assessments, the conscious relation of choice between or adaptation of evaluation methods to their institutional contexts, and the institutional design of the evaluation context as an integral part of the evaluation itself will be increasingly common. We can easily envisage a future in which, with enhanced consciousnes, institutional design will be an integral aspect of good evaluation practice.

A post-modern consciousness is making evaluation part of an openly political and value-related discourse. This does not mean that evaluation should not or cannot be rational. As Lichfield (above) concludes, planning and evaluation will continue to be based on rationality, though a more complex and diverse one than the classic rational model[21]. As we see, the adaptation of evaluation approaches and methods to this reality can and does take different forms, varying under different circumstances: issue or problem type, the intrinsic nature of possible alternatives, the organizational and institutional context, and even social and political cultures.

These forms range from improved versions of appraisal methods that attempt to modify and qualify their basic instrumental rationality, to more comprehensive evaluation frameworks aspiring to substantive rationality. Interactive models of these are being devised and applied in an effort to combine substantive and communicative rationalities, and the institutional design of evaluations and their contexts is receiving growing attention. Finally, the doctrinal, ideological and value-related "frames" that shape evaluations and evaluation methods are also becoming an explicit element of this discourse and of evaluation itself.

Where Do We Go from Here? Above, I have painted a picture of historical and contemporary developments in planning in general and evaluation in particular, and suggested what the future might hold, extrapolating from likely trends. That is not necessarily to conclude: it is so, therefore it is good. Where we are going may not be where we ought to go. Why should we think that?

Three considerations could suggest a clear normative goal: a choice between planning paradigms and an identification of a preferred evaluation approach or "family" of evaluation methods associated with the selected planning paradigm. One could, charitably, be called meta-theoretical, or less charitably: blind faith. The second is frankly ideological, and the third could be called pragmatic.

The first approach believes that one paradigm is superior, and is, or has the potential to be, all encompassing. I call this blind faith, because out-and-out advocates of any single paradigm are dismissing critiques which have identified flaws and limits in each paradigm, and ignoring well-founded admonitions against a single normative-positive theoretical framework. In post-rationalist terms: a "Complete General Theory of Planning" is impossible; in post-modernist-speak: "There is no metanarrative".

The second approach identifies a particular planning paradigm with its ideology, feeling that one model of planning is more compatible with its preferred values. In this sense, practitioners of rational planning identify it with reason and enlightenment, scientific knowledge and a liberal community -- all these are questioned and belittled by its critics. Proponents of communicative practice identify their paradigm with democracy, social pluralism and empowerment -- traits everyone welcomes, but which might be more ideal than real. Coordinative planning promises effectiveness, but threatens goal-displacement. Advocates of "frame-setting" promote an open discourse of ideas in a culturally plural community, but how often is this aspiration constrained by history and power.

The endemic flaw in this approach is that only the normative "ideal" version of each paradigm expresses its ideology, but it is subject to all the limits of any idealised abstraction (as the critics of each have not hesitated to point out). In reality, each paradigm is value neutral. Rational planning can produce Head-Start and concentration camps, communicative interaction can seek consensus or be "power games", effective institutional design can produce the Mondragon community and the Microsoft corporation, and "frame-setting" discourse produces what the actual "community" (with its history, culture, and institutions) "wants", from the ideology you love to the doctrine you hate.

The third approach is the pragmatic approach: one planning paradigm is preferred to others because it works. But this also raises questions: how does it work better? If its claim is as a normative model, that implies regress to the second approach, which has been dismissed above. No claim for "their" paradigm to work better than another as a positive model -- a "truer" description of planning behavior or explanation of its consequences -- has been made by anyone, and if such a claim were asserted, it would be unfounded. Rather, as I have suggested, it is more constructive to regard each of these planning paradigms as a different perspective on a common reality, each as valid and "true" as any other, and each useful for different kinds of analysis.

Consequently, it is difficult, if not impossible, to set any definite normative direction for evaluation: who is to decide, and how, where we ought to go. Perhaps, the answer to: "Where do we go from here?" is: Where we are going. And perhaps "Where do we go from here?" is not the right question, but a simplistic and generalised version of what should be a much more specific and individual question (in fact, of any "good" or "right" action): "What do I do now?"

In the light of the discussion here, this question, in terms of evaluation approaches and methods, can only have contingent answers. Clearly, there is not and cannot be any one "best" way of evaluation. There can only be better or worse ways of evaluating particular kinds of alternatives in addressing various types of problems or issues, in a given context or circumstances. Most of our attention has been devoted to the improvement and refinement of evaluation methods, and extending the repertoire of evaluation approaches available. It would be worth the effort to develop answers to the question: What are the most appropriate evaluation approaches and methods to use, if specified situation characteristics apply. The "Four-fold way" contingent framework of complementary planning paradigms may be a useful basis for such an effort.

5. References

Alexander, E.R. (1998.a) Doing the impossible: Notes for a general theory of planning, *Environment and Planning B: Planning & Design* (forthcoming).

Alexander, E.R. (1998.b) Planning for Interdependence: Linking Amsterdam, its region, and Europe, AME - University of Amsterdam, Amsterdam (forthcoming)

Alexander, E.R. (1998.c) Rationality revisited: Planning paradigms in a post-postmodernist perspective, Presented at the Planning Theory Conference, Oxford 2-4 April, 1998.

Alexander, E.R. (1996) After rationality: Towards a contingency theory for planning, in S.J.Mandelbaum, L.Mazza and R.W.Burchell (eds.), *Explorations in Planning Theory*, Center for Urban Policy Research, Rutgers State Univ. of NJ, New Brunswick, NJ, pp.45-64.

Alexander, E.R. (1995) *How Organizations Act Together: Interorganizational Coordination in Theory and Practice*, Gordon & Breach, Trenton, NJ/Montreaux.

Alexander, E.R. (1994.a) To plan or not to plan, that is the question: Transaction cost theory and its implications for planning, *Environment and Planning B: Planning & Design* 21, 4, 341-352.

Alexander, E.R. (1994.b) The non-Euclidean mode of planning: What is it to be? *Journal of the American Planning Association* 60, 3, 372-376.

Alexander, E.R. (1993) Interorganizational coordination: theory and practice, *Journal of Planning Literature* 7,4, 328-343.

Alexander, E.R. (1992) *Approaches to Planning: Introducing Current Planning Theories, Concepts and Issues* (2nd. Ed.) Gordon & Breach, Philadelphia PA.

Alexander, E.R. (1984) After rationality, what? A review of responses to paradigm breakdown, *J. of the American Planning Association* 50 ,1: 62-69.

Alexander, E.R., R.Alterman and H.Law-Yone (1983) Evaluating plan implementation: The statutory planning system in Israel, *Progress in Planning* 20, 2, 99-172.

Alexander, E.R., and A.Faludi (1996). Planning doctrine: Its uses and implications, *Planning Theory* 16: 11-61.

Allison, G.T. (1971) *Essence of Decision: Explaining the Cuban Missile Crisis*, Little-Brown Boston.

Beauregard, R. (1996) Advocating preeminence: Anthologies as politics." in S.J.Mandelbaum, L.Mazza and R.W.Burchell (eds.), *Explorations in Planning Theory*, Center for Urban Policy Research, Rutgers State Univ. of NJ, New Brunswick, NJ pp. 105-110

Belton, V., F.Ackerman and I.Shepherd (1997) Integrated support from problem structuring to alternative evaluation using COPE and V.I.S.A. *Journal of Multi-Criteria Decision Analysis* 6, 115-136.

Bolan, R. (1996) Planning and institutional design, in S.J.Mandelbaum, L.Mazza and R.W.Burchell (eds.), *Explorations in Planning Theory*, Center for Urban Policy Research, Rutgers Sate Univ. of NJ, New Brunswick, NJ, pp.497-513.

Bryson, J. (1988) *Strategic Planning for Public and Non-profit Organizations*, Jossey-Bass, San Francisco, CA.

Bryson, J., and B.Crosby (1992) *Leadership for the Common Good: How to Tackle Public Problems in a Shared-Power World*, Jossey-Bass, San Francisco, CA.

Dabinett, G. and T.Richardson (1997) Evaluation: Power/knowledge in European spatial planning. Working Paper. 11th. Association of European Schools of Planning Congress, Nijmegen, The Netherlands.

Fischer, F., and J.Forester (eds.) (1993) *The Argumentative Turn in Policy Analysis and Planning*. Durham, NC: Duke University Press.

Faludi, A. (1996) Framing with images, *Environment & Planning B: Planning and Design* 23, 1, 93-108.

Faludi, A., and A. v.d. Valk (1994) *Rule and Order: Dutch Planning Doctrine in the Twentieth Century*, Kluwer, Dordrecht.

Feldman, M. (1981) *Order without Design: Information Production and Policy Making*, Stanford University Press, Stanford, CA.

Flyvberg, B. (1996.a) The dark side of planning: rationality and realrazionalität, in S.J.Mandelbaum, L.Mazza and R.W.Burchell (eds.), *Explorations in Planning Theory*,: Center for Urban Policy Research, Rutgers Sate Univ. of NJ, New Brunswick, NJ pp.383-394

Flyvberg, B. (1996.b) *Rationality and Power*, Avebury, Aldershot, Hants.

Forester, J. (1987) Planning in the face of conflict. *J. of the American Planning Association* 53, 4, 303-314.

Forester, J. (1989) *Planning in the Face of Power*, University of California Press, Berkeley, CA.

Forester, J. (1993) *Critical Theory, Public Policy and Planning Practice*, SUNY Press, Albany, NY.

Forester, J. (1996) Argument, power and passion in planning practice, in S.J.Mandelbaum, L.Mazza and R.W.Burchell (eds.) *Explorations in Planning Theory*, Center for Urban Policy Research, Rutgers State Univ. of NJ, New Brunswick, NJ, pp.241-262.

Friedmann, J. (1973) *Retracking America: A Theory of Transactive Planning*, Anchor/Doubleday, Garden City, NY.

Friedmann, J. (1994) Toward a non-Euclidean mode of planning, *J. of the American Planning Association* 59, 4, 482-5.

Habermas, J. (1984) *The Theory of Communicative Action Vol 1*, Heinemann, London.

Hajer, M. (1995) *The Politics of Environmental Discourse: Ecological Modernisation and the Policy Process*, Oxford University Press, Oxford.

Innes, J. (1995) Planning theory's emerging paradigm: communicative action and interactive practice, *J. of Planning Education & Research* 14, 3, 183-189.

Healey, P. (1992) A planner's day: knowledge and action in communicative practice. *J. of the American Planning Association* 58, 1, 9-20.

Healey, P. (1995) The communicative turn in planning theory and its implications for spatial strategy formation, *Environment and Planning B: Planning & Design* 23, 2, 217-234.

Laumann, E.O., and D.Knoke (1987) *The Organizational State: Social Choice in National Policy Domains*, University of Wisconsin Press, Madison, WI.

Lichfield, N. (1996) *Community Impact Evaluation*, UCL Press, London.

Mandelbaum, S.J. (1979) A Complete general theory of planning is impossible, *Policy Sciences* 11, 1, 59-71.

Mandelbaum, S.J. (1991) Telling stories. *J. of Planning Education & Research* 10, 2, 209-214.

March, J.C. and H.A.Simon, with H.Guetzow (1958) *Organizations*, Wiley, New York.

Marin, B., and R.Mayntz (eds.) (1992) *Policy Networks*, Campus Verl., Frankfurt a. M.

Richardson, T. (1997) Constructing the Policy Process: Creating a new discourse of trans-European transport planning, Working Paper, 11th. Association of European Schools of Planning Congress, Nijmegen, The Netherlands.

Roy, R. (1994) Investment in Transportation Infrastucture: The Recovery in Europe, European Centre for Transportation Studies, Rotterdam.

Sabatier, P.A., and H.Jenkins-Smith (1993) *Policy Change and Learning: An Advocacy Coalition Approach*, Westview Press, Boulder, CO.

Sager, T. (1994) *Communicative Planning Theory*, Avebury, Aldershot, Hants.

Schön, D.A.and M.Rein (1994) *Frame Reflection: Toward the Resolution of Intractable Policy Disputes*, Basic Books, New York.

Throgmorton, J. (1992) Planning as persuasive storytelling about the future: Negotiating an electric power rate settlement in Illinois. *Journal. of Planning Education & Research* **12**, 1, 17-31.

Voogd, H. (1983) *Multicriteria Evaluation for Urban and Regional Planning*, Pion, London.

Wyatt, R. (1996) Transcending the retreat from rationality, *Environment and Planning B: Planning & Design* **23**, 6, 39-654.

1. I am using this term (Healey, 1992; Innes, 1995) to encompass a range of related concepts which have as their common base Habermas' (1984) communicative rationality and his normative theory of communicative action, which apply critical theory to planning (Forester, 1989, 1993) and which use labels ranging from "the communicative-" or "..argumentative turn" (Healey, 1995; Fischer and Forester, 1993) to "communicative planning theory" (Sager, 1994).

[2] This "paradigm" is not limited here to the formal instrumental utilitarian rationality which is the subject of the critiques; it embraces "substantive rationality" which includes systematic goal articulation and prioritization (Alexander, 1992.a: 39-40). Not incidentally, this reflects the realities of planning practice, which employs methods, such as multi-objective decision analysis and community impact analysis (Voogd, 1983; Lichfield, 1996) that are premised on substantive rationality.

[3] As used here, organizations is a somewhat subjective term, and they can range from organizational (sub)units to interorganizational networks (Alexander, 1995.a: 50).

[4] In the area of physical planning the tradtional "Master Plan" was an expression of this aspect of planning; in the policy arena strategic planning also has a "framing" function, as Feldman (1981) discovered when researching interagency policy analyses that were not implemented.

[5] This is what Habermas called "strategic action" (Alexander, 1996: 52).

[6] See, for example, the analysis of Forester's (1996) case of development planning review (Alexander, 1998.a).

[7] A similar contingent framework (of alternative organizational decision-making models) informed Allison's (1971) analysis of the Cuban missile crisis.

[8] A national or regional planning system, for example, can be interpreted as a process of rational deliberation, seeing governments like individuals choosing the best course of action, or as communicative practice, e.g. Israel's national planning described as a strategic "game" (Alexander, Alterman and Law-Yone, 1983). Or it can be viewed as a form of coordinative planning in which the relevant organizations interact - e.g Alexander's (1998.b) description of Amsterdam regional planning, or as a process of frame-setting by the country's planning community, e.g. the evolution of the Netherlands' planning doctrine (Faludi and v.d.Valk, 1994). Each view may be "true", and each may be more or less useful.

[9] In fact, the potential impacts of other institutional arrangements are worth examining, from total privatisation (a la Houston TX) managed by private bargaining and adjudication, through various combinations of planning systems with TDR, to assuring rights to existing property plus planned development rights, only changes from planned development assignable (land use planning, Netherlands/Israel/UK pre1965 -style).

[10] These problems are not unique to Israel; e.g the participative potential of the EU's Strategic Environmental Analysis (SEA), intended for assessing T-TEN proposals, is also not utilised, and this form of multi-criteria evaluation is being applied as a "desk exercise" (Dabinett and Richardson, 1997: 13).

[11] This may be changing, and a growing number of descriptions of such cases is appearing, e.g in the *Journal of Multi-Criteria Decision Analysis*; for an example, see Belton, Ackerman and Shepherd (1997).

[12] My obvious sympathy with Voogd's contingent approach is based on its similarity in principle to my own contingent framework (Alexander, 1998.a).

[13] No pun intended.

[14] For somewhat more extended discussion, see Alexander (1998.a).

[15] The institutional design implications of these proposals are not pursued.

[16] I have suggested that planners pay more attention to institutional design, and advocated the inclusion of more organization theory (which relates to institutional design) in planning curricula (Alexander, 1994.a,b).

[17] This proposal has wide-ranging institutional design implications, which are barely touched upon here.

[18] These limits depend on the arena and topics of the discourse and the nature of the competing paradigms, and have to do with the degree of incommensurability between them. For a more detailed discussion see Alexander and Faludi (1996) on interdoctrinal discourse.

[19] This does not include the actual and potential methodological contributions from related areas, such as group facilitation, conflict-resolution and arbitration.

[20] But it does support the concept of a contingent framework consisting of partial models which are mutually complementary (Alexander, 1998.a). The only exception to the previous statement is the Persian fable of the Shah who wanted a summarised compendium of the world's wisdom, and executed the scholars who failed to produce the book, chapter, or sentence. Finally a hermit came out of the woods to stop the massacre, with the phrase: "This too shall pass."

[21] In fact, in its broad and traditional sense of "reason", rationality includes a wide variety of rational models which can also be associated with the different planning paradigms discussed here (Alexander, 1998.c).

INDEX